Sustainable Development Goals Series

The **Sustainable Development Goals Series** is Springer Nature's inaugural cross-imprint book series that addresses and supports the United Nations' seventeen Sustainable Development Goals. The series fosters comprehensive research focused on these global targets and endeavours to address some of society's greatest grand challenges. The SDGs are inherently multidisciplinary, and they bring people working across different fields together and working towards a common goal. In this spirit, the Sustainable Development Goals series is the first at Springer Nature to publish books under both the Springer and Palgrave Macmillan imprints, bringing the strengths of our imprints together.

The Sustainable Development Goals Series is organized into eighteen subseries: one subseries based around each of the seventeen respective Sustainable Development Goals, and an eighteenth subseries, "Connecting the Goals," which serves as a home for volumes addressing multiple goals or studying the SDGs as a whole. Each subseries is guided by an expert Subseries Advisor with years or decades of experience studying and addressing core components of their respective Goal.

The SDG Series has a remit as broad as the SDGs themselves, and contributions are welcome from scientists, academics, policymakers, and researchers working in fields related to any of the seventeen goals. If you are interested in contributing a monograph or curated volume to the series, please contact the Publishers: Zachary Romano [Springer; zachary.romano@springer.com] and Rachael Ballard [Palgrave Macmillan; rachael.ballard@palgrave.com].

More information about this series at https://link.springer.com/bookseries/15486

Sylvia Croese • Susan Parnell
Editors

Localizing the SDGs in African Cities

Springer

Editors
Sylvia Croese
School of Architecture and Planning
University of the Witwatersrand
Johannesburg, South Africa

African Centre for Cities
University of Cape Town
Cape Town, South Africa

Susan Parnell
Human Geography
University of Bristol
Bristol, UK

African Centre for Cities
University of Cape Town
Cape Town, South Africa

ISSN 2523-3084 ISSN 2523-3092 (electronic)
Sustainable Development Goals Series
ISBN 978-3-030-95978-4 ISBN 978-3-030-95979-1 (eBook)
https://doi.org/10.1007/978-3-030-95979-1

This Springer imprint is published by the registered company Springer Nature Switzerland AG
The registered company address is: Gewerbestrasse 11, 6330 Cham, Switzerland

Foreword

The Sustainable Development Goals (SDGs) provide a framework to address the world's biggest challenges regarding economic, social and environmental development. The challenge of achieving the global goals is increasingly felt in the world's cities. At the same time, many cities and local governments are developing innovative strategies, policies and initiatives to localize the SDGs to achieve the 2030 Agenda, to avoid a reversal of the progress made since 2015 and to support an inclusive, resilient and sustainable post-COVID-19 pandemic recovery.

This book deals with African experiences of SDG localization. It makes an important contribution to the existing body of work and knowledge on the SDGs by bringing grounded academic research together with practice. For an organization such as UN-Habitat, which works to advance the achievement of the 2030 Agenda and the SDGs at the local level according to the New Urban Agenda commitments and principles, this publication provides rich insights that can support normative development as well as inform our operational activities in many cities and communities across the African continent. It highlights African experiences that might inform regional and multilateral reviews and deliberations on the SDGs and profiles innovative cases, practices and processes that might be scaled up to other parts of the continent. The chapters use local examples from different sectors, scales and geographical regions to highlight practical solutions and entry points to SDG localization while acknowledging the complexity of African urbanism.

The book brings home three important messages.

First, the Sustainable Development Goals remain a robust framework—and perhaps our only hope—to advance the transformation towards an inclusive, resilient and more equitable post-pandemic future. However, the successful realization of the SDGs and the 2030 Agenda depends in large measure on the mainstreaming of the goals and targets within local and national visions and the degree to which the goals have been embraced by a wide range of local and national actors.

Second, SDG localization goes beyond the implementation of SDG 11 because all SDGs are relevant to cities. Recognizing the centrality of cities for achieving the SDGs also means gaining better insight into the interlinkages, accelerators and barriers that operate at different scales to impact urban SDG progress.

Third, and finally, the absence of data and the weakness of existing monitoring systems remain as the biggest challenges across African cities, not just in reporting on the SDGs, but also in their achievement. Data gaps, a lack of disaggregation and spatialization, as well as poor data quality and its inaccessibility hamper the systematic identification of priorities, undermine evidence-based policymaking and limit the extent to which precious resources can be directed to the most catalytic and impactful interventions.

This book comes at a crucial moment, when the world is recognizing the importance of SDG localization and its centrality to the Decade of Action. The critical self-reflection on the SDGs offered by leading African thought leaders in this publication is unprecedented. I hope that it will inspire more reflection, more discussion and more collaborative action, which would transform the lives of the most vulnerable people and communities in Africa, by 2030 and beyond.

Global Solutions Division
UN-Habitat,
Nairobi, Kenya

Shipra Narang Suri

Preface

This book project began when its editors were both based at the African Centre for Cities (ACC) at the University of Cape Town (UCT) in South Africa. Between 2018 and 2019, they worked on three different, but interconnected, research projects. The first was the PEAK Urban Research programme on which they both worked as researchers. The second was the fifth Global Report on Decentralization and Local Democracy (GOLD) on the Localization of Global Agendas of the United Cities and Local Governments (UCLG) association, for which they curated the Africa chapter. The third and very different common project was the Leading Integrated Research for Agenda 2030 in Africa (LIRA 2030) programme, where Sue fulfilled the role of Chair of the Advisory Committee and Sylvia was a Principal Investigator.

Taken together, these three different projects built on Sue's involvement in the global debates about the SDGs through the ACC and her work as part of the core team that pushed for the adoption of an urban goal (SDG 11). They also contributed to (funding for) Sylvia's work as an embedded researcher in the City of Cape Town, which started as part of an ACC-hosted international comparative research project of the Mistra Urban Futures network on the implementation of the SDGs and New Urban Agenda, and related research on SDG localization in Cape Town through PEAK Urban, as well as the cities of Luanda and Maputo through the LIRA 2030 programme. Overall, these projects were important for widening the editors' links and connections to other SDG scholars and practitioners across the continent.

This work culminated in the organization of five SDG seminars hosted by the ACC, while an additional two were co-hosted by the UCT Faculty of Economics and the UCT Faculty of Health Sciences over the course of the year 2019. The aim of the seminar series was to bring together a diverse set of actors and stakeholders with different disciplinary and professional backgrounds to share and exchange thoughts and experiences related to SDG localization. The seminars brought about 150 participants together, ranging from researchers and academics to practitioners, representatives from the private sector, non-governmental organizations, as well as national, regional and local government officials and networks, all with a shared interest in the African SDG space. As such, the seminars themselves functioned as important spaces and practices for the co-production and transfer of the knowledge required for meeting the challenges and opportunities of SDG localization, in line with an established practice of transdisciplinary research at ACC.

The idea to turn the inputs of the SDG seminars into a book was made possible with the award of a small Global Challenges Research Fund grant from the University of Bristol. The editors planned to organize a larger workshop and bring prospective contributors to Cape Town, South Africa, in April 2020. However, the onset of the COVID-19 pandemic led to the indefinite postponement of this meeting and forced them to take their work to the virtual space, where all of the final stages of discussion and refinement for the book took place.

During the pandemic months of 2020 and 2021, a broad and unique group of researchers and practitioners from across the African continent came on board to contribute to the project. With the overarching aim of bridging the gap between research and policy with a distinct focus on Africa, the book builds on the work of a total of 32 authors. They include up-and-coming young academics, as well as urban practitioners, including architects, city officials and representatives of city networks, private consultancy firms and international organizations based in cities across South Africa, Uganda, Egypt, Ghana, Burkina Faso, Nigeria, Ethiopia, Zambia and Mozambique.

Taken together, the contributions sit at the interface between urban research and policy, providing insight into the wide range of actors involved in, and necessary for, SDG localization in Africa. Their commitment and insights into what it takes to achieve long-term, inclusive, sustainable and participatory development should be of interest to academics, students, practitioners and policy makers in Africa and beyond.

Johannesburg, South Africa — Sylvia Croese
Cape Town, South Africa
Bristol, UK — Susan Parnell
Cape Town, South Africa
September, 2021

Acknowledgements

Firstly, we thank all of our chapter contributors, who despite the challenging impact of the ongoing COVID-19 pandemic managed to find time to write up their invaluable expertise and experience for this book.

In addition, we would like to thank Lynn Woolfrey and Namhla Mniki-Mangaliso, who presented their work as part of the SDG seminar series and contributed to informing the book's ideas and conception, but whose work for different reasons could not be included as chapters. We also thank all the participants in the seminar series for their presence and inputs. This includes ACC colleagues who supported the seminars in various other ways, including just by being supportive colleagues. These include Rike Sitas, Gareth Haysom, Jane Battersby-Lennard, Vaughn Sadie and Anton Cartwright, as well as Alma Viviers for her assistance with communications and outreach and Liza Rose Cirolia who contributed to our initial work on the GOLD report. There were many other collaborations, including with Ntombini Marrengane on knowledge co-production in the context of the African Urban Research Initiative. In addition, we thank the ACC management staff who supported the project and assisted in managing its administrative and financial side: Edgar Pieterse, Andrew Tucker, Warren Smit, Ithra Najaar, Marlene Joubert and Shakira Jeppie. At Bristol University, Alison Tyers, Simon Glasser, Tiernan Williams, Richard Davis and Tanya Downes provided valued support.

We also thank those who participated in and have become part of a collective SDG journey that began in Cape Town. These include Carol Wright, Natasha Primo and Jameyah Ally-Armien, as well as Gareth Morgan and Cayley Green (formerly) in the City of Cape Town. Other fellow passengers in this journey include Nachi Majoe and Timothy Blatch at ICLEI, Stefanie Chetty at the National Department of Cooperative Governance, Charmaine Smith at the Poverty & Inequality Initiative at the University of Cape Town (UCT) and Clare Jeffrey at the Department of Medicine, Faculty of Health Sciences, UCT. From United Cities and Local Governments (UCLG), they include Edgardo Bilsky, Luc Aldon Duffles and Fernando Santomauro.

Other persons who indirectly contributed to the development of the book through their invitations to teach and present on the SDGs include Saskia Ruijsink and Ore Fika at the Institute for Housing and Urban Development Studies at Erasmus University in Rotterdam, Zarina Patel at the Environmental and Geographical Sciences Department at the UCT and Katsia Paulavets from the International Science Council.

Specifically, we would like to acknowledge the following funders in supporting our various research endeavours: the Swedish International Development Cooperation Agency (Sida) and the Swedish Foundation for Strategic Environmental Research (MISTRA); the PEAK Urban Research programme, funded by the UK Research and Innovation Global Challenges Research Fund, grant reference ES/P011055/1; the LIRA 2030 Africa Programme, implemented by the International Science Council (ISC) in partnership with the Network of African Science Academies (NASAC) with support from Sida, grant number LIRA2030-GR02/18; and QR UK Research and Innovation Global Challenges Research Fund funding through the University of Bristol.

Finally, we would like to thank our publisher Springer for their interest in taking this project on, especially series editor Zachary Romano, Leonie Joubert for her patient and thorough coordination of the editing process and Shipra Narang Suri from UN-Habitat for generously taking the time to engage with our work and writing the inspiring Foreword to this book.

Contents

Contributors

Ernest Nti Acheampong African Technology Policy Studies Network (ATPS), Nairobi, Kenya

Puvendra Akkiah Senior Manager Programmes: Integrated Development Planning, eThekwini Metropolitan Municipality, Durban, South Africa

João Porto de Albuquerque School of Social and Political Sciences, University of Glasgow, Glasgow, UK

Joseph Amikuzuno University of Development Studies (UDS), Tamale, Ghana

Disan Byarugaba Department of Geography, Geo-informatics and Climatic Sciences, Makerere University, Kampala, Uganda

Liza Rose Cirolia African Centre for Cities, University of Cape Town, Cape Town, South Africa

Sylvia Croese School of Architecture and Planning, University of the Witwatersrand, Johannesburg, South Africa

African Centre for Cities, University of Cape Town, Cape Town, South Africa

Sylvestre Da WASCAL, Competence Centre, Ouagadougou, Burkina Faso

Amin El-Didi Cairo Lab for Urban Studies, Training and Environmental Research (CLUSTER), Cairo, Egypt

Peter Elias Department of Geography, University of Lagos, Lagos, Nigeria

Hanaa Gad Cairo Lab for Urban Studies, Training and Environmental Research (CLUSTER), Cairo, Egypt

Bizoola Zinzoola Gandaa University of Development Studies (UDS), Tamale, Ghana

Jennifer van Geesbergen Independent Researcher/Consultant, Paris, France

Marcella O. Guarneri Urban Planning & Design Lab, UN-Habitat, Nairobi, Kenya

Kwame Oppong Hackman WASCAL, Competence Centre, Ouagadougou, Burkina Faso

Buyana Kareem Department of Geography, Geo-informatics and Climatic Sciences, Makerere University, Kampala, Uganda

Pedro Laíce National Association of Mozambican Municipalities (ANAMM), Maputo, Mozambique

Salome Maswime Global Surgery Division, Department of Surgery, Faculty of Health Sciences, University of Cape Town, Cape Town, South Africa

Paul Mukwaya Department of Geography, Geo-informatics and Climatic Sciences, Makerere University, Kampala, Uganda

Omar Nagati Cairo Lab for Urban Studies, Training and Environmental Research (CLUSTER), Cairo, Egypt

Gloria Nsangi Nakyagaba Department of Geography and Environmental Sustainability, University of Oklahoma, Norman, USA

Oblé Neya WASCAL, Competence Centre, Ouagadougou, Burkina Faso

Laura Nkula-Wenz African Centre for Cities, University of Cape Town, Cape Town, South Africa

Kehinde Olufunso Ogunjobi WASCAL, Competence Centre, Ouagadougou, Burkina Faso

Susan Parnell Human Geography, University of Bristol, Bristol, UK

African Centre for Cities, University of Cape Town, Cape Town, South Africa

Thomaz M. T. Ramalho Regional Office for Africa, UN-Habitat, Maputo, Mozambique

Ché L. Reddy Global Surgery Division, Department of Surgery, Faculty of Health Sciences, University of Cape Town, Cape Town, South Africa

Seyni Salack WASCAL, Competence Centre, Ouagadougou, Burkina Faso

Safiétou Sanfo WASCAL, Competence Centre, Ouagadougou, Burkina Faso

Alexis Schäffler-Thomson Pegasys Strategy and Development, Cape Town, South Africa

Gilbert Siame Centre for Urban Research and Planning, Department of Geography and Environmental Studies, University of Zambia, Lusaka, Zambia

Hakim Sseviiri Department of Geography, Geo-informatics and Climatic Sciences, Makerere University, Kampala, Uganda

J. Jacqueline A. Walubwa Department of Geography, Kisii University, Kisii, Kenya

Dieter Zinnbauer Copenhagen Business School, Copenhagen, Denmark

About the Editors

Sylvia Croese is a Senior Researcher at the South African Research Chair in Spatial Analysis and City Planning at the School of Architecture and Planning of the University of the Witwatersrand and Research Associate with the African Centre for Cities (ACC) at the University of Cape Town. She has conducted extensive research on urban politics, policy and governance in African cities through the lens of housing, land, urban infrastructure and mobility. This research is premised on the need for new forms of knowledge that bridge the divide between academic research and (global) urban policy by involving different actors in society through inter- and transdisciplinary research methods of co-production. She has published widely on this work in major international journals as well as co-edited two books: *Refractions of the National, the Popular and the Global in African Cities* (African Minds 2021) and *Reframing the Urban Challenge in Africa: Knowledge Co-production from the South* (Routledge 2021).

Susan Parnell is a Professor in the School of Geography at the University of Bristol and Emeritus Professor at the African Centre for Cities (ACC) at the University of Cape Town. She has held previous academic positions and fellowships at the University of the Witwatersrand in South Africa and the School of Oriental and African Studies (SOAS), University College London (UCL) and London School of Economics (LSE) in the United Kingdom. She has been actively involved in local, national and global urban policy debates around the Sustainable Development Goals and is an advocate for better science policy engagement in cities. She is the author of numerous peer-reviewed publications that document how cities, past and present, respond to policy change. Her most recent books include the co-authored *Building a Capable State: Service Delivery in Post-Apartheid South Africa* (Zed 2017) and the co-edited *Urban Planet: Knowledge Towards Sustainable Cities* (Cambridge 2018).

1 Toward an Embedded, Integrated, and Collaborative Approach to SDG Localization in African Cities

Sylvia Croese

Abstract

The African continent represents the key hub in the global transition to a predominantly urban world. Hence, the fate of African cities—and the global South more generally—will have major implications for the extent to which the world will be able to achieve globally agreed development goals, such as the Sustainable Development Goals (SDGs). Questions about what makes for effective take-up of the SDGs lie at the heart of this book's aim and contribution. A key term in this regard is that of "localization." This chapter makes a case for an approach to SDG localization that is embedded in the continent's diverse local contexts. As an introduction to the book, this chapter departs from integrated perspectives and collaborative methodologies and outlines the ways in which the book's chapters illustrate the need for advancing such an approach in African cities.

Keywords

Sustainable Development Goals (SDGs) · Localization · African cities

1.1 Introduction

The African continent represents the key hub in the global transition to a predominantly urban world, with an urban population that is expected to grow threefold by 2050 (UN-Habitat 2014; UNDESA 2018). Cities such as Lagos, Kinshasa, and Dar es Salaam are predicted to be among the 13 African cities that will represent the world's 20 largest cities with average populations of about 23 million in 2100 (Hoornweg and Pope 2017). This unprecedented scale and speed of urbanization comes with new and intensified challenges at the economic, social, and environmental level, which will greatly determine the ways in which cities will grow and develop (Satterthwaite 2017). Hence, if "cities are where the battle for sustainable development will be won or lost" (UN 2013), the fate of African cities—and the Global South more generally—will have major implications for the extent to which the world will be able to achieve globally agreed development goals, such as the Sustainable Development Goals (SDGs) and related global development agendas, by the year 2030 and beyond.

S. Croese (✉)
School of Architecture and Planning, University of the Witwatersrand, Johannesburg, South Africa

African Centre for Cities, University of Cape Town, Cape Town, South Africa
e-mail: sylvia.croese@wits.ac.za

S. Croese, S. Parnell (eds.), *Localizing the SDGs in African Cities*, Sustainable Development Goals Series, https://doi.org/10.1007/978-3-030-95979-1_1

Implementing the SDGs does not take place in a vacuum, as the way in which global and local processes come together in any location are shaped by local history, geography, and politics, among others. Therefore, a recognition of both the global and local significance of the role of African cities requires a much better understanding of the nature, drivers, and management of urbanization, not just in the continent's megacities but also in its vast number of small- and intermediate-sized towns. It also means interrogating our universally defined understandings of what sustainable development means and the ways in which we think it can be achieved and measured in the African (urban) context. What are the barriers and constraints to achieving sustainable urbanization in urban Africa? What are enabling factors and opportunities? What is the role of local action and actors, and how can they be advanced? Who are the movers and shakers? And what do we know, and what do we still need to know, in order to answer these questions?

Questions about what makes for effective take-up of the SDGs lie at the heart of this book's aim and contribution. A key term in this regard is that of "localization." How can global development goals be adapted to and implemented at the local level in ways that contribute to sustainable urbanization? Our position here is not to promote the uncritical regurgitation of global goals in ways that have no relevance to the local level. In fact, our stance is that global goals can *only* be relevant to the extent that they address and are embedded in complex and diverse local contexts, needs, and realities. At the same time, we recognize and illustrate the significance of having a universally agreed set of global development goals that can be used to galvanize local action. As such, localization goes beyond the mere implementation of global goals and involves a process that covers practices of awareness building, advocacy, adaptation, monitoring, reporting, and evaluation. Importantly, such a process needs to recognize the different ways in which the global goals are interconnected at and across global and local scales, therefore necessitating an integrated perspective. Local or sub-national government actors are key actors in this process, but effective localization requires the involvement of a wide range of actors that are active and have an interest in the local policy and development space in Africa, all the way down to the neighborhood level. For these actors to have maximum impact, a collaborative approach to localization is imperative.

With less than 10 years to go until 2030 at the time of writing, the contributions of this book are important for informing and accelerating local action going forward. At the same time, their relevance extends beyond global reporting time-frames, by locating sustainable development planning and change in the everyday practices of African cities and their residents.

1.2 SDGs and the Global (Urban) Post-2015 Development Agenda

Global development goals are not new, nor is the recognition of the importance of the role of cities as both actors, places, and issues of development (Kosovac and Acuto 2020). But compared to the Millennium Development Goals, which only nominally recognized the urban as manifested in developing countries' slums, the inclusion of a stand-alone urban goal (SDG 11) to make cities "safe, inclusive, resilient, and sustainable" across the world should be seen as "pathbreaking both within the UN system for the acknowledgment it brings of the developmental role of sub-national governments and paradigmatically for global urban policy because it concedes that, in an urban world, cities can be pathways to sustainable development" (Parnell 2016, p. 529).

Globally, a broad coalition of local government representatives, civil society organizations, and scientists has been instrumental in advancing the inclusion of the (recognition of the) role of cities into a set of global development goals and agreements that together make up the post-2015 development agenda (Arajärvi 2019).[1] This

[1] In this book, we see the SDGs as representing an all-encompassing universal development framework, which covers the various elements of economic, social, and

includes the adoption of the Sendai Framework for Disaster Risk Reduction in March 2015, which states that "disaster risk reduction and management depends on coordination mechanisms within and across sectors and with relevant stakeholders at all levels, and it requires the full engagement of *all State institutions of an executive and legislative nature at national and local levels*" (art. 19(2), emphasis added) (UN 2015a). This was followed by the adoption of the Addis Ababa Action Agenda on the financing of development post-2015 in July 2015, which "commit[s] to scaling up international cooperation to strengthen capacities of *municipalities and other local authorities* [...] in implementing resilient and environmentally sound infrastructure, including energy, transport, water and sanitation, and sustainable and resilient buildings using local materials. We will strive to *support local governments* in their efforts to mobilize revenues as appropriate" (par. 34, emphasis added) (UN 2015b). Signatories to the Paris Climate Agreement, adopted in December 2015, in turn "recognize that adaptation is a global challenge *faced by all with local, subnational, national, regional and international dimensions*, and that it is a key component of and makes a contribution to the long-term global response to climate change to protect people, livelihoods and ecosystem [...]" (art. 7(2), emphasis added) (UNFCC 2015). The New Urban Agenda adopted in 2016 represents a culmination of the recognition of the importance of local action, dating back to the first Human Settlements Conference held in 1976 (UN 2016).

Important critiques have been made of the language of universality and inclusivity of the SDGs and related post-2015 development concepts and agreements as masking agendas that privilege global economic and commercial interests and structures while trivializing and depoliticizing local actors and development solutions (Kaika 2017; Weber 2017; Leitner et al. 2018; Macamo 2019). However, actors from the Global South actively played a key role in the preparations for and negotiations around the SDG framework. African policymakers developed the Common African Position, which was adopted by the African Union (AU) as a platform to negotiate for African voices in the SDG framework (AU 2014). African leaders also pushed for goals that were of particular interest to the continent—such as inclusive and sustainable industrialization—in sessions of the Open Working Group on Sustainable Development Goals and other regional and global intergovernmental forums representing Africa and the Global South, such as the UN Group of 77 (Fukuda-Parr and Muchhala 2020). This active engagement shows that these agendas were not imposed on Africa but reflect local development aspirations that are supported and owned at the highest political level. African governments therefore carry responsibility for the design and operationalization as well as, at times, (mis)use of the SDG framework and in that regard should be held to account (Malonza and Ortega 2020; Jönsson and Bexell 2021). But the negotiations also reflect the nature of global development frameworks as the result of highly politicized and at times contested compromises and are therefore, in some way, inherently imperfect (Fukuda-Parr and McNeill 2019).

Nowhere is the paradox inherent in the 2030 vision clearer than in the tension between the universal recognition of cities in the post-2015 development agenda on the one hand and the position of cities in global and national governance frameworks on the other. The Janus-faced nature of the importance of the local scale in international development is particularly acute in the African context where many local governments lack the political, administrative, and fiscal power, mandate, and resources to govern vast areas of settlement, with far-reaching implications for the prospects of effective SDG localization.

1.3 Localizing the SDGs

The need for devolution has been one of the unspoken reasons for the reticence of UN member states to commit to a stand-alone urban

environmental goals included in other development agreements.

SDG. This will be an ongoing political, economic, and fiscal struggle, which will play itself out over the lifetime of the SDGs (UN Open Working Group on SDG 11, in Revi et al. 2014).

Despite the inclusion of a stand-alone urban goal, there has been much less consensus about what safety, inclusivity, resilience, or sustainability mean or look like in the urban context. Relatedly, there is little consensus on how sustainable urbanization can be achieved and measured across different places and scales (Klopp and Petretta 2017). This uneven local vision of the SDGs has resulted in an agenda which, while ground-breaking in the universality of its goals, is much less clear at the level of the targets and indicators developed to measure progress. Moreover, during the first 5 years after the adoption of the global indicator framework for the SDGs, over half of the framework's unique 231 indicators were classified as indicators for which data were not regularly produced by countries (Tier II) or for which there was no internationally established methodology or standard available (Tier III). A monitoring framework for the New Urban Agenda in turn was only launched in 2020. While the absence of comprehensive multi-scale information did not hinder the development of a vast number of different indices, dashboards, and rankings by international or UN-affiliated organizations to track levels of progress toward SDG achievement, most of these nationally produced reports did not include city-level data, as sub-national data and actors are not considered in and of themselves in the follow-up and review frameworks on the SDGs (Dellas et al. 2018).

Local governments and their representatives have actively challenged the deep-seated tension and deficit in global governance frameworks by creating their own spaces and terms for participation. Members of the Global Taskforce of Local and Regional Governments (GTLRG), such as United Cities and Local Governments (UCLG), have played an important role in this regard, by foregrounding the sub-national scale of action and information. The push for building a stronger role for sub-national actors in the multi-lateral system has gone beyond the lobby for the inclusion of a stand-alone urban goal, to advocating for the importance of sub-national action for achieving *all* SDGs. Indeed, even with a stand-alone urban SDG, it will be very difficult to achieve any of the SDGs, ranging from poverty to climate change, without taking the role of cities into account. Moreover, many of the goals are interlinked and cannot be addressed in isolation, while certain goals and targets may function as accelerators and multipliers. From this perspective, "localization" has been understood as "the process of taking into account sub-national contexts in the achievement of the 2030 Agenda, from the setting of goals and targets, to determining the means of implementation and using indicators to measure and monitor progress" (GTLRG 2016, p. 6).

The work of members of the GTLRG has a distinct policy focus and is concerned with offering practical guidance for SDG localization at the local government level (UCLG 2015; Kanuri et al. 2016). A growing number of pioneering cities across the world in turn have started experimenting with their own approaches to SDG localization through the development of Voluntary Local Reviews on the SDGs, which are now recognized by the UN as important contributions to the implementation and monitoring of the SDGs at the national level (UNDESA 2021a). The success of local and regional governments in carving out a global space for collective accounting, reflection, and sharing of best practice is testament to the convening power of the SDGs and the usefulness of the framework as a tool to build and enable local dialogue and planning, as well as city networking and branding (Fox and Macleod 2021). It also provides insight into the kind of factors and actors that enable local action around the SDGs. Importantly, documented experiences of SDG localization show that effective local action goes beyond local governments themselves and includes the role and importance of an active local civil society, city-university partnerships, local political leadership and buy-in, as well as the support of global research and policy organizations, networks, platforms, and communities of practice to help share and disseminate good practices (Valencia et al. 2019; Fox and Macleod 2021; Morales et al. 2021).

Yet, overplaying the success of a few pioneering cities runs the risk of overlooking some of the basic institutional criteria that need to be in place to enable local action. This includes important provisions that need to be in place for local governments to have the financial, technical, and political mandate and capacity to engage with global and local development goals on their own and/or together with other relevant stakeholders. Many of such provisions are not (sufficiently) in place in African cities (UCLG Africa and Cities Alliance 2018). The underlying barriers of highly centralized countries with rapidly evolving urban centers means that many African local governments are not endowed with the political, administrative, and fiscal competencies they require to effectively advance with SDG localization processes in their respective territories.

1.4 SDG Localization in African Cities

The adoption of the SDGs and related development agendas has enjoyed high-level political support in Africa. This support is reflected not just in the widespread participation in the negotiations around the SDGs but also in their follow-up and review. Between 2016 and 2021, 46 out of 54 African countries presented their Voluntary National Reports (VNRs) to the UN High-Level Political Forum, some of them multiple times (UNDESA 2021b).[2]

Africa's commitment to a collective development agenda is also evidenced in the adoption of the continent's own development agenda, Agenda 2063: The Africa We Want. Agenda 2063 represents a homegrown vision of development that is inclusive, holistic, and sustainable. Notably, its adoption preceded that of Agenda 2030, but it covers many similar goals, illustrating the advantages to pursuing these agendas in tandem (UNECA 2017). Moreover, similar to Agenda 2030, Agenda 2063 also recognizes the role of sub-national action in achieving developmental goals. More specifically, Africa's development agenda aspires for "institutions *at all levels of government* [to be] developmental, democratic, and accountable" and for "*cities and other settlements* [to be] hubs of cultural and economic activities, with modernized infrastructure, and [...] access to affordable and decent housing including housing finance together with all the basic necessities of life such as water, sanitation, energy, public transport and ICT" (AU 2015, pp. 2–3; 6, emphasis added).

The recognition of cities as important developmental and economic sites and actors is also evidenced in the development of a growing number of National Urban Plans, which represent a key tool for locating cities in national planning (Cartwright et al. 2018). However, levels of political, administrative, and fiscal decentralization in Africa remain uneven, limiting the capacity of cities to fulfill their roles. A few exceptions include countries such as South Africa, Uganda, Tanzania, and Morocco, and even here, degrees of decentralization are not permanent or linear. On the other hand, countries such as São Tomé and Príncipe, Togo, Comoros, Congo, Democratic Republic of Congo, Equatorial Guinea, Egypt, Mozambique, Liberia, Guinea-Bissau, Central African Republic, Sudan, and Somalia are considered to represent contexts with incomplete decentralization processes (UCLG Africa and Cities Alliance 2018).

The limited extent of formal and functional devolution across Africa is the product of many different reasons, including histories of centralized colonial rule, economic crisis, (ethnic) conflict, and post-colonial political economies and settlements that have favored centralized rule, often—even if inadvertently—reinforced by the support of international donors and organizations. This inter-related set of factors has resulted in partial and complex decentralization reforms, many of which have only been implemented since the 1990s and in many countries are still ongoing. In many countries, large gaps remain between institutional reforms and practice, the effects of which are compounded by rapid urban growth. As a result, local governments face diverse and per-

[2]African countries that had not presented yet as of 2021 were Djibouti, Equatorial Guinea, Eritrea, Gabon, Guinea-Bissau, São Tomé and Príncipe, and South Sudan.

sistent challenges, ranging from an uneven administrative reach over urban territories, central governments reluctant to devolve mandates and power to the local level (especially when opposition parties govern cities), and weak fiscal capability, which limits the capacity of local and regional governments to respond to, engage with, and plan for development in a sustainable and integrated way (UCLG 2019).

The capacity of the sub-national state in Africa to drive SDG localization has taken another hit with the COVID-19 pandemic. Just like in other parts of the world, African cities have been the hotspots of COVID-19 cases. The resultant strain on healthcare systems and economic impact of the global economic slowdown have stretched and diverted already limited fiscal resources, exacerbated by the effects of stringent local lockdowns (Teachout and Zipfel 2020; Morsy et al. 2020). With the most vulnerable in society most affected, the impact of COVID-19 has in many ways slowed down, and in some cases reversed, important progress made on the achievement of global development goals. As a result, the devastating effects of the pandemic have especially been felt in the area of poverty, hunger, and inequality, with detrimental consequences for health, well-being, and economic growth (Sachs et al. 2021). Reflections and debates on the need to "build back better" in post-pandemic times, including a rethinking or recalibration of the SDGs themselves, are therefore timely and appropriate (Nature 2020). But, to a certain extent, debates on the pandemic and its recovery do not recognize that the daily reality of many cities across the world already is, and has been for a long time, that of structural and recurrent crisis. This requires us to challenge some existing assumptions around the workings and scope—including limits—of democracy, economy, and public action.

One of the key examples to illustrate the limits of existing assumptions is the centrality of informality in the African context, as representing the predominant source of economic livelihoods, modes of land tenure and housing construction, actors involved in service delivery, and much more. The prevalence of informality means that the actors and logics underpinning public action as included in global goals such as the SDGs work in very different ways, representing a much wider spectrum of actions and actors than just those of formal government. This spectrum includes non-state actors, ranging from traditional authorities to community-based organizations, religious organizations, neighborhood watch groups, associations, and small businesses, but also includes political parties, large corporations, as well as non-governmental or international organizations, who all play direct or indirect roles in the management and provision of services down to the community and neighborhood level.

The (re)thinking of ways to inform and accelerate local action going forward then urges a reflection on the ways in which these actors can be supported or at least taken into account as also part and parcel of the "city." Such a reflection needs to be accompanied by a better understanding of the roles that they play, the ways and degree to which these roles are regulated, how different actors interact with each other and the state, as well as the role of structural and global forces and influences in shaping them.

1.5 Toward an Approach to Advancing the SDGs Locally

An understanding of the complexity of the different actors and forces that make up African cities needs to be coupled with more sophisticated ways of generating knowledge. Many examples of new epistemological and methodological approaches for thinking and researching "wicked" problems of urban sustainability and development by approaching cities as interconnected systems have emerged in the post-2015 era (Bai et al. 2016; Elmqvist et al. 2018). This emergence involves a (re)appreciation of the role of (urban) science and calls for the need for transdisciplinary approaches to knowledge production that bridge the gap between science, policy, and practice (McPhearson et al. 2016; Acuto et al. 2018). Importantly, these calls have been accompanied by a recognition of the geopolitics around science and knowledge production and efforts to address existing imbalances related to the role and stakes around African urbanization and Africa's role as a knowledge producer (Nagendra et al. 2018; Marrengane and Croese 2021).

This book aims to contribute to this important emerging body of work by straddling both disciplinary and professional boundaries. While there has been a growing academic interest in the challenges and opportunities for the uptake of the SDGs in Africa (e.g., Nagao et al. 2019; Ramutsindela and Mickler 2020; Nhamo 2017; Nhamo et al. 2020, 2021), few studies approach this topic with a transdisciplinary urban lens. This book presents contributions from the city level, by bringing together authors from a wide range of urban actors and practitioners, including researchers, city officials, representatives of consultancy firms, local government associations, and international organizations who have been working at the forefront of SDG localization in Africa. Taken together, the chapters present critical views of existing challenges, but also practical insights into innovative practices, partnerships, and lessons learned. As such, the book speaks both to academic debates on the SDGs and to wider policy debates on the current status and future prospects for SDG localization in Africa.

Geographically, the chapters cover urban research and practice in a variety of African cities, ranging from large metropolitan city regions such as Cairo and Lagos; large cities such as Kampala, Accra, and eThekwini; smaller cities such as Lusaka; as well as towns in Burkina Faso, Mozambique, Guinea-Bissau, and Ethiopia, including the island city of Príncipe Island in São Tomé and Príncipe. Methodologically, the chapters offer local perspectives and experiences of what it takes to effectively localize the SDGs. Taken together, they illustrate the range of different tools and methods that can be employed at different stages of the process of SDG localization while also highlighting the importance of fostering the skills and expertise that are required to apply such methods on the ground.

Examples in the chapters include the creation of Stakeholder Working Groups for Voluntary Reporting on the SDGs (Chap. 2); SDG studios with community actors to generate local understandings of the SDGs (Chap. 3); participatory design and citizen science research as the basis for developing revised indicators to measure and improve urban housing and transport systems (Chap. 4); Learning Alliance Platforms to bring together different actors for participatory inquiry and knowledge production on urban water management (Chap. 5); multi-stakeholder platforms representing actors involved in urban waste recycling and repurposing (Chap. 6); new and cross-cutting disciplinary perspectives for advancing urban health (Chap. 7); local think tank groups to review the availability and management of SDG 11 data as the basis for community data collection (Chap. 8); comparative sub-national fiscal data analysis to assess prospects for financing sustainable development at the city level (Chap. 9); qualitative action experiments with urban planning professionals to address corruption at the local government level (Chap. 10); capacity and awareness-building as well as local data collection for and by city officials as the basis for the development of Voluntary Local Reviews on the SDGs (Chap. 11); the institutionalization of local integrated development planning for the SDGs at the city level (Chap. 12); and Participatory Incremental Urban Planning for the implementation of the SDGs and New Urban Agenda (Chap. 13).

While we have tried to cover a broad range of disciplines, actors, geographical areas, thematic issues, and methods, we recognize that this book is not an exhaustive account and that there are important themes and areas that are not, or not sufficiently, covered. This includes, among others, the area of education, culture, and social development, including the role and inclusion of the most vulnerable in society, such as women, children, and the elderly, but also members of the LGBTQI+ community, migrants, and the differently abled. Other areas that have not been fully addressed are that of the limits and trade-offs of economic development, including the role of private economic actors and technology, the environment and climate change, as well as the role of (party) politics in determining the prospects and dynamics of effective SDG localization. All of these issues require further research and amplification going forward.

1.5.1 Part I: Embedding SDG Localization in Local History, Meanings, and Context

The first part of the book consists of three chapters that all address the importance of the embeddedness of SDG localization in local history, meanings, and context. Chapter 2, in that regard, builds on the South African experience to highlight the importance of contextual knowledge that is necessary to translate the information that is reported at the global level on SDG progress into intelligence that builds on and supports locally available data, planning, and development efforts. Chapter 3 in turn builds on research conducted at the community level in the city of Kampala, Uganda, to call for the importance of the exploration of local meanings of the SDGs that are reflective of the vast diversity of urban contexts on the continent and can represent a critical counterpoint to the SDGs as a universal normative framework. Similarly, Chap. 4 offers a bottom-up approach to SDG localization by conducting grounded research in the housing and transport sectors at the neighborhood level in Cairo, Egypt, to shed light on the intricacies of the workings of informality as the basis for a revised set of SDG targets and indicators that can inform the measurement and improvement of local systems of service provision.

1.5.2 Part II: Integrated Perspectives on Water, Waste, and Health

The second part of the book brings together integrated perspectives on some of the key development sectors in Africa that play a crucial role in addressing a range of interconnected development goals. Chapter 5 does so by unpacking the challenges of water management in Accra, Ghana, and how these can be overcome by supporting more participatory and integrated approaches at the city level. Chapter 6 takes a similar approach by exploring the benefits of multi-stakeholder platforms for more sustainable urban waste management in the context of Ouagadougou, Burkina Faso. Chapter 7 offers unique insight into the importance of integrated approaches to SDG localization by introducing global surgery as a cross-cutting tool and perspective to advancing urban health in Africa in spatially and socially just ways.

1.5.3 Part III: Barriers or Opportunities—Data, Finance, Corruption

The third part of the book sheds light on the cross-cutting challenges for SDG localization in the areas of city-level data, finance, and corruption while outlining important transformative steps to be taken going forward. Chapter 8 does so by presenting an innovative approach for overcoming existing data inequalities reflected in existing urban data collection and management practices through the co-design and production of knowledge at the community level in the cities of Lagos, Nigeria, and Accra, Ghana. Chapter 9 draws on a unique data set on sub-national finance to shed light on the complexity of sub-national fiscal data and systems in order to work toward building public finance systems that can address urban financing needs. Chapter 10 constructively sheds light on the role of corruption in hampering development at the city level and builds on collaborative work with planning professionals in Lusaka, Zambia, to offer insight into prospects for addressing corruption through the promotion of professional integrity and collective accountability mechanisms.

1.5.4 Part IV: Collaborative Experiences from the Frontier of Practice

The final chapters of the book reflect experiences from the frontier of practice, illustrating the diversity of modalities of collaborative action for and around SDG localization across the continent, in spite of complex institutional contexts marked by limited data, resources, and capacity. Chapter 11 does so by providing unique insight into the role of global and national local government associations in supporting SDG localization in contexts of complex and uneven decentralization by building on the experience of the development of a collective Voluntary Local Review in Mozambique. Chapter 12 looks at some of the factors that have shaped the globally renowned approach of the South African city of eThekwini to SDG localization from the perspective of a city official responsible for integrated development planning. Finally, Chap. 13 provides a view of the support provided by UN-Habitat for the implementation of the SDGs and New Urban Agenda through participatory planning in the cities of Bissau, Guinea-Bissau, Príncipe Island, São Tomé e Príncipe, and Hawassa and Bahir Dar in Ethiopia.

References

Acuto M, Parnell S, Seto KC (2018) Building a global urban science. Nature Sustainability 1(1):2–4

African Union (AU) (2014) Common Africa position on the Post-2015 Agenda. https://au.int/sites/default/files/documents/32848-doc-common_african_position.pdf. Accessed 27 July 2021

African Union (AU) (2015) Agenda 2063: the Africa we want. Final popular version. https://au.int/Agenda2063/popular_version. Accessed 27 July 2021

Arajärvi N (2019) Including cities in the 2030 Agenda: a review of the post-2015 process. In: Aust HP, du Plessis A (eds) The globalisation of urban governance: legal perspectives on Sustainable Development Goal 11. Routledge, New York and London, pp 17-37

Bai X, Surveyer A, Elmqvist T, Gatzweiler FW, Güneralp B, Parnell S, Prieur-Richard AH, Shrivastava P et al. (2016) Defining and advancing a systems approach for sustainable cities. Current Opinion in Environmental Sustainability 23:69–78

Cartwright A, Palmer I, Taylor A, Pieterse E, Parnell S, Colenbrander S (2018) Developing prosperous and inclusive cities in Africa—National Urban Policies to the rescue? Coalition for Urban Transitions, London and Washington, DC

Dellas E, Carius A, Beisheim M, Parnell S, Messner D (2018) Local and regional governments in the follow-up and review of global sustainability agendas. Adelphi, Berlin; Cities Alliance, Brussels

Elmqvist T, Bai X, Frantzeskaki N, Griffith C, Maddox D, McPhearson T, Parnell S, Romero-Lankao P et al. (eds) (2018) Urban planet: knowledge towards sustainable cities. Cambridge University Press, Cambridge

Fox S, Macleod A (2021) Localizing the SDGs in cities: reflections from an action research project in Bristol, UK. Urban Geography. doi:https://doi.org/10.1080/02723638.2021.1953286

Fukuda-Parr S, McNeill D (2019) Knowledge and politics in setting and measuring the SDGs: introduction to special issue. Global Policy 10(Suppl.1):5-15

Fukuda-Parr S, Muchhala B (2020) The Southern origins of Sustainable Development Goals: ideas, actors, aspirations. World Development 126:104706

Global Taskforce of Local and Regional Governments (GTLRG) (2016) Roadmap for localising the SDGs: implementation and monitoring at the subnational level. https://www.global-taskforce.org/sites/default/files/2017-06/bfe783_434174b8f26840149c1ed37d8febba6e.pdf. Accessed 2 August 2021

Hoornweg D, Pope K (2017) Population predictions of the world's largest cities in the 21st century. Environment and Urbanization 29(1):195–216

Jönsson K, Bexell M (2021) Localizing the Sustainable Development Goals: the case of Tanzania. Development Policy Review 39:181–196

Kaika M (2017) 'Don't call me resilient again!': the New Urban Agenda as immunology… or … what happens when communities refuse to be vaccinated with 'smart cities' and indicators. Environment and Urbanization 29(1):89–102

Kanuri C, Revi A, Espey J, Kuhle H (2016) Getting started with the SDGs in cities: a guide for stakeholders. United Nations Sustainable Development Solutions Network. https://irp-cdn.multiscreensite.com/be6d1d56/files/uploaded/9.1.8.-Cities-SDG-Guide.pdf. Accessed 27 July 2021

Klopp JM, Petretta DL (2017) The urban Sustainable Development Goal: indicators, complexity and the politics of measuring cities. Cities 63:92–97

Kosovac A, Acuto M (2020) Acknowledging urbanization: a survey of the role of cities in UN frameworks. Global Policy 11(3):293-303

Leitner H, Sheppard E, Webber S, Colve E (2018) Globalizing urban resilience. Urban Geography 39:1276-1284

Macamo E (2019) E se a África rejeitasse os Objectivos de Desenvolvimento Sustentável? O Público, 1 August. https://www.publico.pt/2019/08/01/mundo/opiniao/africa-rejeitasse-objectivos-desenvolvimento-sustentavel-1881616. Accessed on 21 July 2021

Malonza JM, Ortega AA (2020) Fissures in localizing urban sustainability: the case of Rwanda. GeoJournal. doi:https://doi.org/10.1007/s10708-020-10239-8

Marrengane N, Croese S (2021) Reframing the urban challenge in Africa: knowledge co-production from the South. Routledge, London and New York

McPhearson T, Parnell S, Simon D, Gaffney O, Elmqvist T, Bai X, Roberts D, Revi A (2016) Scientists must have a say in the future of cities. Nature 538(7624):165–166

Morales G, Bromaghim E, Kim A, Diamond C, Maggini A, Everhart A, Gruskin S, Chase AT (2021) Classroom walls and city hall: mobilizing local partnerships to advance the Sustainable Development Agenda. Sustainability 13:6173

Morsy H, Balma L, Mukasa AN (2020) 'Not a good time': economic impact of COVID-19 in Africa. Working paper series no. 338. African Development Bank, Abidjan, Côte d'Ivoire

Nagao M, Masinja J, Alhassan A (eds) (2019) Sustainable development in Africa: case studies. Spears Media Press, Denver

Nagendra H, Bai X, Brondizio ES, Lwasa S (2018) The urban south and the predicament of global sustainability. Nature Sustainability 1:341–349

Nature (editorial) (2020) Time to revise the Sustainable Development Goals. Nature 583:331-332

Nhamo G, Togo M, Dube K (eds) (2021) Sustainable Development Goals for society, vol. 1: selected topics of global relevance. Sustainable Development Goals series. Springer Publishing, Cham

Nhamo G, Odularo GOA, Mjimba V (eds) (2020) Scaling up SDGs implementation: emerging cases from state, development and private sectors. Springer Publishing, Berlin

Nhamo G (2017) New global Sustainable Development Agenda: a focus on Africa. Sustainable Development 25:227-241

Parnell S (2016) Defining a global urban development agenda. World Development 78:529-540

Ramutsindela M, Mickler D (eds) (2020) Africa and the Sustainable Development Goals. Sustainable Development Goals series. Springer Publishing, Cham

Revi A, Simon D, Parnell S, Elmqvist T (2014) Consultation with the United Nations Open Working Group on the SDGs' urban SDG goal 11: targets & indicators. Confidential draft for review and discussion, London, 22-24 August 2014

Sachs J, Traub-Schmidt G, Kroll C, Lafortune G, Fuller G (2021) Sustainable development report 2021: the decade of action for the Sustainable Development Goals. https://s3.amazonaws.com/sustainabledevelopment.report/2021/2021-sustainable-development-report.pdf. Accessed 5 August 2021

Satterthwaite D (2017) Will Africa have most of the world's largest cities in 2100? Environment and Urbanization 29(1):217-220

Teachout M, Zipfel C (2020) The economic impact of COVID-19 lockdowns in Sub-Saharan Africa. Policy brief. International Growth Centre. https://www.theigc.org/wp-content/uploads/2020/05/Teachout-and-Zipfel-2020-policy-brief-.pdf. Accessed 5 August 2021

United Cities and Local Governments (UCLG) (2019) Africa: regional chapter. In: The localisation of the global agendas: how local action is transforming territories and communities. Fifth global report on decentralization and local democracy. United Cities and Local Governments, Barcelona. https://www.gold.uclg.org/sites/default/files/goldv_en.pdf. Accessed 15 August 2021

United Cities and Local Governments (UCLG) (2015) The Sustainable Development Goals: what local goverenments need to know. United Cities and Local Governments, Barcelona. https://www.uclg.org/en/media/news/sustainable-development-goals-what-local-governments-need-know. Accessed 27 July 2021

United Cities and Local Governments Africa (UCLG Africa), Cities Alliance (2018) Assessing the institutional environment of local governments in Africa, 3rd edn. United Cities and Local Governments Africa, Rabat; Cities Alliance, Brussels

United Nations (UN) (2016) New Urban Agenda: Quito declaration on sustainable cities and human settlements for all (71/256). United Nations General Assembly, New York

United Nations (UN) (2015a) Sendai Framework for Disaster Risk Reduction 2015–2030. https://www.preventionweb.net/files/43291_sendaiframeworkfordrren.pdf. Accessed 6 August 2021

United Nations (UN) (2015b) Addis Ababa Action Agenda. Third international conference on financing for development, Addis Ababa, 13-16 July 2015. https://www.un.org/esa/ffd/wp-content/uploads/2015/08/AAAA_Outcome.pdf. Accessed 6 August 2021

United Nations (UN) (2013) A new global partnership: eradicate poverty and transform economies through sustainable development. Report of the High-Level Panel of eminent persons on the Post-2015 Development Agenda. United Nations, New York. https://www.un.org/sg/sites/www.un.org.sg/files/files/HLP_P2015_Report.pdf. Accessed 26 July 2021

United Nations Department of Economic and Social Affairs (UNDESA) (2021a) Global guiding elements for Voluntary Local Reviews (VLRs) of SDG implementation. https://sdgs.un.org/sites/default/files/2020-10/GlobalGuidingElementsforVLRs_FINAL.pdf. Accessed 6 August 2021

United Nations Department of Economic and Social Affairs (UNDESA) (2021b) Overview of Voluntary National Reviews and the High-Level Political Forum on sustainable development: evolving in the decade of action and delivery for sustainable development.

United Nations Department of Economic and Social Affairs (UNDESA): Population Division (2018) World urbanization prospects: the 2018 revision (ST/ESA/SER.A/420). https://www.uneca.org/sites/default/files/TCND/ARFSD2021/Presentations/MGoS/Overview%20of%20voluntary%20national%20reviews%20and%20the%20high-level%20political%20forum%20on%20sustainable%20development%20Evolving%20in%20the%20Decade%20of%20Action%20and%20Delivery%20for%20sustainable%20development%20-%20Tonya%20Vaturi.pdf. Accessed 26 July 2021

United Nations Economic Commission for Africa (UNECA) (2017) Africa sustainable development report: tracking progress on Agenda 2063 and the

Sustainable Development Goals. United Nations Economic Commission for Africa, Addis Ababa

United Nations Framework Convention on Climate Change (UNFCC) (2015) Paris Agreement. https://unfccc.int/files/meetings/paris_nov_2015/application/pdf/paris_agreement_english_.pdf. Accessed 6 August 2021

UN-Habitat (2014) The state of African cities: re-imagining sustainable urban transitions. United Nations-Habitat, Nairobi

Valencia SC, Simon D, Croese S, Nordqvist J, Oloko M, Sharma T, Buck NT, Versace I (2019) Adapting the Sustainable Development Goals and the New Urban Agenda to the city level: initial reflections from a comparative research project. International Journal of Urban Sustainable Development 11(1):4-23

Weber H (2017) Politics of 'leaving no one behind': contesting the 2030 Sustainable Development Goals Agenda. Globalizations 14(3):399-414

Part I

History, Ideas, Context

Check for updates

2 The Sustainable Development Goals in South Africa: Transitions Are the Hardest Part

Alexis Schäffler-Thomson

Abstract

Building on the lessons learned by the Millennium Development Goals (MDGs), the Sustainable Development Goals (SDGs) have generated a sophisticated data reporting structure designed to monitor new goals that were introduced, as well as the interactions across goals. The expanded SDG framework requires localization, and the most useful channel for reporting is the Voluntary National Review (VNR), which holds important insights into how countries deal with goal integration and sub-national dynamics. This chapter reflects on how the VNR process unfolded in South Africa for SDG 6, 14, and 15, which are key environmental resources goals that have prominence in the country. These three goals show the extent to which goal interconnections and downscaling issues are understood by the VNR custodians in South Africa and thus highlight what is needed institutionally for achieving this kind of understanding. The chapter argues that the preparation of the VNR highlights the capacity of a government to generate and assess complex data for the VNR, providing a proxy for how well the country can manage the interconnected drivers of transformation across the SDGs.

Keywords

SDG 6 · SDG 14 · SDG 15 · Voluntary National Review (VNR) · SDG reporting · SDG interactions · South Africa

2.1 Introduction

The Sustainable Development Goals (SDGs) expanded on the Millennium Development Goals (MDGs) by increasing the number goals for reporting and emphasizing that the relationships between the 17 goals need to be analyzed in terms of local contexts. The sophisticated data and indicator framework necessitated by the SDGs is ideally matched by a highly consultative Voluntary National Review (VNR) process involving civil society, private sector, and local bodies to drive uptake and context-specific application of the goals by diverse actors and contexts (Klopp and Petretta 2017). The core channel used by the United Nations High-Level Political Forum (UN HLPF) framework for reporting on SDG implementation is the VNR, which holds important insights about a country's ability to apply the SDG framework in local data contexts and understand strategies for achieving the SDG at a local level.

A. Schäffler-Thomson (✉)
Pegasys Strategy and Development,
Cape Town, South Africa
e-mail: alexis@pegasys.co.za

S. Croese, S. Parnell (eds.), *Localizing the SDGs in African Cities*, Sustainable Development Goals Series, https://doi.org/10.1007/978-3-030-95979-1_2

Reflecting on South Africa's first VNR, this chapter interrogates the country's capacity to arrive at a locally framed understanding of the SDGs and of the necessary institutional and reporting infrastructure. Evidence is specifically drawn from the VNR goal reports for SDG 6 (Clean Water and Sanitation), SDG 14 (Life Below Water), and SDG 15 (Life on Land), key environmental resources goals for South Africa. Although not explicitly urban framed, the extent to which the VNR goal reports of SDG 6, SDG 14, and SDG 15 dealt with issues of integration, and made provisions for sub-national contexts, gives important insights into the prospects for SDG localization.

The chapter draws on the author's experience as a consultant in the VNR of SDG 6, 14, and 15, the reporting of which formed part of South Africa's VNR Country Report presentation to the UN HLPF in 2019.[1] The VNR for the SDGs in South Africa was led by Statistics South Africa (Stats SA), the country's data custodian, in partnership with government, academia, civil society, and non-governmental organizations (NGOs).

2.2 The Global SDG Transition in Brief

The SDGs are part of a post-2015 agenda that aims to expand on, and address, the shortcomings of the former UN development era encapsulated by the MDGs. The MDGs were aimed at reducing extreme poverty and improving the lives of the poorest while engendering economic development in a sustainable environment (UNDP 2013). The MDG era was guided by 8 goals, 21 targets, and 63 indicators, established following the Millennium Summit of the United Nations in 2000, as an output of the United Nations Millennium Declaration (Lomazzi et al. 2014, p. 1).

The MDGs are viewed as an important foundation for global development reporting in providing a straightforward framework that distilled many complex on-the-ground issues (Harcourt 2005). The use of simple metrics, such as counting taps and toilets, was seen as providing clear signals to policy makers about the state of basic development metrics, such as water and sanitation (Herrera 2019). The precipitating context of the post-Cold War years saw governments facing increasing scrutiny about failed promises and development seeking a clear purpose to mobilize public support (Harcourt 2005; Fukuda-Parr 2016).

As such, the MDG goals and targets were mostly relevant for developing countries and hence tagged as the "minimum development goals" (Harcourt 2005). Criticism is also pointed at the top-down development agenda embodied by the MDGs, particularly because the goals were drafted with limited consultation of local stakeholders, such as city governments or civil society organizations (Fukuda-Parr 2016; Klopp and Petretta 2017). The extent to which local development priorities were integrated in the MDGs was therefore questioned, and because local governments—who often lacked resources for implementation—were largely missing in the reporting process overall, this highlighted the challenges in appropriate targeting and progress (Klopp and Petretta 2017).

The attraction of the MDGs in being a simple framework was thus also their vice, leading to significant over-aggregation and lack of nuance due to the limited "development" agenda that tended to focus on simplistic poverty eradication metrics, as well as the top-down design of the MDG drafting and reporting process. In terms of the scope limitations, two areas are relevant for the purposes of this chapter. First, the MDGs failed to address environmental issues in any substantive form, barring MDG 7 that obliquely measured progress toward "environmental sustainability." Progress on MDG 7 was assessed in terms of integration of sustainable development into country policies and programs and the reversal of environmental losses (MDG Achievement Fund 2019). The second major scope limitation

[1] The reflections presented here are independent findings, and the author wishes to acknowledge VNR Goal Report co-authors, Amanda Nyingwa (SDG 6) and Hannah Benn (SDG 14), as well as the University of Pretoria (UP) as the overall country report editor.

of the MDGs relates to the lack of clear engagement with cities and the dynamics of urban systems, in terms of social, economic, and environmental functionality (Parnell 2016). Although the MDGs did deal with urban issues, it was from a slum eradication perspective, in order to "[achieve] by 2020 a significant improvement in the lives of at least 100 million slum dwellers" (Klopp and Petretta 2017). As the urban poor were also not involved in interventions designed to assist them (Hasan et al. 2005), the issues around targeting were especially acute and left many gaps in both understanding urban issues and engaging sub-national, and local-level, capacity building.

In the run up to the 2015 MDG deadline, a set of new global processes unfolded, leading to the development of the 2030 Agenda for Sustainable Development and expansion to the 17 SDGs. With origins in the Rio+20 conference, the SDGs made a specific call to countries to reduce inequality and promote prosperity while protecting the planet, which is perhaps the most significant breakaway from the MDGs. Scope was significantly increased by emphasizing that ending poverty must go hand in hand with strategies that build economic growth and address social needs while tackling climate change and environmental protection (UN 2019). The scope expansion took form in the significantly enlarged suite of 17 SDGs, 169 targets, and (initially) 230 unique indicators, with a cross-cutting focus on integrating sustainable development into all country policies and programs. The universally applicable goals are driven by an explicit recognition that sustainable development hinges on climate change, ecological limits, and planetary boundaries (Parnell 2016; Klopp and Petretta 2017).

Greater specificity on the environmental metrics of the SDGs signaled a vastly different approach to their predecessors. SDG 6 advanced from MDG Target 7c to present a thematically broader water framework focused on safe water supply, sanitation and hygiene, degradation of ecosystems, and sustainable water management for social development and economic growth. Where marine and coastal issues had limited representation under MDG 7, SDG 14 explicitly deals with focus areas necessary for addressing ocean-related issues including sustainable fisheries, ocean pollution, and climate change impacts. In terms of ecosystems, SDG 15 also emerged to specifically advance MDG 7 in terms of terrestrial ecosystems and ecosystem services by prioritizing sustainable forest management, combating desertification, land degradation, and biodiversity loss.

Sustainable cities and communities also received a spotlight in SDG 11, which offered for the first time a global urban policy position in dealing with the overall social economic and environmental functionality of cities and the urban system (Parnell 2016). SDG 11 represented the wider shift in development rhetoric, in which urban settlements have a tenuous position, contributing to many of the environmental challenges taken on by the sustainability community while also being key sources of innovation, capacity, and investment necessary for any substantive transition (Swilling and Annecke 2006; Schäffler and Swilling 2013; Pieterse et al. 2018). With the attendant New Urban Agenda (NUA) adopted at Habitat III, SDG 11 thus underscored the important premise that cities and their communities are key arenas of problem-solving and urban movements and processes therefore need to be prioritized as agents in the decision-making process (Revi 2016).

SDG 11 also includes key sustainability metrics such as access to sustainable transport systems and reduction in waste and air pollution, and a number of enabling factors for sustainable human settlements are also included within this SDG. Although the environmental resource metrics of SDG 6, 14, and 15 are not "urban" framed, in terms of SDG 11, the extent to which countries understand such interconnections and the role of local governments for achieving them is a key principle of the SDGs (Sarwar and Nicolai 2018). The United Nations Development Programme (UNDP) developed the Rapid Integrated Assessment (RIA) tool to help countries understand the interlinkages across targets and more broadly determine relevance of the SDGs to their country context at national and sub-national levels (UNDP 2017).

Putting countries—and national and sub-national players in particular—in the driving seat of SDG reporting was a fundamental shift toward localization and was emboldened by Agenda 2030's mantra of "leaving no one behind." Localization is described as "the process of defining, implementing, and monitoring strategies at the local level for achieving global, national, and sub-national sustainable development goals and targets" (UN-Habitat, UNDP and UCLG 2014). A key outcome was that to address the limitations of the MDGs, the *Guidelines to support Country reporting on the Sustainable Development Goals* began from the premise that all reporting needs to acknowledge that countries differ in their priorities and capacities. In addition, it acknowledged that governments and public institutions are to work closely on implementation with regional and local authorities, sub-regional institutions, international institutions, academia, philanthropic organizations, and volunteer groups (UNDP 2017).

In seeking to shape localization efforts, the VNR process was put forward as the channel through which countries track and report on their progress with the SDGs. Countries undertake to present their VNR at the UN HLPF which sits every 2 years, under the auspices of the UN Economic and Social Council (ECOSOC) (IISD 2019). Although country contexts differ, UN member states are advised to prepare their VNRs through an inclusive process that engages entities from government, civil society, the private sector, and academia (Global Alliance 2020). Created as a country-led platform, enabling countries to work on advancing the SDGs in ways that are grounded in national realities, the VNRs are viewed as an "important innovation" of the United Nations in recognizing that "country ownership" is central to implementing Agenda 2030 (Fukuda-Parr et al. 2018).

The VNR process is intended to facilitate learning and is "not an end, but a means" to exchange national experiences and challenges, promote accountability to citizens, and mobilize support to accelerate implementation (IISD 2018). The incentive is also in helping to reveal challenges at the domestic level, enhance cooperation, reduce silos, and identify gaps for support (UN General Assembly 2016). The format of the VNR process consists of national consultation, regional meetings, main messages summarizing countries' key findings, as well as the presentation of the report at the UN HLPF (Fukuda-Parr et al. 2018).

As shown in Fig. 2.1, in order to submit a VNR, National Statistical Offices (NSOs) and their member states need to produce national SDG reports, which undergo several reviews that are in turn overseen by global regional and thematic bodies, as well as goal-specific engagements. While the UN HLPF is the overall global reviewing body, through the UN General Assembly, the lead and convening institutions need to follow the global reporting methodology and undertake internal review protocols for alignment and quality assurance. At the same time, the NSOs and member states need to mobilize stakeholders to ensure "no one is left behind." This includes, for instance, those who hold the primary data or can advise on the appropriate means of domestication, as discussed in Sect. 2.4. At a high level, the reporting process presented in Fig. 2.1 therefore consists of several layers of strategic engagements between political leaders, decision-makers, civil society, academia, and the general public, further to which reporting teams need to engage contextual issues in how the SDGs play out.

Instead of being an accountability mechanism, the VNRs offer a shared learning platform with significant potential for local stakeholders to identify synergies and address difficult issues such as trade-offs (Kindornay 2018). Analyses highlight that the consultative process of VNRs can afford a lot of detail to the scope, length, and quality of the SDG reporting process. In countries such as Ghana, outreach efforts and media campaigning via radio, press, public campaigns, or awareness raising in schools have been core components (UCLG 2019). The creation of Stakeholder Working Group (SWG) platforms and workshops also allows for collaboration between authors (typically technical goal experts) and civil society organizations, organized labor, and the private sector.

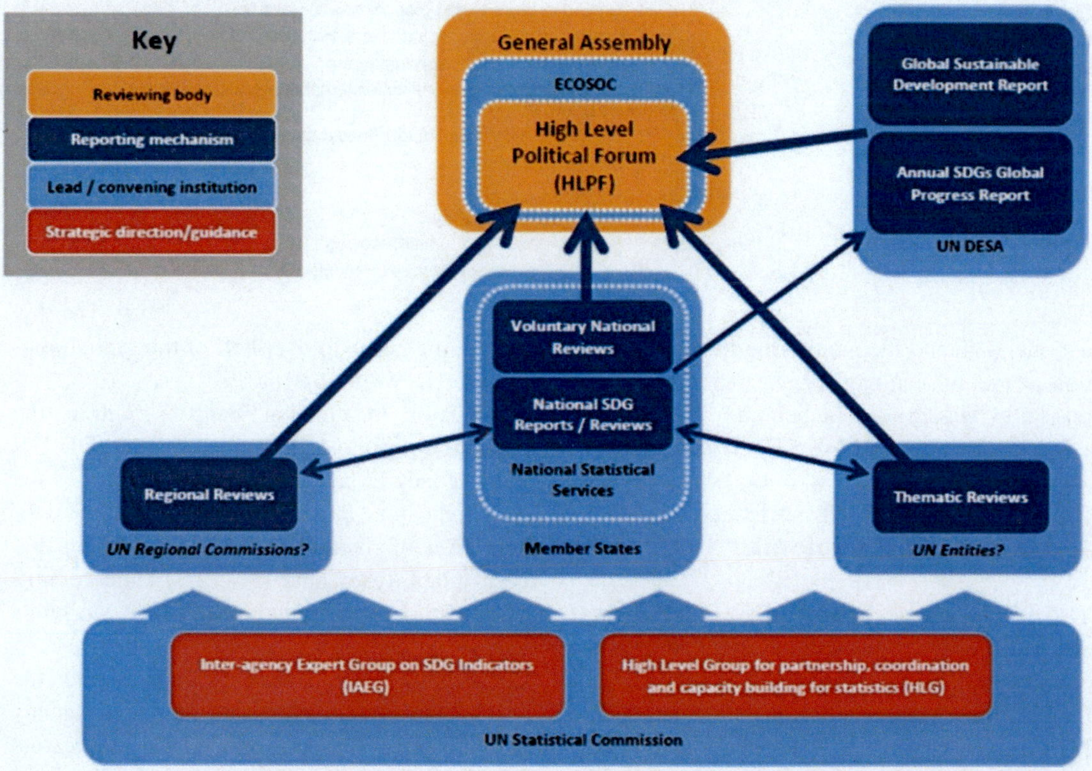

Fig. 2.1 Ultimate reporting into the High-Level Political Forum housed within the UN General Assembly (UN OIOS 2017)

However, whether the SWG workshops are sufficiently attended, and manage to gain the most from the engagement process, is a critical determinant of whether localization efforts are proving successful. Somewhat ironically, the VNRs also tend to be very long processes, and observers note that processes are perhaps unnecessarily detailed (Kindornay 2018). A key challenge of the reporting format is thus how to achieve localization with efficacy, given the extent of stakeholder engagement required.

There are also many concerns about the level of inclusivity of VNR processes, raising questions about the extent to which localization is truly supported by a country VNR. For instance, many countries have lacked clarity about how marginalized groups such as women, youth, and people with disabilities have been effectively engaged in the VNR process, questioning the true extent to which "no one has been left behind" (Sarwar and Nicolai 2018). Criticisms have also pointed to the country-led format which allows countries to present selected results, thus showing an incomplete view of all the SDGs (Leiva and Rahmaty 2019).

In terms of addressing trade-offs, and finding solutions to key sticking points, independent reviewers argue that the VNRs tend to pay little attention to identifying concrete mechanisms to address difficult and often politically sensitive trade-offs and the VNRs may thus simply reflect reporting protocols (Fukuda-Parr et al. 2018). Of South Africa's VNR, observers have pointed to the lack of concrete guidelines or mechanisms to coordinate efforts around SDG planning, monitoring, or reporting at the local government level (Croese 2019). Overall, the participation of local and regional governments in VNR processes across the African continent continues to be uneven at best and lacking at worst (UCLG 2019). Without sufficient local engagement, the VNR will undermine the collection of sub-

Fig. 2.2 Ten principles for global monitoring indicators (UN SDSN 2015a)

Ten principles

1. Limited in number and globally harmonized
2. Simple, single-variable indicators, with straightforward policy implications
3. Allow for high frequency monitoring
4. Consensus based, in line with international standards and system-based information
5. Constructed from well-established data sources
6. Disaggregated
7. Universal
8. Mainly outcome-focused
9. Science-based and forward-looking
10. A proxy for broader issues or conditions

national data and the monitoring of implementation at the sub-national level, areas that are all critical to "leaving no one behind."

2.3 Indicators, Classifications, and Methodologies

The SDGs are accompanied by an Indicator and Monitoring Framework developed to assist countries transform the targets into a management tool for implementation, informed by the MDG shortcomings and lessons. The underpinning methodology sets out to address technical issues in the ways countries report, such as filling gaps in available indicators; harnessing new, innovative sources of data; and moving toward annual monitoring.[2] The principles for global monitoring call on countries to engage with data availability, shared or public access to data, and applicability to context (UN SDSN 2015a).

Overall, the SDG reporting framework represents an attempt to harmonize global data with countries' needs to ensure that data is collected to standard while reflecting local needs and priorities (UN SDSN 2015b). The ability of indicators to be "localized"—within Principle 7 of the monitoring framework (see Fig. 2.2)—is thus central and viewed as a means to encourage active implementation of the agenda within sub-national levels of government, such as cities, comprising home to over half of the global population (UN SDSN 2015a).

Indicator localization happens within the reporting methodology, decided by the SDG UN Inter-Agency Expert Group, which classifies the SDG indicators into three tiers on the basis of their level of methodological development and the availability of data (Fig. 2.3). The country classification is an outcome of close workings between NSOs and statistical agencies of International Organizations (IOs) around the world, as well as members of the Inter-Agency and Expert Group on SDG Indicators (IAEG-SDGs).

The SDG indicators were classified according to whether an indicator is the same as a global indicator. An indicator that is set by the UN is called an unmodified SDG indicator. An indicator accompanied by an uppercase D is domesticated and therefore a national proxy. In other words, a domesticated indicator is aligned to an SDG indicator, but adapted to make sense for the country context. There are further disaggregations that are possible depending on the category. In the case of semi-arid regions in South Africa, for example, forested cover (within SDG 15) is not a relevant measure, and thus a specific matrix was developed as a national proxy to measure those indicators.

In 2020, the UN Statistical Commission (UNSC) met for its 51st session following the Comprehensive Review of the global indicator framework. The Comprehensive Review focuses on refinements, revisions, replacements, additions, and deletions to the indicator framework to guide countries, many of which have advanced national frameworks and reporting platforms for the SDGs (UNSC 2020). The general guiding

[2]Annual monitoring is deemed particularly crucial if the SDG indicators are to serve as a management tool, informing national planning and budgetary processes, as well as global follow-up (UN SDSN 2015a).

Tier I: Indicator is **conceptually clear**, has an internationally **established methodology and standards** are available, and **data are regularly produced** by countries for at least 50% of countries and of population in every region where the indicator is relevant.

Tier II: Indicator is conceptually clear, has an internationally established methodology and standards are available, but **data are not regularly produced by countries.**

Tier III: No internationally established methodology or standards are yet available for the indicator, but methodology/standards are being (or will be) developed or tested.

Fig. 2.3 SDG tier classification as of time of research in January 2019 (Adapted from Indicators and a Monitoring Framework for the Sustainable Development Goals, UN SDSN 2015a)

principles of the Comprehensive Review include the following (UNSC 2020):

- The review needs to take into account investments already made at the national and international levels and should not undermine ongoing efforts.
- The revised framework should not significantly impose an additional burden on national statistical work.
- There should be space for improvements while at the same time ensuring that the changes are limited in scope and the size of the framework remains the same.
- The focus of the group's common work should remain on the implementation of the framework in countries for the achievement of the SDGs.

A key outcome of the examination by the IAEG-SDGs was the elimination of Tier III indicators that were either deleted or promoted to Tier II (UNESCO Institute of Statistics 2020). As of 17 July 2020, the updated tier classification table contained 123 Tier I indicators, 106 Tier II indicators, and 2 indicators that have multiple tiers (different components of the indicator are classified into different tiers) (IAEG-SDGs 2020). The updates sought to address the overburdening of national statistical systems, coordination challenges, and fragile statistical capacities, which are underscored by funding and resource issues (Jütting et al. 2019). Such issues apply most acutely to the African context, which, as a result, still faces significant data gaps that need to be addressed for SDG monitoring and reporting (Borel-Saladin 2017; Woolfrey 2020). While South Africa performs well compared to many other African countries, the budget freeze instituted on Stats SA since is a striking case in point, severely limiting the country's ability to produce accurate official statics and reliable planning data (Stats SA 2020).

The mobilization of stakeholders in the VNR process is therefore critical, not only for country ownership of the process but also to leverage existing capacity to drive participation, support data collection, and engage in constructive science policy dialogues (Fourie 2019). In so doing, the VNR is potentially the most useful reporting platform for showing how effectively governments have created the institutional architecture and platforms required to do justice to SDG integration and localization.

2.4 South Africa's VNR Process

As a signatory to *Transforming Our World: The 2030 Agenda for Sustainable Development*, South Africa committed itself to implement and report on policies and strategies that advance the 2030 Agenda in a manner that is both transformative and sustainable. In 2010, the South African Cabinet endorsed Stats SA as the focal point for reporting and instructed Stats SA to set up and institutionalize the coordinating processes to support the SDG reporting process. Additionally, as a focal point, Stats SA set out to strengthen statistical coordination, as prescribed in the Statistics Act, No. 6 of 1999, besides its usual business of the production of statistics (Stats SA 2019a).

To produce the SDG Country Report, Stats SA opted to produce a variety of goal reports aimed at providing the basis for the synthesis of a Country Report to be presented as part of the VNR described above. It is important to appreciate the SDG Country Report as an outcome of a process that would already have produced a number of goal reports synthesized into a single SDG Country Report. The voluntary reporting process was formally initiated by South Africa in 2017, framed to transition from the MDGs and to ultimately produce the 2019 Country Report of national progress in achieving and reporting on the SDGs at the UN HLPF.

Stats SA engaged several sector, research, and academic experts to assist with the goal report production. The VNR was explicitly framed as a consultative process with private sector, civil society, and other relevant stakeholders to report on the targets and indicators for each SDG. As a feeder into the country report, each goal report set out to cover the following areas:

- Review the current *state of the sector* (goal).
- *Review the policies and intervention* strategies employed or being considered by key government institutions toward achievement of the relevant SDG targets.
- Review the *level of implementation* of the said policies and strategies.
- Review the *key challenges* to the achievement of the SDG targets for each goal.
- Suggest possible or alternative policies and strategies to *fast-track* the achievement of the goals, where necessary.
- Suggest, for each of the goals, *key priority activities* that government needs to pay attention to.
- Suggest formation of *strategic partnerships* with other development stakeholders in the country, including with civil society organizations, organized labor, and private sector.
- Suggest possible strategies for *mainstreaming* SDGs into the national development agenda.

In terms of area (i), the output was an indicator review comprising explanatory target and indicator tables in terms of the global classification and the progress of South Africa on the monitored target. The progress is reported on achievements toward the goal and reporting progress, as shown in Table 2.1.

The indicator review quickly produced insights about goal discrepancies, in terms of both achievement progress and reporting progress. At a high level, reporting on SDG 6 has had a substantial head start, largely due to an emphasis within the MDGs on water and targeted water programs and government activities in South Africa. As a result, South Africa was quickly able to align with the SDG monitoring framework, in terms of developing the appropriate method and collating the appropriate data, which may mean making a provision for domestication or disaggregations. Indicators that did have some, but not all, reported data implied South Africa is making reporting progress. But depending on the outcome and the target period, additional work was necessary to assess the outcome of whether the global umbrella target was in sight.

The data used to review progress on the goals were formally channeled through Stats SA, as the country NSO, in consultation with the Stakeholder Working Group (SWG). All principles of the reporting were followed by Stats SA (see Fig. 2.2), which was responsible for computing, measuring, and sourcing the data. The SWG comprised several stakeholders from government institutions, research bodies, and civil society organizations (CSOs) to ensure the VNR process was consultative and participatory. The SWG consultations took place in five iterations during the VNR process and essentially aimed to vet the data collected and the accuracy of reporting.

While the NSO body is ultimately responsible for all data reporting, the underlying metadata are, in many cases, sourced from government departments and other national bodies. In the case of SDG 15, the South African National Biodiversity Institute (SANBI) had a large role in collecting and collating metadata for various ecosystem indicators, although Stats SA was the main channel for receiving these data. Through

Table 2.1 Target and indicator reporting table—extract from SDG 15 (Adapted from Stats SA 2019b)

Target	SDG indicator	Reported indicator	Indicator type	Tier classification	Reported: status	Not reported: reason
15.3 By 2030, combat desertification; restore degraded land and soil, including land affected by desertification, drought, and floods; and strive to achieve a land degradation-neutral world	15.3.1 Proportion of land that is degraded over total land area		SDG indicator	Tier II	Making progress but further data needed to assess 2030 outcome	
15.4 By 2030, ensure the conservation of mountain ecosystems, including their biodiversity, in order to enhance their capacity to provide benefits that are essential for sustainable development	15.4.1 Coverage by protected areas of important sites for mountain biodiversity	15.4.1D Percentage of mountain ecosystem types that are well-represented in protected areas	Domesticated	Tier I	Some progress but further data needed to assess 2030 outcome	
	15.4.2 Mountain Green Cover Index		Not reported	Tier I	Unable to formally report	No national data available for reporting

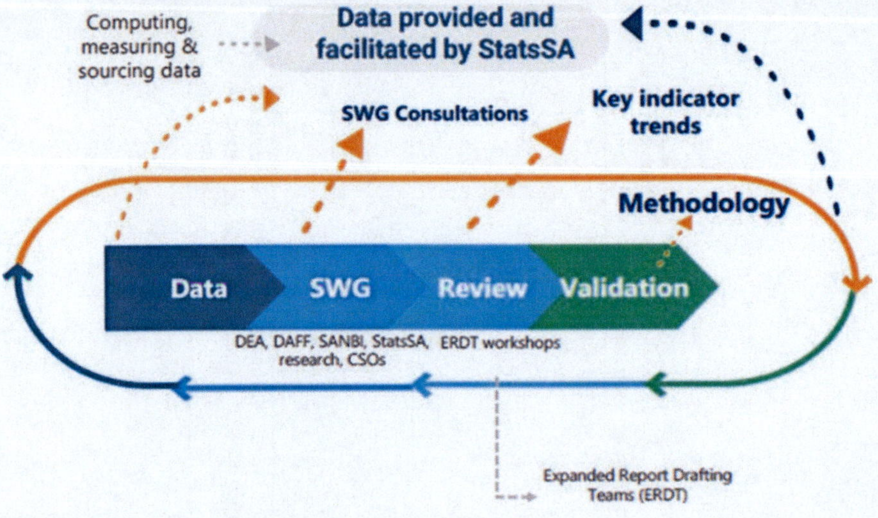

Fig. 2.4 Author reflection of Voluntary National Review (VNR) feedback loops adapted from SDG report writing workshop, 18–19 December 2018 (Stats SA 2018)

the SWG for SDG 15, the report authors were to work closely with SANBI—as the stakeholder closest to the metadata—to better understand key dynamics for the wider national review and in turn to vet the results with other relevant stakeholders.

The Department of Environmental Affairs, Forestry and Fisheries (DEFF [3]) was also closely involved in some metadata, although largely from a policy review angle. As mentioned, it is up to the NSO's discretion as to how the VNR is structured. Stats SA input included in the goal reports a review of the policies and intervention strategies employed or being considered by key government institutions toward achievement of the relevant SDG targets. Barring a few indicators where the DEFF provided data, engagement with the department helped authors to analyze the indicators pertaining to South Africa's enabling conditions or support for terrestrial ecosystem and review the implementation of this support.

The data were reviewed in a sequence of workshops where all SDG report authors come together and assessed the monitoring and reporting of all 17 goals. The monitoring and reporting methodology was then validated by the Expanded Report Drafting Team (ERDT) that includes the NSO, report authors, and an appointed overall editorial author (functioning in an editorial-cum-synthesis-type role) reporting on the results at goal, thematic, and country levels. The upstream synthesis and validation methodology involved analyzing the data according to each goal target and relevant indicators and consulting the relevant sources (government departments, government agencies, research institutions, and CSOs) regarding the validity of the results (Fig. 2.4).

The highly iterative SWG process provided critical feedback loops to available data and essential nodes in goal integration and downscaling. For SDG 15, the collaboration between the metadata providers—SANBI and key government departments, including DEFF—together with public agencies and research partnerships continues and plays a key role in the recent reporting refinements and tier status revision at global level. In particular, SANBI has played a fundamental role, not only in filling baseline data gaps but also in the better use and collation of evolving data such as those collected in the National Biodiversity Assessment (NBA) (Nel and Driver 2012; Sebola 2017; Khatieb 2019).

For SDG 15, the SWG provided an invaluable channel to mediate between the global reporting framework and key data collection processes

[3]Previously, and at the time of the VNR, Department of Environmental Affairs (DEA).

such as the International Union for Conservation of Nature (IUCN) Red List of Threatened Species, known as the Red List Index (RLI) databases, collation of which was underway by SANBI during VNR. Although this meant the RLI was unable to be formally reported upon in the earlier prescribed reporting intervals (around February 2019), set by Stats SA, the report authors leveraged the SWG interaction and were able to engage with SANBI to view the incoming metadata for analysis of the indicator 15.5.1, which is a critical reflection of species trends. Through the work of SANBI, indicator 15.5.1 was ultimately included in the VNR, which demonstrated South Africa is not on track with addressing species loss.

The SWG for SDG 14 is equally informative about the importance of establishing platforms and partnerships for engaging with data for thorough SDG reporting. South Africa was only able to report on two of the ten targets for SDG 14, and thus significant foundational reporting infrastructure was needed. The VNR reporting format guided by Stats SA and the Country Report Office was therefore to focus on those interactions that were either strongly disabling or strongly enabling, drawing some important policy implications. For example, the SWG had to consult on the implications of pursuing certain energy sources on ecosystem goals. The SDG 14 SWG considered the effects of the significant geography of South Africa's coastal zone (more than 150,000 km) under application for prospecting rights after the discovery of natural gas near the coastal town of Mossel Bay. As such, the VNR presented South Africa with the trade-off, where the ocean economy is going to encounter major challenges, not just for reaching the SDGs but also in light of progress of the implementation of national policy frameworks (Benn 2019).

For the SWGs to contribute with this level of analysis, goal authors, together with the Stats SA, the underlying data providers, and key affected groups, needed to be present and engaged in the material. As mentioned above, SDG 6 had a significant head start given the MDGs, although national government was also incredibly proactive in providing data timeously, a significant advantage to the SDG 14 and SDG 15 SWGs.

The workings of South Africa's SWGs for SDG 6, 14, and 15 present lessons about what is required of a localization process and in the context of "leaving no one behind." Localization requires input and participation of those closest to the underlying metadata while needing a degree of independent vetting by sub-national stakeholders. Sub-national governments were not specifically included in the SWG—largely by design of the first VNR—and no platforms or specific mechanisms exist that can functionally share and validate national data at the local level or for reporting sub-national data. As the first VNR, it may be understandable that the focus was national, although future progress will need to work out ways to integrate national and sub-national data points and build on existing efforts by sub-national governments toward localizing the SDGs and knowledge exchanges, as well as theme-specific platforms (Fourie 2019; Croese 2019).

2.5 Key Insights on Data, Indicators, and Reporting

South Africa has made progress in driving the first VNR process for the country and establishing the initial platforms for supporting SDG reporting. The country has adopted the multi-stakeholder approach necessary for an inclusive VNR and, through the SWG, has put in place the foundations toward the supporting reporting architecture. In the VNR, over and above the goal reporting, the report emphasized the value of the partnerships, which were facilitated through the 2019 VNR process, across government, the private sector, CSOs, and development partners on strategic priorities, challenges, lessons, and next steps (Republic of South Africa 2019).

Progress is also evident in that, as of the final 2019 VNR submitted to the UN HLPP, South Africa was able to report on 63% of the 230 SDG indicators, including both Tier I and Tier II indicators, as well as domesticated indicators (Stats SA 2019c). While simply reporting on progress is only a portion of the VNR task and progress may indicate worsening trends (such as the RLI in SDG 15), a fundamental step has been taken to understand and communicate South Africa's sustainable development trajectory. That South Africa's first VNR also prioritized an understanding of the impact of policies and programs toward realizing sustainable development is also notable, although the connection to indicators is often long term and theme-specific achievement is often affected by various actions, actors, and institutions (Fourie 2019).

There are two broad categories of insights and areas for improvement, based on experience drawn from the VNR processes covered in this chapter. The first relates to data availability and measuring, on the one hand. The second relates to multi-stakeholder and multi-scalar institutional mechanisms and processes that are needed to analyze, localize, and use data to inform implementation.

2.5.1 Data Availability and Measuring Progress

An overriding issue is the nature of accessing and integrating baseline data that are correct, consistent, and aligned to the global SDG monitoring methodology. At the outset of the reporting process, a single round of data was earmarked for sharing by December 2019 to ensure a consistent and efficient review. However, multiple rounds of clean data were shared, creating many inefficiencies as every new dataset per indicator required re-analysis. In addition, each iteration of indicator analysis needed to be shared with all the stakeholders in the SWG. While the VNR plans for an iterative SWG and consultations, many more engagements were necessary in practice and created an incredibly cumbersome process. As the last National Coordinating Committee (NCC) did not take place, the validity of the process is also in question. Five months into the goal reporting process, much data was still missing so that the goal authors were unable to formally report on many indicators although the authors were aware of available data.

To account for data issues and indicator misalignment, the goal authors drew heavily on stakeholder consultation, publicly available information, and expert analysis. While these data pursuits were enabled by the SWG, upfront issues with sequencing metadata created ripple effects in numerous follow-up calls and interim reviews, consuming a great deal of the limited resources available. In that regard, the relationship between UN indicators and what is domestically relevant is a core issue. The framing of many of the SDG indicators needs to be reviewed (e.g., 15.6.1), while in other cases government departments have indicated that they have data (e.g., 15.7.1) but have not collated nor developed the methodology to be vetted by Stats SA.

Further, as the SDG indicators reflect an ascending order of whether an indicator is conceptually clear and if the data are readily produced, there are specific advancements that need to be well understood for reporting to proceed. It is fundamentally up to the NSO, together with the primary data collectors to demonstrate data availability and reporting progresses. In the case of SDG 15, SANBI plays a key role in advancing the metadata and quality checks, a process that necessarily takes time and which may not yield data at the set VNR reporting intervals. In future VNRs, consideration needs to be given to whether reporting for the sake of reporting should take precedence over a more prudent approach.

The relevance of many indicators for South Africa needs to be addressed in a manner that is globally aligned and domestically relevant. For instance, South Africa's VNR reporting shows that the country is making progress on specific goals, such as SDG 15.3, which deals with land degradation. However, further analysis is needed to assess the progress toward 2030, and therefore, the supporting indicator data are not sufficient measures of progress. Similarly, target 15.4 in turn measures the conservation of mountain ecosystems in the country, although South Africa uses a domesticated indicator (the percentage of mountain ecosystems that are well protected). The domesticated indicator raises concerns because the approach of domestication may signal progress on the goal, yet 58% of mountain ecosystems in South Africa are not protected or at least not well protected. Every indicator therefore requires sufficient interrogation before a conclusion can be reached as to whether progress is being made or not.

2.5.2 Multi-stakeholder and Multi-scalar Institutional Mechanisms and Processes

The SDGs have shown that governments need not only information but also better intelligence about what is represented by data. The VNR process is the best means of establishing intelligence about sustainable development, and the SWGs established by Stats SA are a foremost platform in that regard. As the VNR process matures, South Africa needs to invest in platforms and mechanisms to achieve better data and the nuance called for by its localization efforts. A key benefit will be in addressing current data gaps of key goals, such as for SDG 14 where very little data were reported, which may be more easily achieved through better use of engaging South Africa's coastal municipalities in data collection.

As other chapters in this book show, multi-stakeholder platforms are also needed to measure sustainable development and local impact beyond quantitative measures. In South Africa's case, local and provincial government have crucial policy insight to assist with measuring implementation challenges, highlighting the value in platforms that also work across scales. Experiments with co-production for SDG localization hold learnings about process and the role of linking VNR reporting to strategic objective reporting, in line with the Integrated Development Plans (IDPs) of municipalities in South Africa (Croese 2019). If strategic local objectives can be broadly linked to SDG reporting, to feed either directly into VNR or through local VLR reporting, greater synergy and overall progress with localization can be achieved.

It is worth noting that globally, the VNRs are driving parallel platforms for debate, knowledge sharing and reciprocal learning, and thematic groups where specialist indicators are reported on by epistemic communities (UN OIOS 2017). The discussions coming out of Partners for Review—a transnational multi-stakeholder network for robust review process of the 2030 Agenda for Sustainable Development—offer nuanced insights and resources to help countries navigate the SDG reporting and review process (Partners for Review 2020). Efforts have begun locally in the establishment of the South African SDG hub, which enables access to relevant, useful, and reliable knowledge resources and policy advice and promotes dialogue through active and diverse partnerships (Future Africa 2019).

As sub-national engagement in the VNR is not yet formalized, working platforms such as the South Africa SDG hub may hold opportunities and continuity for supporting data collection and measuring goal intersections and trade-offs. Local government associations working with municipalities to support efforts toward SDG localization at the city level and reporting and development of VLRs are also significant. In addition to global associations such as United Cities and Local Governments (UCLG), as Chap. 11 shows of Mozambique, there are active efforts in municipalities such as Cape Town and eThekwini (see Chap. 12). Other municipal government associations such as Local Governments for Sustainability (ICLEI) are noteworthy in trying to support local elements of SDG reporting in intermediate cities, with the South African Cities Network (SACN) playing a potentially useful role for metropolitan municipalities. Global platforms such as Partners for Review are similarly useful, but particularly to provide perspective on managing the trade-offs implied in many of the SDGs, in ways that avoid parochialism setting in.

Ways of engaging multiple stakeholders at different scales may also bring some remediation to South Africa's limited statistical capacity, given the recent budget cuts. Although municipalities are also fiscally constrained, partnerships between sub-national governments and their local universities or research institutions, such as the Gauteng City-Region Observatory (GCRO), show significant potential for circumventing capacity constraints. As nuanced SDG reporting requires not only strong statistical skills but also analytical and development context, platforms that engage wider stakeholders such as research bodies and citizens can contribute to the depth of technical work needed.

2.6 Conclusion

This chapter has shown that South Africa's initiative regarding the VNR is critical to leverage for further progress on SDG reporting and implementation. Such work is supported by a robust international methodology, but needs to be aligned to contextual knowledge. The follow-through might be challenging and cumbersome and will require a degree of patience. The reality is that many of the targets will not be met and, at the same time, reporting progress will likely improve as VNR processes mature. It is fundamental that baseline time series data become more efficient across the board and sub-national tiers of government will likely be key to reporting improvements.

Going forward, it will be useful to follow what will happen with the dedicated national coordinating vehicle for SDG monitoring and implementation that was adopted in March 2019. The nature of that interaction will be a key factor in how the SDG process unfolds relative to valuable sub-national experiments and multi-stakeholder platforms also underway.

References

Benn H (2019) Review of SDG 14 and policy implications. Unpublished presentation notes

Borel-Saladin J (2017) Data dilemmas: availability, access and applicability for analysis in sub-Saharan African cities. Urban Forum 28:333–343

Croese S (2019) Localisation of the 2030 agenda and its Sustainable Development Goals in Cape Town. https://www.mistraurbanfutures.org/sites/mistraurbanfutures.org/files/cape_town-_final_city_report-_sdgs_project-_nov_2019-_croese_0.pdf. Accessed 13 July 2021

Fourie W (2019) South Africa: promoting constructive science-policy dialogue for the SDG review. Partners for Review. https://www.partners-for-review.de/wp-content/uploads/2020/08/SouthAfrica.pdf. Accessed 13 July 2021

Fukuda-Parr S (2016) From the Millennium Development Goals to the Sustainable Development Goals: shifts in purpose, concept, and politics of global goal setting for development. Gender and Development 24(1):43–52

Fukuda-Parr S, Bruckner M, Hegestad T, Kuehner M, Tavares M (2018) Voluntary National Review reports—what do they report? Committee for Development Policy, United Nations. http://www.partners-for-review.de/wp-content/uploads/2018/07/CDP_BP46_June_2018.pdf. Accessed 13 July 2021

Future Africa (2019) South Africa SDG hub. https://www.futureafrica.science/index.php/hub/sasdkh. Accessed 13 July 2021

Global Alliance (2020) Annual analysis of Voluntary National Reviews: promoting evidence-based action on SDGs and demonstrating how business strengthens the international system. https://www.un-globalalliance.org/additional-resources. Accessed 15 November 2020

Harcourt W (2005) The Millennium Development Goals: a missed opportunity? Development 48(1):1–4

Hasan A, Patel S, Satterthwaite D (2005) How to meet the Millennium Development Goals (MDGS) in urban areas. Environment and Urbanization 17(1):3–19

Herrera V (2019) Reconciling global aspirations and local realities: challenges facing the Sustainable Development Goals for water and sanitation. World Development 118:106–117

Inter-Agency and Expert Group on SDG Indicators (IAEG-SDGs) (2020) Tier classification for global SDG indicators. https://unstats.un.org/sdgs/files/Tier%20Classification%20of%20SDG%20Indicators_31%20December%202018_web.pdf. Accessed 8 November 2020

International Institute for Sustainable Development (IISD) (2018) DESA issues handbook to guide preparation of 2019 VNRs. SDG Knowledge Hub. https://sdg.iisd.org/news/desa-issues-handbook-to-guide-preparation-of-2019-vnrs. Accessed 8 November 2020

International Institute for Sustainable Development (IISD) (2019) DESA issues handbook for countries preparing 2020 VNRs. SDG Knowledge Hub. https://sdg.iisd.org/news/desa-issues-handbook-for-countries-preparing-2020-vnrs. Accessed 8 November 2020

Jütting J, Avendano R, Kuhn M (2019) Counting the invisible: three priorities for strengthening statistical capacities in the SDG era. OECD Development Matters. https://oecd-development-matters.org/2019/12/06/counting-the-invisible-three-priorities-for-strengthening-statistical-capacities-in-the-sdg-era. Accessed 29 November 2020

Khatieb S (2019) Understanding 15.4.1D updated personal communication. South African National Biodiversity Institute (Personal Communication)

Kindornay S (2018) Progressing national SDGs implementation: an independent assessment of the Voluntary National Review reports submitted to the United Nations High-level Political Forum on Sustainable Development in 2018. Canadian Council for International Co-operation, Ottawa. https://www.

partners-for-review.de/wp-content/uploads/2020/03/Progressing-national-SDG-implementation2018_3rd-Edition.pdf. Accessed 29 November 2020
Klopp J, Petretta DL (2017) The urban Sustainable Development Goal: indicators, complexity and the politics of measuring cities. Cities 63:92–91
Leiva J, Rahmaty M (2019) Four questions on how HLPF 2019 can improve SDG implementation. IPI Global Observatory. https://theglobalobservatory.org/2019/05/four-questions-how-hlpf-2019-improve-sdg-implementation. Accessed 29 November 2020
Lomazzi M, Borisch B, Laaser U (2014) The Millennium Development Goals: experiences, achievements and what's next. Global Health Action 7(1):1–9
MDG Achievement Fund (2019) Millennium Development Goals. http://www.mdgfund.org/node/922. Accessed 8 November 2020
Nel JL, Driver A (2012) National biodiversity assessment 201: technical report volume 2: freshwater component http://biodiversityadvisor.sanbi.org/wp-content/uploads/2016/07/NBA2011_TechnicalReport_Vol2Freshwater.pdf. Accessed 17 August 2021
Parnell S (2016) Defining a global urban development agenda. World Development 78:529–540
Partners for Review (2020) Partners for Review: a transnational multi-stakeholder network for a robust review process of the 2030 Agenda for Sustainable Development. https://www.partners-for-review.de. Accessed 6 December 2020
Pieterse E, Parnell S, Haysom G (2018) African dreams: locating urban infrastructure in the 2030 Sustainable Developmental Agenda. Area Development and Policy 3(2):149–169
Republic of South Africa (2019) South Africa's Voluntary National Review (VNR) report 2019: empowering people and ensuring inclusiveness and equality. https://sustainabledevelopment.un.org/content/documents/23402RSA_Voluntary_National_Review_Report__9_July_2019.pdf. Accessed 29 November 2020
Revi A (2016) Afterwards: habitat III and the Sustainable Development Goals. Urbanisation 1(2):x–xiv
Sarwar MB, Nicolai S (2018) Briefing note: what do analyses of Voluntary National Reviews for Sustainable Development Goals tell us about "leave no one behind"? Overseas Development Institute. https://www.odi.org/publications/11147-what-do-analyses-voluntary-national-reviews-sustainable-development-goals-tell-us-about-leave-no-one. Accessed 29 November 2020
Schäffler A, Swilling M (2013) Valuing green infrastructure in an urban environment under pressure—the Johannesburg case. Ecological Economics 86:246–257
Sebola RJ (2017) SANBI' s contribution to national development goals and UN Sustainable Development Goals. http://biodiversityadvisor.sanbi.org/wp-content/uploads/2017/12/2.-Mandivenyi-Biodiversity-SDG-agenda.pdf. Accessed 29 November 2020
Statistics South Africa (Stats SA) (2018) SDG report writing initiation workshop (unpublished notes)
Statistics South Africa (Stats SA) (2019a) Reporting drafting team TOR goals 1-17:1–7 (unpublished notes)
Statistics South Africa (Stats SA) (2019b) South Africa SDG 15 report 2019, final version. doi:10.3143/geriatrics.56.contents1
Statistics South Africa (Stats SA) (2019c) Sustainable Development Goals (SDGs) country report. http://www.statssa.gov.za/MDG/SDGs_Country_Report_2019_South_Africa.pdf. Accessed 29 November 2020
Statistics South Africa (Stats SA) (2020) Statement from the South African Statistics Council on Stats SA funding. http://www.statssa.gov.za/?p=12992. Accessed 29 November 2020
Swilling M, Annecke E (2006) Building sustainable neighbourhoods in South Africa: learning from the Lynedoch case. Environment and Urbanization 18(2):315–332
United Cities and Local Governments (UCLG) (2019) Local and regional government's report to the 2019 UN HLPF: towards the localization of the SDGs. https://www.uclg.org/sites/default/files/towards_the_localization_of_the_sdgs_0.pdf. Accessed 29 November 2020
United Nations (UN) (2019) The Sustainable Development Agenda. https://www.un.org/sustainabledevelopment/development-agenda. Accessed 8 February 2019
United Nations Development Programme (UNDP) (2013) Accelerating progress, sustaining results: the MDGs to 2015 and beyond. https://www.undp.org/content/undp/en/home/librarypage/mdg/accelerating-progress—sustaining-results. Accessed 29 January 2021
United Nations Development Programme (UNDP) (2017) Rapid integrated assessment (RIA) to facilitate mainstreaming of SDGs. United Nations Development Programme, New York
United Nations Educational, Scientific and Cultural Organization (UNESCO): Institute of Statistics (2020) Report of the IAEG-SDG: 2020 comprehensive report. Technical Cooperation Group on the Indicators for SDG 4. http://tcg.uis.unesco.org/iaeg-sdg-2020-comprehensive-report. Accessed 7 November 2020
United Nations (UN) General Assembly (2016) Critical milestones towards coherent, efficient and inclusive follow up and review at the global level. https://sustainabledevelopment.un.org/content/documents/8917SAMOA(2).pdf. Accessed 7 November 2020
UN-Habitat, United Nations Development Programme (UNDP), United Cities and Local Governments (UCLG) (2014) Roadmap for localizing the SDGs: implementation and monitoring at subnational level. https://www.uclg.org/sites/default/files/roadmap_for_localizing_the_sdgs_0.pdf. Accessed 29 November 2020
United Nations Office of International Oversight Services (UN OIOS) (2017) Follow-up and review of the SDGs: advice for the UN Secretariat. www.unevaluation.org. Accessed 13 July 2021

United Nations Statistical Commission (UNSC) (2020) IAEG-SDGs 2020 comprehensive review process. https://unstats.un.org/sdgs/iaeg-sdgs/2020-comp-rev. Accessed 7 November 2020

United Nations Sustainable Development Solutions Network (UN SDSN) (2015a) Indicators and a monitoring framework for Sustainable Development Goals: launching a data revolution for the SDGs. Report to the Secretary-General of the United Nations by the Leadership Council of the Sustainable Development Solutions Network, 12 June 2015. https://sustainabledevelopment.un.org/content/documents/2013150612-FINAL-SDSN-Indicator-Report1.pdf. Accessed 7 November 2020

United Nations Sustainable Development Solutions Network (UN SDSN) (2015b) Data for development: a needs assessment for SDG monitoring and statistical capacity development. https://sustainabledevelopment.un.org/content/documents/2017Data-for-Development-Full-Report.pdf. Accessed 7 November 2020

Woolfrey L (2020) How data-ready are governments to monitor SDG progress? An assessment of socio-economic data for SDG planning and reporting in Egypt, Zambia and Zimbabwe. Paper submitted for presentation at the Economic Research Forum's 26th Annual Conference, 29-31 March 2020. https://erf.org.eg/publications/how-data-ready-are-governments-to-monitor-sdg-progress-an-assessment-of-socioeconomic-data-for-sdg-planning-and-reporting-in-egypt-zambia-and-zimbabwe/. Accessed 7 November 2020

3 Global Norms, African Contexts: A Framework for Localizing SDGs in Cities

Kareem Buyana, J. Jacqueline A. Walubwa, Paul Mukwaya, Hakim Sseviiri, Disan Byarugaba, and Gloria Nsangi Nakyagaba

Abstract

The Sustainable Development Goals (SDGs) set out a normative agenda that offers opportunity for cities to steer profound change globally. But if cities are to play an effective role in localizing the 2030 Agenda, there is a need for systematic engagement with the conundrums presented by the normative dimensions of the SDGs. We argue that African cities offer unique contextual insights into the linkages and overlaps among SDG norms, due to their immensely diverse nature and historically distinct drivers of urbanization. Set against African urban realities, the chapter presents an analytical framework that construes the linkages between global normative statements and local SDG meanings as mainstream and counter-mainstream interpretations. The framework was built by exploring alternative ways of localizing SDGs, through a transdisciplinary waste-to-energy research project in Kampala, Uganda. Based on the locally experienced tensions within the SDGs, we argue for a flexible approach to localization.

Keywords

Sustainable Development Goals (SDGs) · Cities · Localization · Africa · Kampala · Energy

3.1 Introduction

As sustainability became recognized as the global challenge of our time, the United Nations adopted 17 Sustainable Development Goals (SDGs) as an international framework for moving toward more equitable, peaceful, resilient, and prosperous societies and specifically within the limits of what nature can offer (UN 2015). Since the functions of cities intersect with most of the normative dimensions of SDGs, an integrated approach to urbanization is articulated under SDG 11, which stipulates objectives, numerical targets, and indicators that couple the global imperative of sustainability to safety, inclusiveness, and resilience in cities. However,

K. Buyana (✉) · P. Mukwaya · H. Sseviiri
D. Byarugaba
Department of Geography, Geo-informatics and Climatic Sciences, Makerere University, Kampala, Uganda
e-mail: kbuyana@gmail.com; pmukwaya@gmail.com; hsseviiri@gmail.com; disanbyarugaba@gmail.com

J. J. A. Walubwa
Department of Geography, Kisii University, Kisii, Kenya
e-mail: walubwajacque@gmail.com

G. N. Nakyagaba
Department of Geography and Environmental Sustainability, University of Oklahoma, Norman, OK, USA
e-mail: gloria.n.nakyagaba-1@ou.edu

S. Croese, S. Parnell (eds.), *Localizing the SDGs in African Cities*, Sustainable Development Goals Series, https://doi.org/10.1007/978-3-030-95979-1_3

taken as a whole, the SDG norms preclude agreement on a precise meaning at whatever scale and are therefore open to not only different but also fluid interpretations across places and societies.

Whereas globally it is estimated that 23% of the SDG indicators have an urban component and can be measured using statistical evidence from municipalities (UN-Habitat 2018), the attainment of SDGs in cities differs among regions—Europe, the Americas, Africa, Antarctica, and Asia. This is partly why so many scholars and practitioners have called for tenable and flexibly deployable approaches to localization, if progress toward the SDGs is to be visible by 2030 (Parnell 2016; Patel et al. 2017; Arfvidsson et al. 2017; Davidson et al. 2019; Croese et al. 2020; Malonza and Ortega 2020).

Set against the vast diversity of the African urban context, this chapter presents an analytical framework that construes the linkages between global normative statements and local SDG meanings as mainstream and counter-mainstream interpretations. The mainstream interpretations allow discernible connections to be made between local SDG meanings and the universal tenets for SDG implementation—the "5 Ps": people, planet, prosperity, peace, and partnerships. Counter-mainstream interpretations represent double-bind relationships between local SDG meanings and global normative statements. Although not all interactions among SDG norms can fall neatly into mainstream or counter-mainstream interpretations, the framework organizes empirical evidence from a transdisciplinary research project on SDG implementation at local scale in the city of Kampala, Uganda, into a coherent whole. The project involved the scaling up of the adoption of energy briquettes as alternative cooking fuel for low-income households and, at the same time, addressing the health and environmental consequences of indiscriminate waste dumping. Building on this work, the framework shows how diverse social constructs can be translated into normative codes, thus making it possible to discern the (dis)connections between local and global measurements of sustainable cities.

3.2 The Peculiar Nature of African Cities in the Context of SDGs

Urban societies in Africa are enormously varied with regard to history, demographic transitions, cultures, and governance arrangements. The informality of settlements, multiplicity of ecologies, and intersecting inequalities add to the complex realities in which the SDGs must be grounded. The pursuit of policy coherence, as a universal standard for implementing SDGs in an integrated manner (Tosun and Leininger 2017), is not only subject to divergent interpretations, but the goals may also remain largely abstract due to stark urban differentiations across Africa. The range and scale of different urban realities mean that internationally and nationally defined SDG priority targets may not necessarily reflect differences in urban realities across municipalities in a typical African city.

More than most global regions, African cities are heterogeneous. The continent is not only large and hugely physically varied; it also has highly ethnically diverse populations. Kampala, Addis Ababa, Mombasa, Johannesburg, and Lagos are very cosmopolitan, while other African cities are quite low in ethnic diversity, like Maseru, Bujumbura, Banjul, Tunis, Cairo, and Algiers (UNDESA 2014). How African cities are physically structured and administratively configured varies too. The history of urbanization in Africa is traceable to differing colonial footprints of administration, as opposed to industrial and technological drivers in most parts of Europe, North America, and developed Asia. For instance, British colonial officials chose to possess land for town planning from ethnic groups, such as the Igbo in Nigeria and the Baganda in Uganda, who have a history of cultural institutions that hinge on effective approaches to indirect rule (Anderson and Rathbone 2000; Boone 2007; Sikor and Lund 2009). Studies have revealed that there is a tight relationship between contemporary urban governance strategies in South African cities and the apartheid practices by which urban privilege was enforced by the British (Miraftab 2012). The planning ideals that informed the construction of railways, housing estates, seaports, and industrial towns during the post-World War II colonial era still illustrate the rationality of invoking a generic urban form that prioritizes western architecture while obscuring Africa's traditional influences on socio-spatial ordering of cities (Byerley 2013; Martin and Bezemer 2020). Cities like Luxor in Egypt are characterized by historical archaeological sites that have for long influenced their physical expansion and natural growth (Mahmoud et al. 2019). Similarly, the tribal communities of Bakgatla-ba-Kgafela in Mochudi, Botswana, and Moruleng, South Africa, have conveyed the significance of conserving architectural heritage in the use of urban spaces (Mwale and Lintonbon 2020).

Notwithstanding important variation, Africa is seen as a critical focus of SDG implementation. How Africa is targeted for SDG implementation has to take account of the diversity of its urban conditions. Although Africa has the fastest urbanization rates in the world, it has not yet reached the level of 50% urbanization (Parnell and Pieterse 2014). But in the next 30 years, urban dwellers will outweigh rural residents for the first time on the continent. This is because seven out of the ten new megacities anticipated by 2030 across the globe will be in Africa. These include Cairo (Egypt), Accra (Ghana), Johannesburg (South Africa), Khartoum (Sudan), Kinshasa (Democratic Republic of the Congo), Lagos (Nigeria), and Nairobi (Kenya) (UN-Habitat 2016).

However, despite their growth, urbanization processes in Africa remain differentiated across regions. Whereas Northern Africa is highly urbanized, with most of its cities unevenly spread along its Mediterranean coastline and the Nile Valley and Delta, West African cities have a long history of cross-border mobility linked to factors such as long-distance trade (UN-Habitat 2016). In East Africa, rural-urban migration is a salient component of rural households' strategies for income diversification, coupled to flows of refugees and internally displaced persons that create multi-national borders and in-land city neighborhoods (Büscher 2018). The mining economy and racial segregation that shapes the sub-regional character of South African cities cannot be underestimated, coupled with geographical variations in natural resource endowments across urban areas in neighboring regions such as Gaza and Cabo Delgado in Mozambique (Collier 2017). The urbanization patterns of cities like Bujumbura, Mogadishu, Kinshasa, Khartoum, Juba, Monrovia, and Freetown have been partly shaped by the dynamics of prolonged civil strife and drastic political transitions (Bakonyi et al. 2019). This implies that attempts around the development of SDG localization mechanisms need to be explicitly cognizant of the nuances in the scale and speed of urbanization across the continent.

African cities are enormously varied in their population structure. This raises the question as to how SDG 10, which aims to reduce inequalities and support the universal SDG principle of "leaving no one behind," can match the interests of multiple vulnerable urban sub-populations, whose needs differ across cities. The proportion of cross-border communities, people with disabilities, those living with HIV/AIDS, refugees, street families, out-of-school adolescents, out-of-work adults, elderly citizens, ethnic minorities, and educated and non-educated slum entrepreneurs, is increasingly visible in African urban populations (Gibbs et al. 2020; Yeboah 2020). Meanwhile, there is little scientific knowledge of which urban sub-populations have been left behind, which remain furthest behind, the reasons as to why they have been left behind, and what kind of urban policies and sector-specific programs can potentially reach those left behind (Kabeer 2016; Cetrulo et al. 2020). The urban youth bulge, which coincides with intersections of economic deprivation and spatial disadvantage (Urdal and Hoelscher 2009; Ulbrich et al. 2019), pertains to the capacity of SDGs to promote inclusive urban development. The emergence of COVID-19 as a global public health and economic crisis has added new layers of intersecting inequalities. Informal settlements can only continue to practice prudence around social distancing, if urban services are channeled to them either free of charge or at a much-reduced cost (Corburn et al. 2020). The closure of schools has challenged low-income families that have no history of home-schooling or virtual education, leading to social fatigue over containment strategies (Armitage and Nellums 2020). Refugees and undocumented workers, who fear deportation and retaliation by employers, have little incentive to embrace testing at designated health units (Mukumbang et al. 2020). While the New Urban Agenda (UN 2016) commits to the provision of inclusive and safe streets that are free from crime and violence, violence against women has been accentuated by confined living conditions in both low- and high-income families (Maclin et al. 2020; Wright and Skubak Tillyer 2020). Therefore, if the principle of "leaving no one behind" is to be operational in African cities, there are different levels of demographic specificities required in the analysis and in assessing the various SDG implementation mechanisms. To capture the overall progress on implementing the SDGs across each goal, for different places and for multiple constituencies, it is ideal to have sectoral and socially interoperable and, ideally, spatialized data.

As shown in Table 3.1, urban differentiations in Africa include demographic and other parameters that are critical for context-sensitive mapping of the SDGs and their linkages in urban contexts. For instance, within the framework of reporting progress through the Voluntary National Reviews by member states at the United Nations (Persson et al. 2016; Sebestyén et al. 2020), it is possible to have an open data initiative on land ownership by type of tenure system, which would avail data on the injustices harbored by the proliferation of informal land markets (Porsani et al. 2017; Steel et al. 2020) as per SDG 16 on peace and human rights. It could also use such data to measure the prospects of property tax, as a source of domestic financing for SDG implementation (Gambetta et al. 2019; Goodfellow and Owen 2020). Africa also holds enormous potential for using spatial media technologies, for example, to digitally map and visualize risk-prone livelihoods and infrastructure, as one of the approaches to picturing the status and trends in SDG indicators (Luque-Ayala and Neves Maia 2018; Kovacic et al. 2019; Zhilin et al. 2020).

But data openness between governments and citizens is often restricted to particular datasets, as an opportunity for advancing a political discourse around service provision to serve specific stakeholder interests (Cinnamon 2020). Insufficient civic engagement and disaggregation of data is also attributable to models which originate from the Global North and are uncritically replicated in Africa. For example, climate models, such as the Global Protocol for Community-Scale Greenhouse Gas Emissions, can be faulted on account of the difficulties to represent such emissions in Africa, where aggregates as pertinent as the number of vehicles registered and imported fossil fuels can be inadequate (UNECA 2017). Project-based statistical endeavors are also common in Africa, when donors prioritize SDG indicators for national statistical offices, but with conditions and limited follow-up or effort to build on previous investments and pilot projects (Weber et al. 2018). Resolving these challenges across the data space in Africa is critical for ensuring not only equity and accountability but also brevity in SDG monitoring.

Although Table 3.1 is not a comprehensive mapping of all possibilities of data and SDG linkages, it does highlight the wealth of analytical opportunity. However, for Africans, there are some clear omissions in the data that distort the priorities afforded to SDG implementation activities. For example, evidence shows that pro-private sector discourses within approaches to SDG financing are yet to include possible strategies for leveraging the ingenuity of informal economic clusters, which form around peri-urban spaces, transport corridors, mining, and border cities (Barua 2020). According to the UN Global Compact, more than 80% of its 9500 corporate members have committed to advancing one or more of the SDGs (Carby-Hall 2020; Martínez-Ferrero et al. 2020). However, it remains largely unclear how SDG 8 on decent work and economic growth, SDG 9 on innovation, SDG 12 on sustainable production and consumption, and SDG 15 on life on land can be aligned to business operations that are spatially uneven with alterna-

Table 3.1 Summary of linkages between SDG norms and the contextual features of African cities (Authors' aggregation of cited literature)

SDG norms	Linkages with the contextual features of African cities (positive, negative, or ambivalent)
Policy coherence and integration	**Negative linkage**: Policy coherence and integration cannot be easily pursued and may remain largely elusive. This is because African cities are enormously varied in terms of historical and current drivers of the scale and speed of urbanization, at city, country, and regional level (West; East; South; Central; and Northern Africa). Series of country-specific, inter- and intra-city consensus-building meetings with different stakeholders would have to be done, if nationally defined SDG priority targets are to guide the actualization of policy coherence and reflect differences in urban realities across city regions and municipalities in a given country
Leaving no one behind	**Ambivalent linkage:** SDG 10 on reduced inequalities and the universal SDG principle of "leaving no one behind" may match the interests of some but not all types of urban sub-populations in African cities. There is possible exclusion of urban sub-populations that include cross-border communities, people with disabilities, those living with HIV/AIDS, refugees, street families, out-of-school adolescents, out-of-work adults, elderly citizens, ethnic minorities, disadvantaged women, and educated and non-educated slum entrepreneurs. African cities have also been confronted by intra-urban inequalities that stem from intersections of structural, social, economic, and spatial disadvantage
Pro-private sector engagement and innovative financing strategies to mobilize domestic resources	**Ambivalent linkage:** The private sector in Africa cities comprises both large formal corporations and unincorporated small and medium informal enterprises. The customer base and supply chain linkages of both categories are dominated by entrepreneurs and employees in business dealings that are shrouded in accepted informalities. The ambivalence is such that actors in Africa's urban informal economy, and their contacts and informal trust networks in municipal authorities and national business associations, can conditionally support the role of large foreign formal corporations in implementing SDGs, or simply embrace what are presented as practices of corporate philanthropy, or even favor anti-private sector discourses
National commitment to follow-up and review systems	**Ambivalent linkage:** It is possible to report progress on SDG implementation in African cities through the Voluntary National Reviews by member states at the United Nations and visualize the trends in informal settlements using spatial media technologies and open data initiatives. But data openness between governments and citizens is restricted to particular datasets, which obscures the unique and complex realities in Africa. Insufficient disaggregation of data is also attributable to models which originate from the Global North and are uncritically replicated in Africa, as well as donor-driven prioritization of SDG indicators in collaborations with national statistical offices
Global solidarity and inclusive partnerships	**Ambivalent linkage**: The landscape of actors in formal and informal sectors of African cities, and the parallel nature of their operations beyond the purview of state authority, can bring about interdependencies but also unrealized synergies, thus falling short on the universal standard on inclusive partnerships during SDG implementation at global, regional, national, and local levels
Placing people, prosperity, and the planet at the center of SDG implementation	**Ambivalent linkage:** In African cities, there are various trade-offs among SDGs that particularly require the protection of nature while calling for inclusive economic growth and sustainable livelihood improvement. For example, urban informal settlements with low-quality housing situated in floodplains and wetland areas often adopt green roofing and water harvesting technologies that offer opportunities for improved storm-water management, better regulation of building temperatures, and reduced urban heat-island effects. This pulls in the question of what people and what parts of the planet ought to be prioritized, if SDG 1 (no poverty), SDG 14 (life under water), SDG 15 (life on land), and SDG 11 (sustainable cities) are to be addressed in an integrated manner

Key:
Positive linkage: the SDG norm is explicitly linked to the contextual features of African cities
Negative linkage: the SDG norm is explicitly delinked from the contextual features of African cities
Ambivalent linkage: the SDG norm is intricately linked to the contextual features of African cities

tive interpretations of what is formal or informal employment.

Meanwhile, in a further illustration of the unintended outcomes of evidence-led SDG implementation, the description of urban Africa as under-serviced may have nudged urban managers into large-scale infrastructure for industrial development and information technologies, assuming this will interlink SDG 11 with SDG 8 and 9 (Pieterse et al. 2018; Matamanda and Nel 2020). But pursuing global norms of consolidating safe and decent working conditions though formalization can bring about cognitively dissonant messaging in African urban communities, where everyday business dealings are shrouded in uncertainty and accepted informalities (Hummel 2017; Thieme 2018; Grossman 2020). Indeed, in African cities, it is not always clear what the best pathways are toward decent work. In African cities where natural resources exist in abundance, the desire to create employment in the face of disappointing government policies often leads people to mine illegally on dump sites left behind by formal mining operations (Hoffman 2007; Mususa 2012; Stewart et al. 2020; Antwi-Boateng and Akudugu 2020; Makhetha and Maliehe 2020). In other cities, employment in the informal transport sector, locally known as boda-bodas and tuk-tuk rides and other unincorporated businesses, has an influence on collaborations between municipal actors and informal trust networks within national business associations (Goodfellow 2017; Ezeibe et al. 2017; Mohan and Tan-Mullins 2019).

Urban Africa is a combination of coastal, inland, highland, and arid cities, which means that pathways to urban resilience not only differ on account of multiple ecologies but also illustrate the trade-offs among SDGs that require the protection of nature while calling for inclusive economic growth. An estimated 54 million Africans live in vulnerable Low-Elevation Coast Zones (LECZ)—defined as areas that are 10 m or less above sea level—and this figure is projected to rise to over 100 million by 2030 (Becker et al. 2019). This raises double-bind or contradictory relationships between SDG 11 indicators, such as ratio of land consumption rate to population growth rate, and SDG 8 indicators, such as annual growth rate of real GDP per capita. Cities with greater exposure to excessive heat are located in Western and Central Africa, although East African cities indicate an increase of more than 2000 times the current level by 2090 (Rohat et al. 2019). Although this may align to SDG 13 indicators, such as number of countries with national and local disaster risk reduction strategies, urban ecologies in Africa harbor resilience pathways that raise the question of what people and what parts of the planet ought to be prioritized, if SDG 1 (no poverty), SDG 14 (life under water), SDG 7 (sustainable energy), and SDG 11 are to be addressed in an integrated manner. For example, urban informal settlements with low-quality housing situated in floodplains often adopt green roofing and water harvesting technologies that offer opportunities for improved storm-water management, better regulation of building temperatures, and reduced urban heat-island effects (Buyana et al. 2020). Although policymakers have linked the collective ingenuities of informal urban dwellers to illegality in slum upgrading projects, local community networks in unplanned settlements have set in motion urban resilience pathways, such as the uptake of solar electrification for water disinfection, through slum-dweller associations and social enterprises (Thorn et al. 2015). This means that effective implementation of SDGs in African cities will not be simple. To be effective in the complexity of the African city, SDG implementation calls for alternative approaches that are grounded in in-depth knowledge that sensitively reflects the different challenges and solutions deployed across varied populations, ecologies, and human settlement forms.

3.3 The Approach to Localizing SDGs in Kampala City, Uganda

While actors from civil society, academia, government, and the private sector are indispensable players in defining target areas for implementing SDGs at city level, the research project in Kampala, called Localized Norms for Sustainable Energy in Kampala (LONSEK), followed the premise that localization needs to consider initiatives outside hierarchical and highly formalized contexts, other than international and inter-municipal coalitions like the Compact of Mayors, the Cities Climate Leadership Group (C40), and the United Cities and Local Governments network (UCLG) (Stafford-Smith et al. 2017; Bowen et al. 2017). On the other hand, if locally grounded meanings of SDG norms are to be obtained, there is a need for transdisciplinary research approaches that bolster deliberative learning processes with societal actors that are immersed in a specific situation, to not only shape the identification of problems and scalable local solutions but also permit broader joint reflections on alternative urban development visions (Schneider et al. 2019; Buyana 2020).

The LONSEK project explored opportunities in the informal urban waste sector, particularly the instances through which individual households and community ingenuities relate with transitions to affordable and clean energy that supports the most vulnerable groups (SDG 7), while leapfrogging the interdependent outcomes of healthy and sanitized urban environments (SDG 11 and 3), together with ending poverty and reducing inequalities (SDG 1 and 10). The project was undertaken in Kasubi-Kawaala Parish, which borders Kampala Central Division in the western part of the city (Fig. 3.1). Although drainage and road conditions have improved, most households in Kasubi-Kawaala rely on charcoal as a source of cooking energy and can barely afford formal waste collection services. Households therefore resort to illegal waste dumping that brings about environmental health burdens, including contamination of air and water sources. Lighting is from energy-saving bulbs and candles, coupled to adjustments in energy-use practices, for example, abandoning boiling of water and foregoing hot water baths and forsaking foods that require long hours of preparation, alongside illegal tapping of electricity from Umeme, the dominant hydro-power distribution company in the country.

The household health and energy coping mechanisms in Kasubi-Kawaala have emerged from the waste-to-energy sector. In order to investigate the nature of locally available technologies and capabilities among local actors in the waste sector, the LONSEK team from Makerere University worked with Kasubi Local Community Development Association (KALOCODE) and undertook key informant

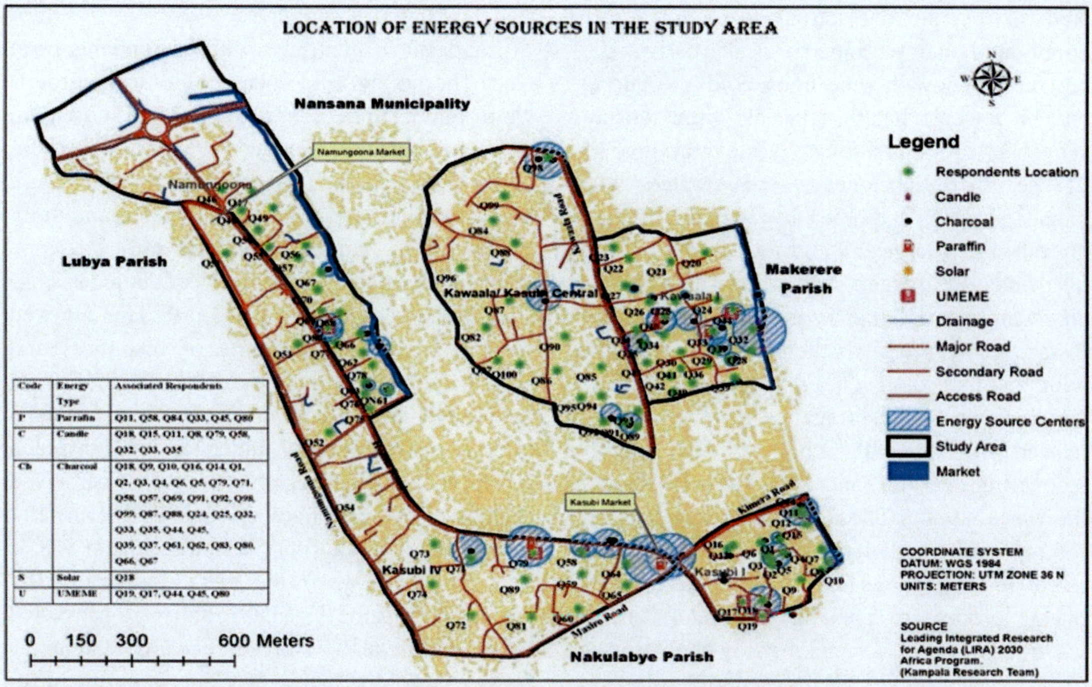

Fig. 3.1 Map of project area and location of energy sources. (Source: authors)

interviews among purposively selected respondents from 100 households. The representatives from KALOCODE coordinated the process of establishing contacts with individual households and community groups of energy-briquette producers. The research team then followed up the contacts for interviews, using questions pertaining to practice-based challenges for transformation of waste management in the neighborhood. The questions were centered on how to enable transition of the current micro-scale interventions of energy briquettes to meso- and macro-scale, within the context of localizing SDGs in Kampala. This question became the boundary subject for social engagement and learning with the individual household respondents and community groups of energy-briquette producers. Interview data from households and energy-briquette producers was collected using the biographical method that took the form of a life narrative (Zinn 2005). Hence, the research project assumed that it is possible to reconstruct the individual self and identity by putting respondents in the context of their life course, as the act of telling a story that is linked to the main question. Through respondents' narratives, it was possible to recreate the actions taken toward the adoption of energy briquettes as alternative cooking fuel for low-income households, to address energy, poverty, health, and environmental consequences of indiscriminate waste dumping.

The energy briquettes are created when banana peelings and other dried organic material are put into a large bin and then burned at high heat and low oxygen, which creates a kind of charcoal material that is made out of garbage instead of trees. This is then crushed and mixed with clay and cassava flour, as a glue, and rolled into balls to create briquettes that can be used instead of charcoal. Other activities involve recovering reusable and recyclable items from the waste stream. These include polythene bags for growing mushrooms; banana, cassava, and sweet potato peelings and cow dung for compost; plastic bottles for packing juice and drinking water; newspapers for making tray eggs; tins and mineral water bottles for making shoe soles; bottle straws for knitting baskets; charcoal and saw dust for reducing odor from latrines; oily milk packages used as fuel for cooking; and discarded cardboard serving as walls and roofs of houses for a cool indoor climate.

These life-course actions of energy-briquette producers as well as users at household level were transcribed and reconstructed as storylines and visuals of lived experiences for an engaged, contextually rich, and nuanced interpretation of linkages between the emerging waste sector and SDGs. The LONSEK project organized SDG studios as seminar workshops for researchers, policy officials from Kampala Capital City Authority (KCCA), and local community actors from KALOCODE, to mediate discussions on how to depict the local context in relation to the SDGs through visuals and storylines. The SDG studios were premised on the principles of visual ethnography, which entails a situational combination of field techniques for exploring how we understand and learn about social phenomenon, using note taking, audio-visual recording, interviews, examination of local relative to policy and academic knowledge, and observation that is rooted in the ideal of participant observation to live, to some extent, as the briquette-energy producers do (Falzon 2016).

In addition, interviews with respondents from research, policy, and local community groups were conducted in order to gather their individual views, alongside the collective discussions held in the SDG studios. Thirty (30) respondents were interviewed in total, 12 from local community groups, 8 researchers, and 10 policy actors. Analysis of interview data was conducted as conversations were being carried out. This allowed for the immediate grouping of responses and their triangulation with the data obtained from the group-based interactions in SDG studios. The interviews were key for understanding how each actor in the research process perceived sustainability challenges at a local scale in relation to the global normative statements on SDGs.

The discussions in the SDG studios initially focused on all the 17 SDGs that were presented by the researchers, basing on the value statements framework of the United Nations Department for Economic and Social Affairs (UNDESA 2014).

But, in an example of how to localize the emphasis and nature of SDG interactions, those preferred for further discussion by local community actors and policy officials included SDG 11 (cities), SDG 7 (clean and affordable energy), SDG 1 (no poverty), SDG 2 (zero hunger), SDG 6 (clean water and sanitation), SDG 8 (decent work and economic growth), SDG 9 (innovation), SDG 10 (reduced inequalities), SDG 12 (responsible consumption and production), and SDG 13 (climate action).

The researchers developed a five-step process for deriving the local applications of SDGs, as shown in Table 3.2.

This five-step process of collective debate was key to mediating learning on the local appropriateness of SDG norms. It also nurtured a participatory process for identifying linkages and overlaps across different SDGs and how these are compatible or incompatible with the lived urban realities, exhibited in the form of visuals and storylines from energy-briquette producers. The challenge, though, was that it was a recursive process. Policy officials and local community actors had not widely engaged with the value statements of SDGs, even though there is a national SDG prioritization framework for Uganda, a national SDG Roadmap and Coordination Framework, as presented in the country's second Voluntary National Review on the implementation of Agenda 2030 (Republic of Uganda 2020). This prioritization framework was developed by the National Planning Authority with support from the United Nations Development Programme (UNDP), but local actors were not included. Similarly, there was lack of research on Agenda 2030 at a municipal level in Kampala, making it hard to identify barriers to municipalities localizing the nationally defined priority SDG targets.

Table 3.2 The five steps for relating the SDGs to local contexts

1. Categorize or group applicable and inapplicable SDG norms
2. Assess revisions required, based on the original intent of the SDG norm
3. Revise, replace, or alter the norm language as locally appropriate
4. Develop new norms to align with local context
5. Validate proposed norms to existing local and scalable solutions in the community

These limitations gave birth to the idea of designing Local Agenda 2030, as a visual aggregation of local versus global SDG meanings, in the form of a chart. The policy officials involved in the SDG studios argued that SDGs required a cross-sectional way of working within urban residents through collaborations with private and public stakeholders, but also research institutions for data and public education activities. Besides, municipal authorities like KCCA are still organized by sector, and such an organizational structure poses challenges to collaborative mechanisms for localizing SDGs (Caprotti et al. 2017; Valencia et al. 2019). The research team was able to ensure that local community actors are actively engaged in appropriating local meanings to the SDGs, in line with the visuals and storylines about the emerging waste sector, resulting in the launch of Local Agenda 2030-Kampala City (Fig. 3.2).

The analytical framework (Fig. 3.3) is a conceptually dynamic diagram that construes the linkages between global normative statements and local meanings of SDGs as mainstream and counter-mainstream interpretations, based on the findings and learnings from Kampala. The mainstream interpretations link local SDG meanings with the "5 Ps," people, planet, prosperity, peace, and partnerships, which are universally central to SDG implementation. Counter-mainstream interpretations represent double-bind relationships between local SDG meanings and global normative statements. Although the framework was developed based on SDG localization efforts in Kampala, the iterative methods used for its formulation (SDG studios, individual interviews across different actors, and selection of mainstream and counter-mainstream interpretations) can facilitate application in other city contexts. Beyond the methods and tools used for the participatory formulation of the framework, the process was a combination of representation, deliberation on alternative urban sustainability visions and balancing acts between individual and collective views. Those most affected by

Fig. 3.2 Launch of Local Agenda 2030-Kampala City. (Source: authors)

local sustainability challenges were represented by local community groups and were also given a chance to speak out so that their real-world challenges and perspectives on potential solutions could influence the design of the framework. This did not only enhance bottom-up participation in shaping the design and content of the framework but also required balancing acts by the researchers. The researchers who moderated the discussions frequently recognized that individual participants could shift between representative roles, at one moment seeming to speak for larger groups they represent and at other times expressing more personal or individual views. These dynamics of stakeholder inclusion made the framework a combination of pragmatic actions at the local level on the one hand and normative ambitions pertaining to translating the SDGs into local meanings on the other.

The local SDG meanings presented in the analytical framework indicate that turning global normative statements into socially relevant constructs of sustainability is largely dependent on the strategy chosen for interpretative analysis and the situation in which implementation can occur on the ground. For example, SDG 9 on inclusive and sustainable industrialization and innovation is locally appropriated to what actors in the informal urban waste sector are immersed into, thus the local meaning of "recycle old materials." Similarly, SDG 2 on food security and improved nutrition was translated as "avoiding food wastage" during the processes of harvesting, transportation, sale, preparation, and consumption of food. This local narrative is quite delinked from global statistics and policy responses, indicating that food insecurity is not just a product of a lack of food but is associated with rising urbanization, decreasing arable land, and weather extremes due to climate change (Battersby 2017; Barthel et al. 2019; Brunori et al. 2020). There is a contradiction here: whereas global normative statements on SDG 2 emerge from vulnerability assessment exercises of global food systems, the local interpretation is skewed toward eliminating wastage in local food value chain activities.

Mainstream interpretations

SDGs	SDG 6	SDG 7	SDG 11	SDG 13
Global normative statements	Ensure availability and sustainable management of water and sanitation for all	Ensure affordable, reliable, sustainable and modern energy for all	Make cities and human settlements inclusive, safe, resilient and sustainable	Take urgent action to combat climate change and its impacts
Local SDG meanings	Ensure protection of water sources and safe reuse of water	Always save energy	Keep cities clean and pollution free	Educate people on climate change

Counter-mainstream interpretations

SDGs	SDG 1	SDG 2	SDG 9	SDG 10	SDG 12
Global normative statements	End poverty in all its forms everywhere	End hunger, achieve food security and improved nutrition and promote sustainable agriculture	Build resilient infrastructure, promote inclusive and sustainable industrialization and foster innovation	Reduce inequality within and among countries	Ensure sustainable consumption and production patterns
Local SDG meanings	Donate what you don't use	Avoid throwing away food	Recycle old materials	Don't discriminate	Use only what you need

Fig. 3.3 Analytical framework for localizing SDGs in cities

The other notable double-bind relationship is between the local meanings of SDG 11 and the images presented by local community actors. Whereas SDG 11 global normative statements look to a reduction in the proportion of urban population living in slums and informal settlements, the images taken reflected a mix of semi-permanent and permanent roof, wall, and floor materials, with local actors arguing that the measure for decent housing should focus on the number of iron sheets needed, affordability of materials required, public service delivery points (water, health, and electricity in the neighborhood), and tenure status. Therefore, mainstream and counter-mainstream interpretations demonstrate how the framework in Fig. 3.3 can be used to mediate discourses on the integrated nature of SDGs while discerning mechanisms for engaging citizens on the (dis)connections between locally embedded visions and global measurements of sustainable, inclusive, safe, and resilient cities.

3.4 Limitations of the Analytical Framework

The analytical framework adopted in Kampala utilizes a critical-interpretive approach to explore and engage with the local meanings of SDGs (Glaser and Strauss 1967). However, there are limitations associated with a critical interpretive synthesis of global normative statements versus local SDG meanings. This approach can interfere with synergies among local, national, and global measurements of sustainability. The normative dimensions of SDGs, and their linkages with local contexts, need to demonstrate that it is possible to have both universal and geographic targeting, as well as a more nuanced view of the interlinkages, so that interpretations can move beyond simplistic notions of trade-offs and synergies (ISC 2017; Schmieg et al. 2018).

As an example, women in Kasubi-Kawaala Parish have primary responsibility for domestic tasks, such as the preparation of meals and boiling water. These tasks require the transportation and use of charcoal and are therefore energy and transport intensive. Though growth in ownership of personal motorized transport among women can symbolize improved standard of living at a local level, it can conflict with global efforts around the reduction of greenhouse gas emissions and the contamination of air and water sources. In the future, however, the transition toward zero-emission cars fueled by renewable electricity may remove this trade-off. Similarly, establishing a local seed fund for the provision of clean cooking stoves to women may serve to reinforce the disproportionate burden they bear of unpaid care work in the household, including child and elder care, and provisioning of food, fuel, and water (Abdelnour et al. 2020).

Hence, not all interactions between SDGs and the contextual features of cities fall neatly into the mainstream or counter-mainstream side of the analytical framework. However, if deployed, the framework can allow for broad multi-disciplinary and multi-sectoral conversations, translate diverse social constructs into normative codes, and organize empirical evidence on SDG implementation at local scale into a coherent whole. This makes it possible to synthesize knowledge and provide concrete clusters of normative statements about synergies and trade-offs, endorsed and debated by stakeholders from science, policy, and local communities.

3.5 Conclusion

Although the SDGs have demonstrated the potential to contribute to the transition toward more sustainable, inclusive, and resilient cities, many of the localization efforts are upstream in nature, with city-level aggregates that stem from a selected set of indicators, which construe the city as a measurable entity of data streams, thus obscuring vast urban differentiations on the ground. This is why there is a need for systematic and context-sensitive engagement with the normative dimensions and indicator targets of SDGs, if cities are to take on an effective role in global efforts around localization. We conclude by re-echoing our argument that African cities offer unique and comprehensive contextual accounts of the linkages and overlaps among SDG norms, due to the immensely diverse nature of their settlement patterns, population structure, ecologies, cultures, and the historically distinct drivers of urbanization across the continent.

Acknowledgments We are grateful to the support received from the International Science Council (ISC), under the Leading Integrated Research on Agenda 2030 (LIRA), GRANT NUMBER: LIRA 2030-GR10/18.

References

Abdelnour S, Pemberton-Pigott C, Deichmann D (2020) Clean cooking interventions: towards user-centred contexts of use design. Energy Research & Social Science 70:101758

Anderson M, Rathbone R (2000) Africa's urban past. James Currey, Oxford

Antwi-Boateng O, Akudugu MA (2020) Golden migrants: the rise and impact of illegal Chinese small-scale mining in Ghana. Politics & Policy 48(1):135-167

Arfvidsson H, Simon D, Oloko M, Moodley N (2017) Engaging with and measuring informality in the proposed urban Sustainable Development Goal. African Geographical Review 36(1):100-114

Armitage R, Nellums LB (2020) Considering inequalities in the school closure response to COVID-19. The Lancet Global Health 8(5):e644

Bakonyi J, Chonka P, Stuvøy K (2019) War and city-making in Somalia: property, power and disposable lives. Political Geography 73:82-91

Barthel S, Isendahl C, Vis BN, Drescher A, Evans DL, van Timmeren A (2019) Global urbanization and food production in direct competition for land: leverage places to mitigate impacts on SDG2 and on the Earth System. The Anthropocene Review 6(1-2):71-97

Barua S (2020) Financing Sustainable Development Goals: a review of challenges and mitigation strategies. Business Strategy & Development 3(3):277-293

Battersby J (2017) MDGs to SDGs—new goals, same gaps: the continued absence of urban food security in the Post-2015 Global Development Agenda. African Geographical Review 36(1):115-129

Becker M, Karpytchev M, Papa F (2019) Hotspots of relative sea level rise in the tropics. In: Venugopal V, Sukhatme, J, Murtugudde, R, Roca, R (eds) Tropical Extremes: Natural Variability and Trends. Elsevier Publishing, pp 203-262

Boone C (2007) Property and constitutional order: land tenure reform and the future of the African state. African Affairs 106(425):557-586

Bowen KJ, Cradock-Henry NA, Koch F, Patterson J, Häyhä T, Vogt J, Barbi F (2017) Implementing the "Sustainable Development Goals": towards addressing three key governance challenges—collective action, trade-offs, and accountability. Current Opinion in Environmental Sustainability 26-27:90-96

Brunori G, Avermaete T, Bartolini F, Brzezina N, Marsden T, Mathijs E, Moragues-Faus A, Sonnino R (2020) The vulnerability of food systems. In: Brunori G, Grando S (eds) Innovation for sustainability. Research in rural sociology and development, vol 25. Emerald Publishing Limited, Bingley, UK, pp 69-105

Büscher K (2018) African cities and violent conflict: the urban dimension of conflict and post conflict dynamics in Central and Eastern Africa. Journal of Eastern African Studies 12(2):193-210

Buyana K (2020) Keeping the doors open: experimenting science–policy–practice interfaces in Africa for sustainable urban development. Journal of Housing and the Built Environment 35:539–554

Buyana K, Lwasa S, Tugume D, Mukwaya P, Walubwa J, Owuor S, Kasaija P, Sseviiri H et al. (2020) Pathways for resilience to climate change in African cities. Environmental Research Letters 15(7):073002

Byerley A (2013) Displacements in the name of (re) development: the contested rise and contested demise of colonial 'African' housing estates in Kampala and Jinja. Planning Perspectives 28(4):547-570

Caprotti F, Cowley R, Datta A, Broto VC, Gao E, Georgeson L, Herrick C, Odendaal N et al. (2017) The New Urban Agenda: key opportunities and challenges for policy and practice. Urban Research & Practice 10(3):367-378

Carby-Hall J (2020) Multinationals, SMEs and non-profit organisations participating in the UN Global Compact. Lex Social: Revista de Derechos Sociales 10(2):130-173

Cetrulo TB, Marques RC, Malheiros TF, Cetrulo NM (2020) Monitoring inequality in water access: challenges for the 2030 Agenda for Sustainable Development. Science of the Total Environment 727:138746

Cinnamon J (2020) Attack the data: agency, power, and technopolitics in South African data activism. Annals of the American Association of Geographers 110(3):623-639

Collier P (2017) Africa's prospective urban transition. Journal of Demographic Economics 83(1):3-11

Corburn J, Vlahov D, Mberu B, Riley L, Caiaffa WT, Rashid SF, Ko A, Patel S et al. (2020) Slum health: arresting COVID-19 and improving well-being in urban informal settlements. Journal of Urban Health 97(3):348–357

Croese S, Green C, Morgan G (2020) Localizing the Sustainable Development Goals through the lens of urban resilience: lessons and learnings from 100 resilient cities and Cape Town. Sustainability 12(2):550

Davidson K, Coenen L, Acuto M, Gleeson B (2019) Reconfiguring urban governance in an age of rising city networks: a research agenda. Urban Studies 56(16):3540-3555

Ezeibe CC, Nzeadibe TC, Ali AN, Udeogu CU, Nwankwo CF, Ogbodo C (2017) Work on wheels: collective organising of motorcycle taxis in Nigerian cities. International Development Planning Review 39(3):249-273

Falzon MA (2016) Introduction: multi-sited ethnography: theory, praxis and locality in contemporary research. In: Falzon MA (ed) Multi-sited ethnography: theory, praxis and locality in contemporary research. Routledge, London, pp 15-38

Gambetta N, Azadian P, Hourcade V, Reyes ME (2019) The financing framework for sustainable development in emerging economies: the case of Uruguay. Sustainability 11(4):1059

Gibbs A, Dunkle K, Washington L, Sikweyiya Y, Willan S, Shai N, Jewkes R (2020) Factors associated with young people's attendance at an IPV prevention intervention in informal settlements in South Africa: a prospective analysis. Global Public Health 15(2):161-172

Glaser BG, Strauss AL (1967) Discovery of grounded theory: strategies for qualitative research. Routledge; Aldine de Gruyter, New York

Goodfellow T (2017) Urban fortunes and skeleton cityscapes: real estate and late urbanization in Kigali and Addis Ababa. International Journal of Urban and Regional Research 41(5):786-803

Goodfellow T, Owen O (2020) Thick claims and thin rights: taxation and the construction of analogue property rights in Lagos. Economy and Society 49(3):406-432

Grossman S (2020) The politics of order in informal markets: evidence from Lagos. World Politics 72(1):47-79

Hoffman D (2007) The city as barracks: Freetown, Monrovia, and the organization of violence in postcolonial African cities. Cultural Anthropology 22(3):400-428

Hummel C (2017) Disobedient markets: street vendors, enforcement, and state intervention in collective action. Comparative Political Studies 50(11):524-555

International Science Council (ISC) (2017) A guide to SDG interactions: from science to implementation. In: Griggs DJ, Nilsson M, Stevance A, McCollum D (eds) International Council for Science, Paris

Kabeer N (2016) Leaving no one behind: the challenge of intersecting inequalities. In: World Social Science Report 2016—challenging inequalities: pathways to a just world. International Social Science Council, Paris; Institute of Development Studies, University

of Sussex; United Nations Educational, Scientific and Cultural Organisation (UNESCO), Paris, pp 55-58
Kovacic Z, Musango JK, Ambole LA, Buyana K, Smit S, Anditi C, Mwau B, Ogot M et al. (2019) Interrogating differences: a comparative analysis of Africa's informal settlements. World Development 122:614-627
Luque-Ayala A, Neves Maia F (2018) Digital territories: Google maps as a political technique in the re-making of urban informality. Environment and Planning D: Society and Space 37(3):449–467
Maclin BJ, Bustamante ND, Wild H, Patel RB (2020) To minimise that risk, there are some costs we incur: examining the impact of gender-based violence on the urban poor. Global Public Health 15(5):734-748
Mahmoud H, Alfons R, Reffat RM (2019) Analysis of the driving forces of urban expansion in Luxor City by remote sensing monitoring. International Journal of Integrated Engineering 11(6):296-307
Makhetha E, Maliehe S (2020) 'A concealed economy': artisanal diamond mining in Butha-Buthe district, Lesotho. The Extractive Industries and Society 7(3):975-981
Malonza JM, Ortega AA (2020) Fissures in localizing urban sustainability: the case of Rwanda. GeoJournal, 1-20
Martin AM, Bezemer PM (2020) The concept and planning of public native housing estates in Nairobi/Kenya, 1918–1948. Planning Perspectives 35(4):609-634
Martínez-Ferrero J, Eryilmaz M, Colakoglu N (2020) How does board gender diversity influence the likelihood of becoming a UN Global Compact signatory? The mediating effect of the CSR committee. Sustainability 12(10):4329
Matamanda AR, Nel V (2020) Sustainable urbanization in Africa: the critical enablers and disablers. In: Leal FW, Azul A, Brandli L, Özuvar P, Wall T (eds) Sustainable cities and communities: UN Encyclopedia of the UN Sustainable Development Goals. Springer Publishing, Cham., pp 738-751
Miraftab F (2012) Colonial present: legacies of the past in contemporary urban practices in Cape Town, South Africa. Journal of Planning History 11(4):283-307
Mohan G, Tan-Mullins M (2019) The geopolitics of South–South infrastructure development: Chinese-financed energy projects in the global South. Urban Studies 56(7):1368-1385
Mukumbang FC, Ambe AN, Adebiyi BO (2020) Unspoken inequality: how COVID-19 has exacerbated existing vulnerabilities of asylum-seekers, refugees, and undocumented migrants in South Africa. International Journal for Equity in Health 19(1):1-7
Mususa P (2012) Mining, welfare and urbanisation: the wavering urban character of Zambia's Copperbelt. Journal of Contemporary African Studies 30(4):571-587
Mwale KP, Lintonbon J (2020) Heritage, identity and the politics of representation in tribal spaces: an examination of architectural approaches in Mochudi, Botswana and Moruleng, South Africa. International Journal of Heritage Studies 26(3):281-298
Parnell S (2016) Defining a global urban development agenda. World Development 78:529-540
Parnell S, Pieterse E (2014) Africa's urban revolution. Zed Books Limited, London
Patel Z, Greyling S, Simon D, Arfvidsson H, Moodley N, Primo N, Wright C (2017) Local responses to global sustainability agendas: learning from experimenting with the urban Sustainable Development Goal in Cape Town. Sustainability Science 12(5):785-797
Persson Å, Weitz N, Nilsson M (2016) Follow-up and review of the Sustainable Development Goals: alignment vs. internalization. Review of European, Comparative & International Environmental Law 25(1):59-68
Pieterse E, Parnell S, Haysom G (2018) African dreams: locating urban infrastructure in the 2030 Sustainable Developmental Agenda. Area Development and Policy 3(2):149-169
Porsani J, Börjeson L, Lehtilä K (2017) Land concessions and rural livelihoods in Mozambique: the gap between anticipated and real benefits of a Chinese investment in the Limpopo valley. Journal of Southern African Studies 43(6):1181-1198
Republic of Uganda (2020) The second Voluntary National Review report on the implementation of the 2030 Agenda for Sustainable Development. Office of the Prime Minister, Kampala. https://sustainabledevelopment.un.org/content/documents/26354VNR_2020_Uganda_Report.pdf. Accessed 19 July 2021
Rohat G, Flacke J, Dosio A, Dao H, van Maarseveen M (2019) Projections of human exposure to dangerous heat in African cities under multiple socioeconomic and climate scenarios. Earth's Future 7(5):528-546
Schmieg G, Meyer E, Schrickel I, Herberg J, Caniglia G, Vilsmaier U, Laubichler M, Hörl E et al. (2018) Modeling normativity in sustainability: a comparison of the Sustainable Development Goals, the Paris Agreement, and the papal encyclical. Sustainability Science 13(3):785–796
Schneider F, Kläy A, Zimmermann AB, Buser T, Ingalls M, Messerli P (2019) How can science support the 2030 Agenda for Sustainable Development? Four tasks to tackle the normative dimension of sustainability. Sustainability Science 14(6):1593-1604
Sebestyén V, Domokos E, Abonyi J (2020) Focal points for sustainable development strategies—text mining-based comparative analysis of Voluntary National Reviews. Journal of Environmental Management 263:110414
Sikor T, Lund C (2009) Access and property: a question of power and authority. Development and Change 40(1):1-22
Stafford-Smith M, Griggs D, Gaffney O, Ullah F, Reyers B, Kanie N, Stigson B, Shrivastava P et al. (2017) Integration: the key to implementing the Sustainable Development Goals. Sustainability Science 12(6):911-919
Steel G, Abukashawa S, Hussein MO (2020) Urban transformations and land governance in peri-

urban Khartoum: the case of Soba. Tijdschrift Voor Economische en Sociale Geografie 111(1):45-59

Stewart P, Bezuidenhout A, Bischoff C (2020) Safety and health before and after Marikana: subcontracting, illegal mining and trade union rivalry in the South African mining industry. Review of African Political Economy 47(163):27-44

Thieme TA (2018) The hustle economy: informality, uncertainty and the geographies of getting by. Progress in Human Geography 42(4):529-548

Thorn J, Thornton TF, Helfgott A (2015) Autonomous adaptation to global environmental change in peri-urban settlements: evidence of a growing culture of innovation and revitalization in Mathare Valley Slums, Nairobi. Global Environmental Change 31:121-131

Tosun J, Leininger J (2017) Governing the interlinkages between the Sustainable Development Goals: approaches to attain policy integration. Global Challenges 1(9):1700036

Ulbrich P, de Albuquerque JP, Coaffee J (2019) The impact of urban inequalities on monitoring progress towards the Sustainable Development Goals: methodological considerations. ISPRS International Journal of Geo-Information 8(1):6

United Nations (UN) (2015) Transforming our world: the 2030 Agenda for Sustainable Development (70/1). United Nations General Assembly, New York. https://sustainabledevelopment.un.org/post2015/transformingourworld. Accessed 19 July 2021

United Nations (UN) (2016) New Urban Agenda: Quito declaration on sustainable cities and human settlements for all (71/256). United Nations General Assembly, New York

United Nations Department of Economic and Social Affairs (UNDESA) (2014) World urbanization prospects: the 2014 revision, highlights. United Nations Department of Economic and Social Affairs, New York

United Nations Economic Commission for Africa (UNECA) (2017) Urbanisation and industrialisation for Africa's transformation: economic report for Africa. United Nations, Addis Ababa, Ethiopia. https://www.uneca.org/sites/default/files/fullpublicationfiles/era-2017_en_fin_jun2017.pdf. Accessed 19 July 2021

UN-Habitat (2016) World cities report 2016: urbanization and development—emerging futures. United Nations Habitat, Nairobi. https://unhabitat.org/sites/default/files/download-manager-files/WCR-2016-WEB.pdf. Accessed 19 July 2021

UN-Habitat (2018) SDG 11 synthesis report 2018 on sustainable cities and communities. United Nations Habitat, Nairobi

Urdal H, Hoelscher K (2009) Urban youth bulges and social disorder: an empirical study of Asian and sub-Saharan African cities. Policy research working paper no. WPS 5110. World Bank, Washington, DC

Valencia SC, Simon D, Croese S, Nordqvist J, Oloko M, Sharma T, Taylor Buck N, Versace I (2019) Adapting the Sustainable Development Goals and the New Urban Agenda to the city level: initial reflections from a comparative research project. International Journal of Urban Sustainable Development 11(1):4-23

Weber C, McCollum DL, Edmonds J, Faria P, Pyanet A, Rogelj J, Tavoni M, Thoma J et al. (2018) Mitigation scenarios must cater to new users. Nature Climate Change 8(10):845-848

Wright EM, Skubak Tillyer M (2020) Neighborhoods and intimate partner violence against women: the direct and interactive effects of social ties and collective efficacy. Journal of Interpersonal Violence 35(19-20):3913-3938

Yeboah T (2020) Future aspirations of rural-urban young migrants in Accra, Ghana. Children's Geographies 19(2):1-14

Zhilin LI, Gong X, Chen J, Mills J, Songnian LI, Zhu XU, Peng TI, Hao WU (2020) Functional requirements of systems for visualization of Sustainable Development Goal (SDG) indicators. Journal of Geovisualization and Spatial Analysis 4(1):1-10

Zinn J (2005) The biographical approach: a better way to understand behaviour in health and illness. Health, Risk & Society 7(1):1-9

Check for updates

Localizing the SDGs Through the Formal-Informal Interface: The Case of Ard al-Liwa, Cairo

4

Omar Nagati, Hanaa Gad, and Amin El-Didi

Abstract

The inclusion of a standalone urban Sustainable Development Goal (SDG) reflects a recognition of the importance of cities for achieving the United Nations Agenda 2030. While the goal and its targets represent a more comprehensive view of the urban, compared with the Millennium Development Goals (MDGs), taking into account issues such as mobility, heritage, waste management, and public spaces, the universal nature of the SDGs limits their applicability to urban contexts in the Global South. They fall particularly short when it comes to engaging with the workings of urban informality, which is central to cities in the Global South and especially Africa, thereby reinforcing existing policies that tend to stigmatize informal practices. This chapter draws on research conducted at the neighborhood level in Cairo, Egypt, to illustrate the limitations of the SDG framework and current government approaches to informality by showing that informality exists on a spectrum that ranges from the formal to the informal. The actors, practices, and processes that exist along this spectrum discredit the mainstream dualistic understanding of formality and informality. They also offer alternative visions of sustainable development, as well as productive suggestions for more localized understandings of, and approaches to, the urban that are needed if Africa is to achieve the SDGs by 2030.

Keywords

SDG 11 · SDG 6 · Policy-practice · Formal-informal interface · SDGs · Housing · Water access · Transportation · Cairo

4.1 Introduction

This chapter lies at the intersection between two prevailing debates on African cities and the Global South in general. The first addresses the localization of the Sustainable Development Goals (SDGs) within Agenda 2030. The second engages with the elusive condition of informality in general and in particular with the perception that there is a formal-informal binary. In doing so, the chapter builds on research conducted in Cairo, Egypt, by Cairo Lab for Urban Studies, Training and Environmental Research (CLUSTER) on the urban formal-informal interface and SDG localization. CLUSTER was founded in 2011 in downtown Cairo as a platform

O. Nagati (✉) · H. Gad · A. El-Didi
Cairo Lab for Urban Studies, Training and Environmental Research (CLUSTER), Cairo, Egypt
e-mail: onagati@clustercairo.org; hanaagad@clustercairo.org; amin.aldidi@clustercairo.org

S. Croese, S. Parnell (eds.), *Localizing the SDGs in African Cities*, Sustainable Development Goals Series, https://doi.org/10.1007/978-3-030-95979-1_4

for urban research, architecture, art, and design initiatives. CLUSTER engages critical discourses while being grounded in professional practice, with an emphasis on participatory design processes.

CLUSTER's work on SDG localization has been conducted in collaboration with Ardhi University in Dar es Salaam, Tanzania. The aim was to develop a bottom-up approach and framework to SDG localization in African cities in order to mediate the top-down policies by national and local governments and grounded practices, through a set of revised indicators and measures to improve local practices. The focus was on SDG 11.1 and 11.2, on housing and transportation, respectively, and SDG 6.1, on access to water. CLUSTER's work on the formal-informal interface in turn is part of broader research conducted through the African Urban Research Initiative (AURI), an Africa-based research network, to inform and enhance the policy actors and networks responsible for sustainable urban policy and management in different African contexts. Research in three Egyptian cities, as well as comparative research across four African cities, showed that the formal-informal interface is viewed as a spectrum or a continuum rather than a binary structure. It offers a generative framework to understand a complex interconnection between formal policies and informal practices, using themes such as transportation, food supply, and water infrastructure as sites for investigation (Nagati and Stryker 2021).

The SDG framework has many shortcomings, many of which are products of the development and export of models without a grounded reflection of the complex realities of the Global South, which often means that the SDGs are inapplicable or incomplete. Localization processes need to make up for such insufficiencies. On the other hand, current policy around urban informality in Africa is largely based on a characterization of informality as abnormal and inherently negative. More so, it tends to treat informal and formal sites and practices as discrete and often homogenous entities, thus reproducing the segregation, marginalization, and stigmatization of informal practices.

Building on the case of Cairo, this chapter aims to expose an inherent contradiction in government policy toward informality. This is an explicit policy to formalize and integrate the informal into the formal system while maintaining its "illegal" status due to its inability to pursue this agenda and exacting fines in the meantime as a source of revenue. Such a paradox of delegitimization-as-extraction, this chapter argues, has become a de facto policy approach toward informality in Egypt. This maintains a gray zone, or interface, between the formal and informal, as a result of the state's incapacity to fully incorporate informal practices. Based on a review of such a policy and practices, and their working and results at the neighborhood level, the chapter argues that appropriately addressing the complexity of informality involves engaging with the gray zone between the formal and the informal and using this as an alternative framework for understanding local practices as well as developing more grounded measures for SDG monitoring and localization.

This chapter builds on previous papers developed in collaboration with Ardhi University in Dar es Salaam, Tanzania, for the Council for the Development of Social Science Research in Africa (CODESRIA) (Nagati et al. 2021a, b). The research and data gathering involved a series of field research workshops in the neighborhood of Ard al-Liwa in Cairo in 2019 and 2020. Empirical data collection, participant observation, and interviews were complemented with archival research of previous studies, published reports, and media analysis. Data analysis, including the development of a visualization strategy, was mostly conducted during the pandemic lockdown in 2020, when access to the field was limited. The overall framework for this case study hinged on developing a middle ground between SDGs targets and state policies, on the one hand, and performance-based standards in informal areas, on the other, for the purpose of articulating local indicators for the monitoring and implementation of universal goals.

4.2 SDG Localization, Urban (In) formality, and Scale

Each SDG is designed with a monitoring and assessment framework of specific targets and measurable indicators. While this framework seeks to provide a comprehensive system for defining and measuring progress toward achieving the goals at the global and national scales, the effectiveness of the practical application of these goals has been heavily scrutinized (Simon et al. 2016; Arfvidsson et al. 2017; Patel et al. 2017). Advocates of SDG localization have primarily argued in favor of support for the inclusion of the sub-national scales of government in the SDG process and effectively cascading programs and projects down to different scales of government, institutions, and organizations within a given country (and in cooperation with other countries). Localizing, which is a process by which the SDG agenda is translated across these scales of government through, and improving the mechanisms through which the principles of the SDGs become effectual, entails incorporating and aligning local priorities and opportunities in the national design, implementation, and monitoring process (Moallemi et al. 2019, p. 1).

While efforts toward localization are appreciated, our research shows that the SDG framework itself is not sufficiently informed by local experience for it to be effective and applicable. New methodologies are needed for localizing the SDG framework, especially in Global South contexts, while acknowledging the universal scope and nature of the SDGs at a global level (Nagati et al. 2021a).

The main shortcomings of the SDG framework are predominantly linked to the prioritization of the standardization and harmonization of development standards at a universal level, rather than ensuring relevance and applicability in defined local settings. As such, context-sensitive specificities are under- or misrepresented, obscured, or entirely missed. Through the metrification of results, local conditions and solutions became unrooted, abstracted, and decontextualized. We find that within the context of the Global South, particular challenges with appropriately grasping, measuring, and representing local conditions arise due to the lack of baseline data, as well as issues with the streamlined monitoring of results and the transparent communication thereof.

A number of studies have been produced that evaluate how local actors can be equipped and enabled in order to localize the SDGs, as well as the challenges of such work (Kanuri et al. 2016). Researchers have pointed toward the challenge of applying universally proclaimed and harmonized development goals, standards, and ideals toward contexts that have vastly different socioeconomic and political structures, institutional capacities, and resources, as well as cultural settings. Caprotti and colleagues have closely looked at this point and demonstrate—on the example of the urban SDG—the following (Caprotti et al. 2017, p. 370):

> For one thing, while there may be an obvious logic behind codifying and standardizing information and knowledge about urban development (to enable shared practice learning, scale up innovation and improve benchmarking), at the same time this comes at the risk of decontextualizing and devaluing the intrinsically local and social urban realities.

From this perspective, the post-2015 agenda (which includes the New Urban Agenda) serves as a technocratic development tool that is based on a smart-city understanding of measurable socioeconomic (urban) settings (Broome and Quirk 2015; Fukuda-Parr 2016). This is further illustrated by looking at SDG 11, with its first indicator on housing designed to measure "the proportion of the urban population living in slums, informal settlements or inadequate housing." This indicator is very telling and symptomatic of the SDGs' design in general and the conception of the indicators.

First, the indicator subsumes dramatically different standards, conditions, and quality levels of architectures and settlement structures. In most urban contexts in the world, informal housing solutions have developed to meet market demands when government or private sector solutions have failed to meet the needs of the local population. Yet, even though this urban condition is shaping the reality of most urbanites all over the planet, it is dramatically different in terms of its functionality, its legal setup, its efficiency, and its adequateness. Further, the definition of what counts as slum or informal housing settlement drastically varies from one place to the next. The indicator, however, cannot reflect these individualities and this diversity. It simply makes a statistical assumption upon its proclaimed scale. Whether an informal settlement is appropriate, affordable, connected, sustainable, safe, and environmentally friendly or not is not being displayed and measured based on what the indicator allows. In this sense, the SDG target on housing obscures this urban condition behind statistical "facts."

Caprotti et al. (2017) argue that developmental tools such as the SDGs and the New Urban Agenda follow a statistical ideal that works along big data analytics and modeling tools to guide policymakers and planners. However, to numerically measure and thus reduce a local infrastructure or service in this way excludes a rich diversity of local (urban) realities and, hence, decontextualizes local development standards. Following their line of thought, Agenda 2030 is aimed at standardizing and harmonizing efforts toward improving local services and networks. But by leaving out the specificities of a local context, its goals can become redundant for some while missing large parts of locally available services and networks that the standardized, government-focused approach of the SDGs cannot measure.

The critiques of the SDG framework thus relate to the particular challenges that (local) governments face with its implementation, as well as the very design of the SDGs, its targets, and indicators. We find that Agenda 2030 tends to represent and favor certain ideals of development that come out of the Global Northern context, leading to the problematization or misrepresentation of local conditions beyond the West. These issues are particularly exacerbated where informal urbanism is the norm, undermining the ambitions and benefit of the SDGs as a globally inclusive framework for development. On several levels, the existing SDGs lack a grounded understanding of cities in the Global South and the practical ways in which such an agenda could be implemented. This ties into the broader problem of eurocentricity in urban studies, where the experiences of cities in Western Europe and North America are taken to be the norm or a default global standard which is compounded by a relative lack of Global South knowledge production (Robin and Acuto 2018).

The universally formulated SDGs have largely been uncritically adopted and translated into national development agendas in Africa, often with support of UN agencies and other international organizations. These are overlaid on a specific development trajectory, which includes a neo-liberal turn following a post-colonial legacy resulting in rapid urbanization. A failed post-independence state led to an erosion of welfare programs and urban services, resulting in divergent urban conditions across cities on the continent including a deteriorating urban core and sprawling informality alongside neo-liberal exclusive (gated) development. Most of the targets and indicators formulated in SDG 11 but also other urban-related SDGs, such as SDG 6, largely fail to recognize these specific conditions and thus overlook not only the possibility but rather the necessity to work with informality as a modus operandi.

The paradoxical but pervasive position of formal structures versus informal urban practices raises important questions around existing systems of political and urban governance in the African context. Localizing the SDGs in line with complex local realities requires governance systems that channel the needs and priorities of local communities into policies, programs, and projects on the one hand while offering mechanisms for the implementation, monitoring, and evaluation of these policies through an intricate system of local governance. In many cases, the absence of participatory modes of governance does not only impede a measurable implementation of SDGs or the New Urban Agenda but also results in the appropriation of the discourse of global development agendas to justify lucrative or politically motivated projects under the guise of sustainability (Malonza and Ortega 2020). Examples from Egypt include the demolition of the historical neighborhood of Maspero in the city center and the expropriation of private properties in al-Warraq "rural" Nile island as potential sites for urban development (Zaazaa et al. 2015).

This leads to a final question, which is that of scale. How can a third position be mediated between the universal and the particular, both conceptually and operatively? In this research, we chose the city and district level as a middle ground between supra-national and local contexts, by engaging the neighborhood of Ard al-Liwa in Cairo and its complex formal-informal interrelations as the site of our empirical analysis.

4.3 The Formal-Informal Interface in Egypt

Ard al-Liwa, with a population of approximately 150,000[1] citizens, is located within the informal belt surrounding the city of Giza from the west and north and split from it by the railway and al-Zomor canal that extends from Imbaba and Bashtil to the north and Umraniya to the south (see Fig. 4.1). Ard al-Liwa occupies a focal point in the western part of Cairo's informal built environment, as it is surrounded by major axes on the city scale while being a smaller and denser area compared to the broader surrounding neighborhoods. Its administrative dependency changed from being an extension of Kirdasa (rural village) to being part of the Aguza district (predominantly formal). As it is informally built on top of agricultural land, Ard al-Liwa is considered an unplanned area where its lack of regulations and local services are its primary characteristics. The history, size, and regional context are purposely engaged as a site and ecosystem to test our methodological toolkit for SDG localization empirically. As such, rather than proposing an alternative norm that implies universal pretensions, on the one hand, or immersing into one specific locale with all its particularities and idiosyncrasies, on the other, a city- or district-level scale may present a middle ground that would potentially expand the applicability of this research methodology to other cities in Africa, ranging from small towns to large urban neighborhoods.

In Egypt, as in many other cities in Africa and the wider Global South, informal urban practices emerged as a response to neo-liberal policies during the last decades of the twentieth century. As the state gradually withdrew from its role as a provider of social and urban services, communities developed creative solutions to fill this gap, in housing and infrastructure, health and education, as well as transportation and mobility. While the state does not recognize informal housing as legal, it does provide access to infrastructure, such as water, electricity, and natural gas. This is at a cost, technically under a formal umbrella, such as through installation fees, water meters, and electricity bills. Further, the past decade has witnessed attempts to formalize informal practices (e.g., Standardized Housing Law, Reconciliation Law, and Financial Inclusion policy) to exact fines, fees, and taxes.[2] In other areas, the state maintains the informal as illegal to exact fines, as exemplified in street vendors and informal modes of transportation (auto-rickshaw, microbus, mini-van taxi). The interface is thus maintained as a middle ground for the state's varying interpretations, and, as will be further explained in the case studies below, this principle of delegitimization-as-extraction can be seen as de facto policy toward informality through persistent legislative inaction.

The empirical analysis in this chapter is organized into three main sections. The first section offers a background to the evolution of policies toward urban informality in Egypt. The second section presents the formal-informal interface through an analysis of policies and practices in

[1] According to different publications of various census and development agencies, the population of Ard al-Liwa differs. The estimation changes according to a number of criteria, one of them being that the boundaries of the area are not defined clearly. After reviewing various statistics, we refer to the latest national census, CAPMAS: 142,601 in 2017. According to Egypt's General Organization for Physical Planning (GOPP), the population grew from 101,177 in 2006 to 145,811 in 2012.

[2] See law descriptions in the relevant empirical research sections below.

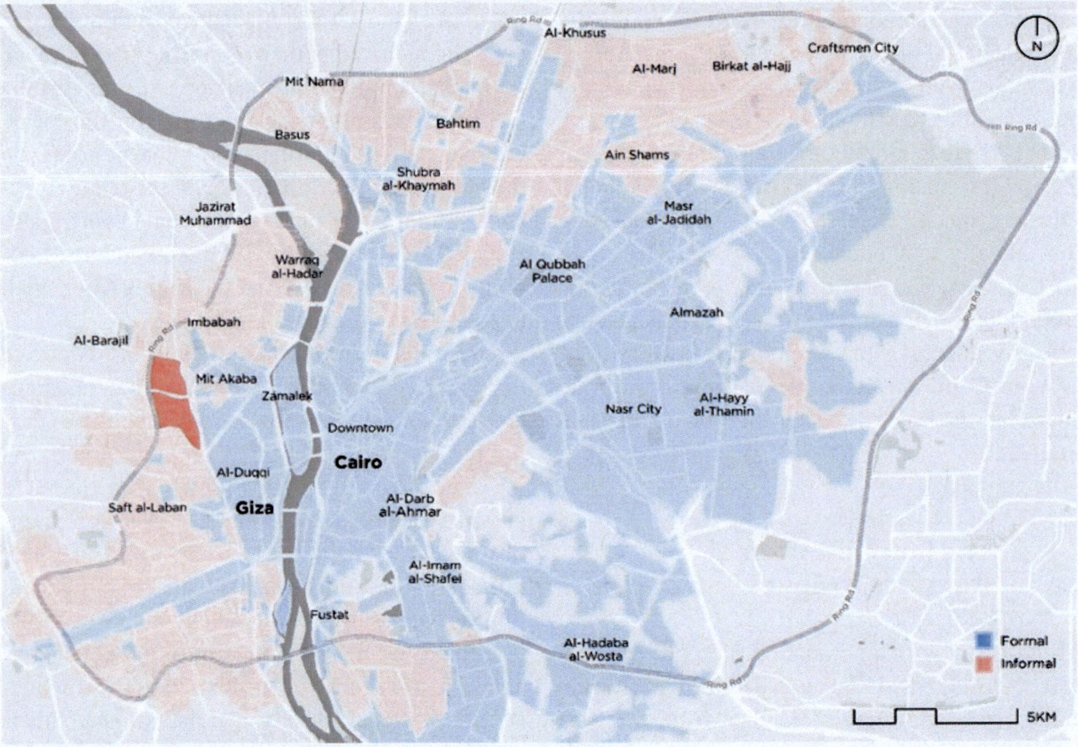

Fig. 4.1 Ard al-Liwa (in bright red), in the context of Greater Cairo (formal areas shown in blue, informal areas in red), bordered by the ring road surrounding the city's periphery (Map by authors, adapted from CLUSTER 2018, based on GCR Google satellite map, 2021)

the neighborhood of Ard al-Liwa in Cairo around three themes: housing, transportation, and water infrastructure. Lastly, a third section revisits these themes to unpack the above contradictions while addressing the problematic approach to SDG localization in Egypt and African cities more generally.

4.3.1 Background

Egypt constructs around one million housing units a year, a significant number when population growth is taken into account. This housing has been provided by three primary sectors since the 1950s: the informal private sector, which mainly builds on private agricultural land without permits; the formal private sector, which builds on government-sanctioned subdivisions; and the public sector, through a number of state-owned agencies and companies. Over the past decade, the vast majority of all housing has contributed to construction that is seen by the government as informal. The Ministry of Housing reports that 40% of the population lives in informal settlements in Cairo, the city where the size of the informal areas is best known. A census data comparison between 1996 and 2006 places the figure at 70% (Sims 2012), with the percentage rising after the revolution of 2011 (for further up-to-date analysis, see Shawkat 2020).

The entity put in place in the effort to address informality in a strategic manner, led by a central government rather than remaining a local governance issue, is the Informal Settlements Development Facility (ISDF). The ISDF was established in 2008, under the prime minister and mandated to deal with unsafe housing conditions. Considering that the ISDF taxonomy of informal settlements covers an array of conditions, based on a UN-Habitat classification of slums, the outlying ISDF strategy emphasizes that upgrading informal areas in situ should be the main approach, with the exception of displacement when it is the only option to eliminate life threat (American University in Cairo Center for Sustainable Development 2014). These intentions, nevertheless, did not yet scale up or receive support from relevant stakeholders. In addition, the ISDF strategy for dealing with informality is a topic that remains up for debate, from an effectiveness point of view, as well as a rights-based and sustainability perspective. The debate on the current strategy and practices remains confined to professional circles or governmental bodies, rather than a societal dialogue on the means with which to approach informality. Furthermore, the ISDF adopted a model for using the land occupied by informal settlements as an asset, covering the cost of redevelopment through land sale or investment. This impelled the ISDF to focus on areas with the potential for redevelopment and investment, further convoluting what on-ground definitions of informal areas are in use by state bodies.

Recently, Egypt has gone through major political, social, and economic turmoil. The Arab Spring and the Egyptian revolution helped change the long-term prevailing political status quo in the country, resulting in a new national constitution and the country's first presidential elections held in 2013. The new constitution was approved through a national referendum, and it is said to be committed toward sustainable development. It was adopted in January 2014, and, according to governmental papers, it marks a significant improvement in securing citizens' rights to education, health, social protection, and development while being aligned with many of the 17 SDGs (Arab Republic of Egypt 2016a). In addition to the national constitution, the *Sustainable Development Strategy: Egypt Vision 2030* is the most important umbrella document for governmental planning, policy, and budgeting on sustainable development in the country until 2030. This policy roadmap was formulated and endorsed by the cabinet of current president Abdel Fattah el-Sisi who took on the presidential seat in 2014. It is the major reference document through which the SDGs will be implemented at the national level, and it states explicitly that it seeks to "[a]lign SDS (Sustainable Development Strategy) objectives with those of the post-2015 United Nations Sustainable Development Goals (SDGs) and the Sustainable Development Strategy for Africa 2063" (Arab Republic of Egypt 2016b).

Each dimension of the agenda includes several "pillars," one of which is urban development. The Key Performance Indicators (KPIs) of Vision 2030 with regard to urban development rest on two interlinked major concerns. One is the elimination of informal (and insecure) areas. The second is the expansion of new urban communities. New urban communities are satellite cities, towns, or entire cities constructed either by private large-scale real estate developers or through state-led initiatives. The KPIs focus on building up new urban communities, many of them in the Greater Cairo region, and settling urban dwellers in these new areas. This goes hand in hand with the elimination of "slums," as well as the eviction of people from areas that are declared "unsafe."

4.3.2 Three Case Studies from Ard al-Liwa

In the following sections, three case studies are presented from the neighborhood of Ard al-Liwa, to problematize and unpack the binary approach to informality as encapsulated in three selected SDG targets and their respective indicators (see Fig. 4.2). Through detailed process descriptions of local practices related to the provision of urban housing, water, and transportation, the case studies provide insight into the formal-informal interface as a continuum, offering an alternative framework for localizing strategies.

For the purpose of urban development on a neighborhood level, there are two types of land division that can be distinguished. The first is the official system of zoning and land allocation in new satellite desert cities planned by the state, Egypt's largest landowner, through its various organs. The other is an informal pattern, whereby small owners of agricultural land, especially those bordering cities, control plot divisions

Fig. 4.2 Analysis framework of the urban fabric, modes of transport, and water provision present in Ard al-Liwa (Nagati et al. 2021b)

(Ramadan 2020). The rapid growth of informal areas should be seen in the light of the failure of the new desert cities to attract population (Raggal 2020). As such, the state is left with the paradox of its inability to cope with an exponential demand for housing and its unwillingness to legitimize informal urban development, hence the contradictions apparent in the various urban themes presented here.

Housing (SDG 11.1)

Considering the SDG agenda and its classifications of informality, as exemplified by the SDG 11 target on housing and its corresponding indicator (see Table 4.1), there remains an indistinctness and a disconnect from the reality of housing provision. This is specifically the case in Global South cities, where measures of safety, adequateness, and affordability vary. This blanket terminology carries over to government development agendas such as Egypt's Vision 2030, which describes informality as one monolith that is to be "eliminated," resulting in policies of exclusion, dislocation, and inequality.

As urban expansion continues to be mainly informally driven in Cairo, it is necessary to understand its processes, and its complex interactions with the formal system in an exemplary context such as that of Ard al-Liwa, in order to inform a localized framework which acknowledges all practices on the ground, and their role in inclusive and participatory urbanism.

A typical land division process includes an agriculture basin of one *feddan* (approx. 1 acre or 0.4 hectare) that is sold to a local developer and in turn subdivided into individual plots; each is one *qirat* (one *feddan* = 24 *qirat*). Each plot, approximately 175 m^2, is required (by the developer) to leave at least 3 m on the front side, forming a street width of 6–8 m. While the land purchase is registered and thus integrated into the formal system, plot subdivision is entirely informal following customized codes and regulations, including plot sizes, street width, and building height. Housing units on these plots are generally organized around three typologies, a single-family of one or two stories (*bayt*), a multi-family building of three to six stories (*imara*), and a tower building (*burj*) of more than six stories (Angélil and Malterre-Barthes 2016, p. 82). The plot size has evolved over time. While, historically, a typical building's footprint for family units varied from 75 to 125 m^2, with one or two small apartments per floor, more recently, the land division shifted toward a bigger footprint for high-rises (locally referred to as "towers"), reaching up to 250–450 m^2, with larger apartments and more units per floor (ibid).

Table 4.1 SDG 11 and its target and measuring indicator specific to the issue of housing

SDG 11: Make cities and human settlements inclusive, safe, resilient, and sustainable	**Target 11.1:** By 2030, ensure access for all to adequate, safe, and affordable housing and basic services and upgrade slums	**Indicator 11.1.1:** Proportion of urban population living in slums, informal settlements, or inadequate housing

The resulting urban fabric, including street width, allows for limited choices for transportation, limiting large vehicles from maneuvering, even on main streets. This poses difficulties for mini-buses, cars, and taxis while becoming more conducive for smaller modes of transport such as the auto-rickshaw (locally known as tuktuk), which in turn affects the capacity of how many cars can pass. This pattern of growth contributes to particular modes of mobility and transportation networks, in terms of people's commutes and traffic flow, including an informal system that helps to fill the gap where the formal transportation network falls short.

The construction process (see Fig. 4.3) demonstrates basic knowledge gained from working in formal areas, which in practice is being filtered through a complex process of trial and error over millions of reiterations, stripping buildings of all superfluous elements to its bare essentials. In the absence of engineering drawings and calculations, local developers tend to overestimate the code for steel reinforcement and concrete mix as a safety precaution, reflecting their interest to secure their investment. This "over-code," in terms of structural elements, is contrasted with the minimal spatial requirements used for rooms,

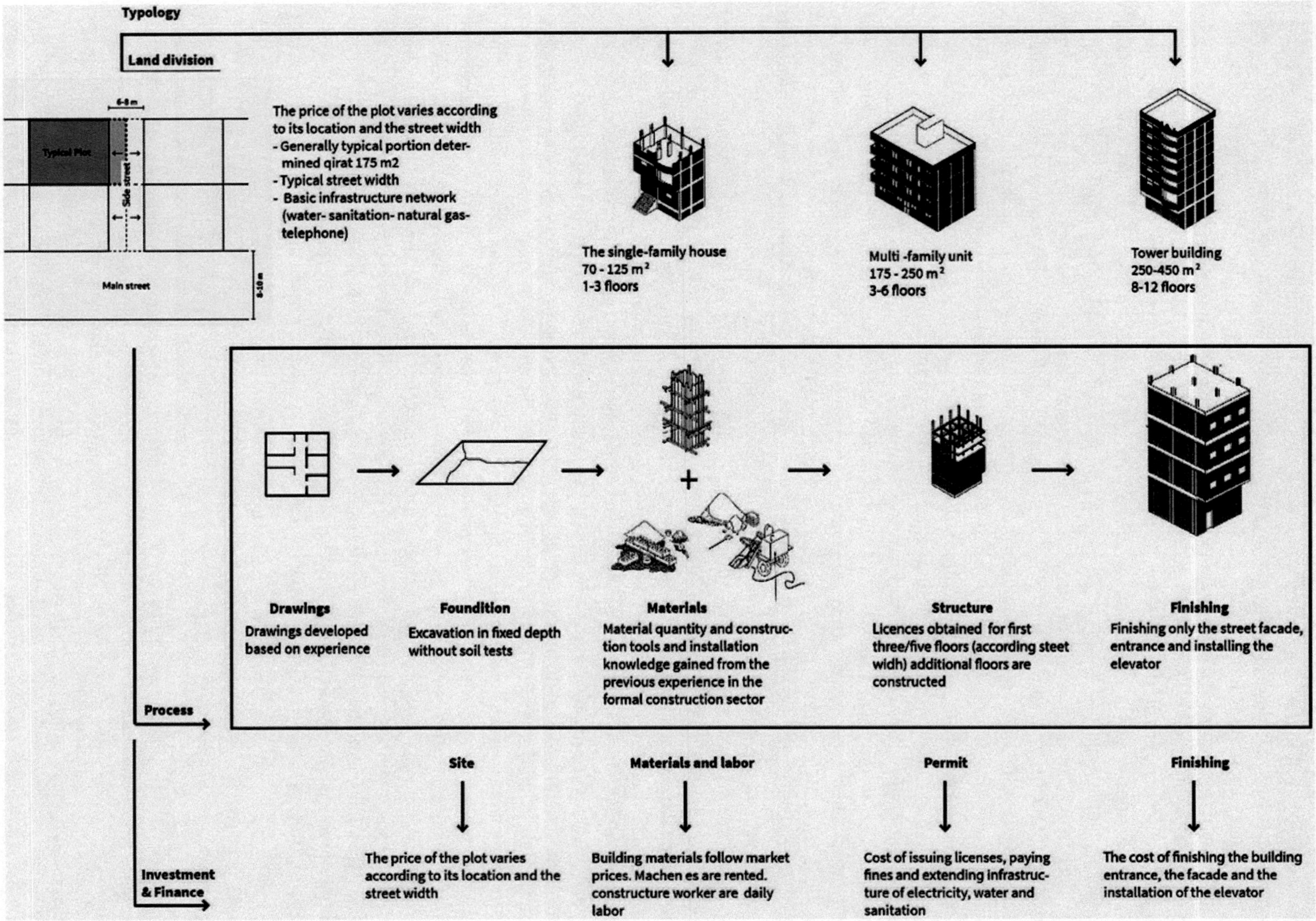

Fig. 4.3 Housing typology and construction process (Diagram adapted from *Housing Cairo: the Informal Response*, Angélil and Malterre-Barthes 2016)

stairs, and ventilation shafts, in a clear non-compliance with the formal building code. Some of the raw materials are purchased from the formal market, including tax receipts, such as cement and re-bars, while other materials are often supplied by local traders, such as sand, bricks, and wood as part of the parallel informal economy. Construction workers are largely recruited informally as daily wage laborers, lacking tax records or insurance.

Financing a building starts from the plot purchase, where price per square meter varies according to location, on a main or sub street, corner plot versus inner plot, and the proximity to general amenities, and access points to the neighborhood at large. The construction cost includes materials and labor, often commissioned to one general contractor. It involves the foundation work, main structure and infrastructure, as well as inside partition walls (core and shell). But it excludes the finishing which is often left to the future owner of each individual unit to complete. The only exception to the finishing phase is the front façade, with the main entrance and the installation of an elevator, all being the responsibility of the plot owner-cum-developer. Once the building is completed, the process of connecting it to the formal infrastructure network, water, electricity, and natural gas, reintroduces the interface as a negotiable framework between formal and informal processes. The partial formalization, including installation assessment fees, actual conduits and meters, as well as monthly bills, are all paid to the respective authorities with official receipts, even if the entire building lacks a permit and thus legitimate formal registration.

Water Access (SDG 6.1)

Regarding access to drinking water, SDG 6.1 aims for safe and affordable drinking water for all, regardless of housing status, by measuring the proportion of the population using safely managed water services (see Table 4.2). To that end, the provision of clean water as an essential need for citizens is recognized in the Egyptian Constitution, Article 79, which states: "Each citizen has the right to healthy, sufficient amounts of food and clean water." The right of the public to supervise and monitor services to ensure their quality is also stipulated in the World Charter to the Right to the City (WCRC) (Tadamun 2014).

Table 4.2 SDG 6 and its target and measuring indicator specific to the issue of water access

SDG 6: Ensure availability and sustainable management of water and sanitation for all	**Target 6.1:** By 2030, achieve universal and equitable access to safe and affordable drinking water for all	**Indicator 6.1.1:** Proportion of population using safely managed drinking water services

Numerous official regulators manage and monitor water quality, including the Cairo Alexandria Potable Water Organization (CAPWO) and the Egyptian Water Regulatory Agency (EWRA), as well as representative bodies of planning and quality control. The official body to which citizens can directly submit applications or complaints is the Holding Company for Water and Wastewater (HCWW), which was created in 2004 by a presidential decree, in an attempt to restructure the existing regional water and sanitation companies into subsidiary entities beneath it (Farmer 2014). On a district level, a water delivery certificate is presented to HCWW. Finally, the Consumer Protection Agency (CPA) receives water complaints as a commodity.

Similar to the housing construction process, getting a water connection to a building or an apartment reveals the various gray areas between formal and informal. The following process outlines the requirements to obtain a water connection for the building. First, an application is submitted to the district, including a valid identity document, a copy of the ownership contract, and the construction license, which leads to obtaining a letter addressed to the relevant service center (if the building was built without license, then an electricity receipt from the neighbor should be provided). This is followed by paying the required fees for inspection and cost assessment for a plot connection to the public network to be implemented. If the water connection already exists, then a reconciliation process has to be undertaken through the same procedures, but without going to the district first. If the existing connection does not meet the criteria, a request is made for a new one. In the case of reconciliation, previously consumed water is estimated retroactively (Tadamun 2015).

Although the process of applying for a new connection to a public water network is almost identical with that of a reconciliation of informal connections, in addition to the high cost of the retroactive estimation, the latter option is more common in informal areas, i.e., making informal connections and restoring the street pavement to its original state and then reconciling. This evasive tactic toward formal connection can be explained by a number of factors as perceived by local residents, such as the lack of transparency and reliability, the limited access to information, and the time it takes for an application to reach the cost assessment stage. Most importantly, if a formal request for a water connection to an informal building is submitted, the building may be rendered as "unauthorized," resulting in its possible forced removal (Tadamun 2016).

In terms of access to water, it is important to note that a connection to the formal networks does not necessarily reflect water availability, hence the widespread use of water pumps in order to address the issue of frequent water unavailability and low pressure. In other words, the official expression of resource availability through networks carries a conceptual flaw.

The above outline of the procedures to access a formal infrastructure network from an informally developed urban system exposes the fact that, despite the hardship and complexity to access the formal network, formal connection does not necessarily mean water availability, hence the prevailing reliance on informal modes to compensate for this absence. Further, the above process illustrates the mistrust by local communities toward the formal option as it involves further risks to jeopardize their already precarious status as informal, hence illegal. The formalization process embodies in itself a potential criminalization, thus perpetuating informality as a byproduct.

Transportation (SDG 11.2)

SDG 11.2 specifies the need for access to safe, affordable, accessible, and sustainable transport modes, prioritizing inclusive public transport (see Table 4.3). As local development agendas focus their attention on expanding the formal transport network (metro, buses) connecting formal satellite neighborhoods, little consideration is given to connecting informal settlements to the rest of the city or the modes of transport that have been appropriated to adapt to the urban fabric of such areas. However, the complex mobility systems in such neighborhoods, and their entanglement with the formal, present an opportunity to inform a development framework aimed at integration.

The transport system in Ard al-Liwa has many formal and informal alternatives and combinations thereof. The difference between formal, semi-formal, and informal modes, i.e., the degree of informality, depends on a set of variables such as actors, economy, and structure (Nagati et al. 2021b). They may include undefined routes or schedules, the absence of fixed stations, unregulated fares, and number of passengers. Some vehicles operate without registration or driving licenses, including underage drivers inside informal areas.

At one end of the spectrum are the formal modes of the transportation system—metro, bus, tram, and taxi—almost all of which (except some

Table 4.3 SDG 11 and its target and measuring indicator specific to the issue of transportation availability in cities

SDG 11: Make cities and human settlements inclusive, safe, resilient, and sustainable	**Target 11.2:** By 2030, provide access to safe, affordable, accessible, and sustainable transport systems for all, improving road safety, notably by expanding public transport, with special attention to the needs of those in vulnerable situations, women, children, persons with disabilities, and older persons	**Indicator 11.2.1:** Proportion of population that has convenient access to public transport, by sex, age, and persons with disabilities

taxis) only connect to the entrance of the informal neighborhoods, but do not enter them. At the other end of the spectrum is the tuktuk, an iconic vehicle associated with informal growth that is most compatible with narrow streets and the absence of formal traffic regulations. The tuktuk-based system is almost entirely unregulated or licensed in Egypt, lacking most of the formal factors described above, such as stations, licenses, routes, or pricing.

In between these two extremes lies the microbus, a ubiquitous example of semi-regulated means of transportation, which may best illustrate the blurred boundaries of the two systems. These are licensed when routes are within the formal city, but not in informal areas. This is the only mode that traverses the formal-informal boundaries. Their main stops are often located at this interface. As such, researchers, planners, and policymakers often use them to measure the extent of informal sprawl around the existing urban agglomeration (El-Khateeb 2017). Other forms of semi-regulated modes include private cars operating as taxis, either independently or affiliated with e-hailing services such as Uber, which, despite having licenses for drivers and vehicles, are still outside the commercial taxation system and thus subject to occasional busting by the traffic police (Wadi 2017).

The tuktuk thus remains a key example while examining attempts to regulate informal modes of transport. The authorities' decision to completely ban the tuktuk from new satellite cities and replace it with the mini-van taxi in order to maintain a "civilized appearance" may be contrasted with the improbability of elimination in informal areas due to their narrow street network (Ashwal and Harbi 2019). From the official perspective, the regulation of the tuktuk is linked to multiple issues. Economically, tuktuks should be integrated in the formal economy, including through taxes and fees, in addition to being one of the major employment suppliers in the informal sector for craftsmen and low-skilled laborers gaining more income as tuktuk drivers. Its formalization, however, requires a number of regulating measures, including licensing and monitoring of its routes, stops, and fares. Some local initiatives have attempted to address these regulatory measures, including registration and color-coding, as well as designation of specific routes (Saad 2020).

According to estimates by the Central Agency for Public Mobilization and Statistics (CAPMAS) in 2018, there are three million tuktuks operating in Egypt, of which 260,000 are licensed vehicles. When comparing the average price of tuktuks to the price of a mini-van (the alternative to the tuktuk), the former could be estimated at EGP 20,000, while the mini-van average is EGP 120,000. Around 30 million citizens depend on tuktuks as a means of transportation. Tuktuks provides three million job opportunities in addition to indirect jobs, such as maintenance workshops and spare parts markets (Ramadan 2019).

The above considerations may explain the reluctance to integrate the tuktuk into the formal system despite the explicit regulations to do so.[3] Such a paradox facing regulators, of formalization versus economic gain, may be further demonstrated by recent decisions to ban tuktuks from driving on main streets while collecting fines from (and occasional confiscation of vehicles of) violators, who may resume their trips on main streets once they show the receipts of the paid charges.

Similar to housing and water infrastructure, transportation systems in informal areas expose an inherent contradiction in government policies toward informality, an official condemnation and delegitimization, while tacitly exploiting it as an economic resource. The following section analyzes some of these contradictions across the above themes, highlighting the gray zone, or the formal-informal interface as an entry point for SDG localization.

[3]Traffic Law No. 121 of 2008, amended by Law No. 54 in 2014, explicitly states that the tuktuk has to be licensed (Salah 2015).

4.4 Policy-Practice Gap

Current policies to legitimize urban informality, whether directed toward housing, or services such as water and electricity, as well as those related to transport and mobility, have been criticized as being an attempt to collect funds from the great pool of the unregistered informal economy, viewed as dead capital (De Soto 1997). Such a broader agenda of formalizing the informal economy (Raggal 2020), which seems to be in direct contradiction with the announced "slum eradication" projects, poses several questions concerning the legitimacy of informal practices and the state's interpretation of the SDGs and their implementation (Ramadan 2020).

The above case studies reveal a recurring theme of contradictions between an acknowledgment of informal practices and explicit intentions to regulate and formalize them, on the one hand, and maintaining their illegal status in order to exact fees, taxes, and fines as a source of revenue, on the other. This section further illustrates the gap between top-down policies and local practices within informal areas. This gap requires close interrogation in order to be effectively bridged through the use of localized approaches to the monitoring and implementation of urban development goals.

In 2019, a revised Reconciliation Law (Law 17/2019)[4] was introduced to settle all building violations in informal areas and was put into action toward the beginning of 2020 (Shawkat 2019a). Unless the building is constructed on state land or is encroaching on an infrastructure or public domain, its illegal status could be changed into a licensed real estate property in exchange for a certain fine to be calculated based on its location, number of floors, or other criteria. The law was applied retroactively, even for buildings or additions made as far back as 1986, effectively impacting around 2.8 million buildings, including 20 million units all over Egypt (Khalafawi 2020). This caused an uproar among millions of families and residents, many of whom have purchased their units from a developer whom, they consider, should be the party held liable (Shawkat 2019b). The issue of exacting fines three or four decades after the "violation" raises a few questions. The first is about responsibility, given that the state's inability to provide housing has forced individuals to develop alternatives, while local authorities turned a blind eye concerning these "violations" for decades, often involving some "special deals" (Ramadan 2020). Secondly, many of these "illegal" structures have been provided with water, electricity, and natural gas from respective state authorities, technically having a legitimate authorization (Raggal 2020). The final question relates to the practicality of implementation. If the developer can no longer be traced, who is responsible for paying the reconciliation fine? Should families be evicted and buildings demolished, or does this law only apply to uninhabited structures?

When it comes to urban transportation, formal traffic regulation presence is rare in informal areas and lacking on-site monitoring and regulations by local authorities. This reflects the state's inability or unwillingness to engage local practices and negotiate measures to improve informal standards. In Ard al-Liwa, community groups have been lobbying local MPs to locate a traffic unit inside the neighborhood, which was met with a request by the relevant authority "to provide an empty plot."[5] In contrast, the state has invested in a giant network of traffic corridors and arteries connecting new urban centers while bypassing informal areas, thus avoiding addressing relevant traffic and mobility issues for local communities. Moreover, these infrastructure projects often create problems for local communities as they dissect their neighborhoods or are

[4] The text of the first paragraph of Article 1 has been amended, which stipulates that it is permissible to reconcile and legalize situations in actions committed in violation of the provisions of the Building Law promulgated by Law No. 119 of 2008, and the amendment came by adding the phrase: "the law regulating building and the previous special laws regulating construction." to incorporate all the violations that occurred before the issuance of the Reconciliation Law within the framework of reconciliation in accordance with the law (Arab Republic of Egypt 2019).

[5] Interview by the research team with a local Member of Parliament, Mona Monir, November 2019.

Fig. 4.4 A plot area close to Cairo Ring Road and Ard al-Liwa, being excavated in preparation for construction (CLUSTER 2018)

constructed in immediate proximity to adjacent buildings. An example of the latter is the al-Zomor urban corridor.

The al-Zomor elevated corridor, under construction since 2019, links southern to northern districts of Giza city and provides an alternative connection along the Cairo Ring Road. It follows the former route of the now partially tunneled al-Zomor canal, traversing a number of major informal neighborhoods to the west and southwest of Giza (see Fig. 4.4). Since this new project was planned decades after these neighborhoods had been informally developed, it faces, as similar urban corridors do, obstacles of existing "unplanned" structures and inconsistent street width, forcing road planners and traffic engineers to develop creative solutions beyond the rigid international code. Al-Zomor corridor raised negative media coverage due to the fact that parts of its elevated sections were constructed immediately adjacent to residential buildings on its side, with less than 1 m separating the freeway from the residents' balconies (see Fig. 4.5). The authorities' reaction was to blame informal-cum-illegal encroachments on the building line, while residents responded by showing approved permits of heights and balcony projections. The project reveals both the complexity of the formal-informal interface, in a literal concrete sense in this case, and a hypocrisy in the official narrative given the loose interpretation of formal codes to construct this edifice while condemning informal housing units, some of which already hold permits (Gaber 2020).

In contrast to the gap between top-down policies and infrastructure projects and local context, which at times were seemingly widening, the following example demonstrates an opposite, bottom-up approach. This is where community groups engaged a state water network project to help ground its implementation. In response to a general plan to extend water pipes in the neighborhood, the group began by assessing local needs, helping file official complaints and petitions to change the proposed route, and following up on the implementation process to ensure its compliance with their proposed adjustments.[6] Such practices provide a useful basis for the development of an alternative framework which

[6]Based on interviews by the research team with the local community group.

Fig. 4.5 Construction of al-Zomor elevated corridor passing through Ard al-Liwa (Photo credit: Jakob Lindfors)

begins by reviewing local needs and practices and proposing improvement measures. A similar approach is applicable to issues such as housing and mobility, as a basis for the adaptation of existing indicator systems to better fit local needs.

Grounded fieldwork, informed by the participatory tools of citizen science, is needed to adapt and revise development indicators, in order to ensure their efficacy. For example, our research shows the potential for a quantifiable assessment of indicators, through which the tuktuk as a mode of informal transportation might be measured against other modes of transport and not be dismissed due to its informal status. Once these gaps are identified, possible policies could then be proposed to narrow these gaps and bring informal processes closer to the normative criteria. Key here is the improvement and enhancement through upgrading local systems, rather than substituting them with imposed global or universal norms (Nagati et al. 2021a).

4.5 Conclusions and Lessons Learned

One of the key critiques of the SDGs and national development plans that are aligned to the SDGs, such as Egypt's Vision 2030, is that they do not address the complexity of informal conditions, most prevalent in urban contexts in Africa and in the Global South more generally. This chapter attempts to unravel this complexity by unpacking the inherent contradictions in government policy toward informality while revealing the gray zone—the formal-informal interface—as a reality and thus as an opportunity to develop more grounded measures for localizing SDGs.

While the SDGs offer an expanded understanding of the urban, through the inclusion of SDG 11 compared to its precursor MDG target 7.D aimed at improving "slum" dwellers' lives, it continues to be guided by binary understandings of the formal and informal. Hence, just as governments justified slum clearance through the use of the MDG slogan "Cities without Slums" (Huchzermeyer 2013), governments continue to (mis)appropriate global development goals to demolish or expropriate properties of prime real estate value under the pretext of advancing and implementing the SGDs and their targets. The case of Egypt shows how the interpretation of the SDGs by governments of the Global South is framed by urban development strategies, priorities, and biases within a neo-liberal paradigm. This is manifest in the allocation of resources to grand urban schemes and traffic corridors with little benefit to dwellers of informal areas while occasionally resulting in their eviction and relocation to make room for these projects. The language of the SDGs is often appropriated to justify further marginalization and the exclusion of more vulnerable communities.

While officially condemning informality as slums, and adopting policies of eradication, the state lacks both the economic capacity and political will to substitute informal housing, amounting to up to 70% in Egypt (Sims 2012). In practice, urban policies therefore deal with informality as reality, with varying conciliatory measures of regulating and upgrading, while extracting revenues through fines and fees.

This chapter aims to identify a middle ground between universal SDGs and local practices, building on grounded research as a basis for developing revised criteria, targets, and indicators to localize the SGDs while improving local practices. Case studies on the provision of housing, water infrastructure, and transportation demonstrate how the formal and the informal have become inseparable, often depending on one another, and in some cases indistinguishable. This highlights the need to recognize the formal-informal interface as an intermediate zone, or spectrum, of urban practice beyond the reductionist formulation and constructed polarization of formal versus informal conditions. Based on this analysis, the chapter suggests the formal-informal interface as a generative space and more conductive framework to experiment with grounded, inclusive, and participatory governance and research modalities, including citizen science, in order to localize the SDGs informed by local necessities.

References

American University in Cairo (AUC) Center for Sustainable Development (2014) Input for Egypt's strategy for dealing with slums. The American University in Cairo, Cairo

Angélil M, Malterre-Barthes C (2016) Housing Cairo: the informal response. Ruby Press, Berlin

Arab Republic of Egypt (2019) Reconciliation Law no. 17 of 2019. Ministry of Housing, Utilities and Urban Communities. admin.mhuc.gov.eg//Dynamic_Page/636990508748531932.pdf. Accessed 30 October 2020

Arab Republic of Egypt (2016a) Egypt National Review report for input to the 2016 HLPF. Ministry of International Cooperation. sustainabledevelopment.un.org/content/documents/10738egypt.pdf. Accessed 10 June 2020

Arab Republic of Egypt (2016b) Why do we need the Sustainable Development Strategy (SDS)? In: 2030 Egypt Vision. Arab Development Portal, p 3. Ministry of Planning and Economic Development. arabdevelopmentportal.com/sites/default/files/publication/sds_egypt_vision_2030.pdf. Accessed 10 June 2020

Arfvidsson H, Simon D, Oloko M, Moodley N (2017) Engaging with and measuring informality in the proposed urban Sustainable Development Goal. African Geographical Review 36:100–114

Ashwal I, Harbi S (2019) Local development: starting the codification of "van" against the random of "tuktuks". Shorouk. shorouknews.com/news/view.aspx?cdate=11082019&id=61afb838-fd82-4fa7-a226-eeaa364ebf8d. Accessed 30 October 2020

Broome A, Quirk J (2015) The politics of numbers: the normative agendas of global benchmarking. Review of International Studies 41:813-818

Caprotti F, Cowley R, Datta A, Broto VC, Gao E, Georgeson L, Herrick C, Odendaal N et al. (2017) The New Urban Agenda: key opportunities and challenges for policy and practice. Urban Research & Practice 10(3):367-378

CLUSTER (2018). Formal-informal interface: Comparative analysis between three Egyptian cities. Cairo Lab for Urban Studies, Training and Environmental Research, Cairo. Unpublished report

De Soto H (1997) Dead capital and the poor in Egypt. Egyptian Center for Economic Studies, Cairo

El-Khateeb M (2017) Urban mobility in greater Cairo: a history of patchwork solutions. Tadamun. tadamun.co/urban-mobility-greater-cairo-history-patchwork-solutions/?lang=en#.X4MAO\WgzbIV. Accessed 30 October 2020

Farmer TR (2014) Cairo ecologies: water in social and material cycles. Dissertation. hdl.handle.net/2152/25228. Accessed 30 October 2020

Fukuda-Parr S (2016) From the Millennium Development Goals to the Sustainable Development Goals: shifts in purpose, concept, and politics of global goal setting for development. Gender and Development 24(1):43–52

Gaber M (2020) This is how the scene of the Zomor Bridge looks like from the balcony. Youm7. youm7.com/story/2020/5/14/اذكه-ودبي-دهشم-يربوك-رمزلا-نم-ةنوكلبلا/4774008. Accessed 30 October 2020

Huchzermeyer M (2013) 'Slum' upgrading or 'slum' eradication? The mixed message of the MDGs. In: Langford M, Sumner A, Yamin AE (eds) Millennium Development Goals and Human Rights: Past, present and future. Cambridge University Press, Cambridge, pp 295–315

Kanuri C, Revi A, Espey J, Kuhle H (2016) Getting started with the SDGs in cities: a guide for stakeholders. irp-cdn.multiscreensite.com/6f2c9f57/files/uploaded/Cities-SDG-Guide.pdf. Accessed 30 October 2020

Khalafawi S (2020) Learn about the number of buildings in violation and their types in Egypt. Youm7. youm7.com/story/2020/9/13/فرعت-ىلع-ددع-اتاراقعلا-ةفلاخملا-اهعاونأو-رصم-ىف/4964723. Accessed 30 October 2020

Malonza JM, Ortega AA (2020) Fissures in localizing urban sustainability: the case of Rwanda. GeoJournal. doi:https://doi.org/10.1007/s10708-020-10239-8

Moallemi EA, Malekpour S, Hadjikakou M, Raven R, Szetey K, Moghadam MM, Bryan BA (2019) Local Agenda 2030 for Sustainable Development. The Lancet Planetary Health 3(6):e240-e241

Nagati O, El-Didi A, Gad H, Kihila J, Mbuya E, Njavike E (2021a) Towards a bottom-up approach for localizing SDGs in African cities. Council for the Development of Social Science Research in Africa, Dakar (in press)

Nagati O, Jean-Baptiste N, Siame G, Vokouma Boussari J (2021b) Formal-informal interface: comparative framework across four African cities. Council for the Development of Social Science Research in Africa, Dakar (in press)

Nagati O, Stryker B (2021) Formal-informal interface: comparative analysis between three Egyptian cities. In: Marrengane N, Croese S (eds) Reframing the urban challenge in Africa: knowledge co-production from the South. Routledge, London

Patel Z, Greyling S, Simon D, Arfvidsson H, Moodley N, Primo N, Wright C (2017) Local responses to global sustainability agendas: learning from experimenting with the urban Sustainable Development Goal in Cape Town. Sustainability Science 12:785–797

Raggal A (2020) Urban and violence: from Mubarak's deal to Sisi's plan. Mada Masr. www.madamasr.com/ar/2020/09/30/opinion/u/نارمعلا-فنعلاو-نم-ةقفص-كرابم-ىلإ-ةطخ-لا/. Accessed 25 October 2021

Ramadan H (2019) After 14 years on the streets of Egypt, how much does it cost to replace a tuktuk with a minivan?. Masrawy. masrawy.com/news/news_various/details/2019/9/12/1633746/دعب-14-ةنس-يف-عراوش-رصم-كم-فلكي-لادبتسا-لاتوكتوك-ب-لاميني-فان-. Accessed 30 October 2020

Ramadan M (2020) That is how the state forced peasants to cultivate cement. Mada Masr. madamasr.com/ar/2020/10/01/feature/داصتقا/اذكه-تربجأ-ةلودلا-نيحالفلا-ىلع-ةعارز/. Accessed 12 October 2020

Robin E, Acuto M (2018) Global urban policy and the geopolitics of urban data. Political Geography 66:76–87

Saad A (2020) Awsim begins to regulate tuktuks: the green is ours and the red is for outsiders. Elwatan News. alwan.elwatannews.com/news/details/4999081/ميسوأ-أدبت-تقنين-لاتوكتوك-بملصق-رقمي-الأخضر-منا-والأحمر-من-برة. Accessed 30 October 2020

Salah A (2015) The first official study: 30 million citizens use "tuktuk" daily. Elwatan News. elwatannews.com/news/details/787436. Accessed 30 October 2020

Shawkat Y (2019a) Egypt's Construction Violations Reconciliation Law: all you need to know. Built Environment Observatory (Marsad al-Umran). marsadomran.info/en/policy_analysis/2020/01/1880. Accessed 15 October 2020

Shawkat Y (2019b) Ten most important questions around the Reconciliation Law concerning building violations. Built Environment Observatory (Marsad al-Umran). marsadomran.info/policy_analysis/2020/08/2008/#1. Accessed 15 October 2020

Shawkat Y (2020) Housing in Egypt. The American University in Cairo Press, Cairo

Simon D, Arfvidsson H, Anand G, Bazaz A, Fenna G, Foster K, Jain G, Hansson S et al. (2016) Developing and testing the urban Sustainable Development Goals targets and indicators—a five-city study. Environment and Urbanization 28:49–63

Sims D (2012) Understanding Cairo: the logic of a city out of control. The American University in Cairo Press, Cairo and New York

Tadamun (2016) Coming up short: Egyptian government approaches to informal areas. Tadamun. tadamun.co/coming-short-government-approaches-informal--areas/?lang=en. Accessed 30 October 2020

Tadamun (2015) A guide for the greater Cairo residents to drinking water. Tadamun. issuu.com/87709/docs/tadamun_gcr_local_services_guide_wa. Accessed 15 October 2020

Tadamun (2014) The right to water and domestic and urban services in the Egyptian constitution. Tadamun. tadamun.co/the-right-to-water-and-domestic-and-urban-services-in-the-egyptian-constitution/?lang=en#.X4L_5GgzbIV. Accessed 12 October 2020

Wadi D (2017) The drivers of Uber and Careem between reconciling their legal status and living conditions. Dotmsr. dotmsr.com/news/200/772807/سائقي-أوبر-وكريم-بين-أوضاعهم-القانونية-والمعيشية. Accessed 30 October 2020

Zaazaa A, Borham A, Abotera M, Muhammad A, Abutawila A, ElMansy A, El Marsafy N, Mourad R et al. (2015) Maspero parallel participatory project. Madd Platform. https://issuu.com/maddplatform/docs/maspero_parallel_participatory_proj. Accessed 23 August 2021

Part II

Integrated Perspectives on Water, Waste and Health

5 Toward an Integrated Approach: Water Management Dynamics in the City of Accra

Ernest Nti Acheampong

Abstract

Water plays a crucial role in addressing an interconnected set of Sustainable Development Goals (SDGs), including urban health, food security, and climate resilience, as well as unlocking opportunities for economic growth and the promotion of sustainable consumption norms in African cities. However, in many cities, water management represents a complex multifaceted challenge characterized by political, economic, and social interactions. This chapter builds on the case of Accra, Ghana, to illustrate the challenges of urban water management in Africa under conditions of rapid urban growth, urban sprawl, and industrial development. These issues are compounded by the impact of climate change on water supply, pollution, and urban health. By bringing together a range of different stakeholders involved in urban water management through the use of a transdisciplinary research approach, the chapter argues in favor of more participatory and integrated approaches to urban water management. These include greater emphasis on the importance of investment in water infrastructure and appropriate technologies to increase efficiency, promote wastewater recycling and reuse, and improve the governance of urban water flows.

Keywords

Urban water management · Accra · Transdisciplinary research · Learning Alliance Platform (LAP)

5.1 Introduction

The need to achieve universal and equitable access to safe and affordable drinking water for all is not just one of the key targets among the Sustainable Development Goals (SDGs) but is also key to addressing an interconnected set of related goals, including urban health, food security, and climate resilience. It is also central to unlocking opportunities for decent work and economic growth and the promotion of sustainable consumption norms in African cities. However, water managers are experiencing intense pressure to efficiently manage water supply to meet growing water demands in the face of rapid urban growth and the increasing threat of climate change. Compounding these challenges is the strong influence by many different water stakeholders with varied interests in water consumption, management perspectives, and sustainability goals at multiple levels (Head et al. 2016). Such disparity increases sociopolitical risks associated

E. N. Acheampong (✉)
African Technology Policy Studies Network (ATPS), Nairobi, Kenya
e-mail: enacheampong@gmail.com

S. Croese, S. Parnell (eds.), *Localizing the SDGs in African Cities*, Sustainable Development Goals Series, https://doi.org/10.1007/978-3-030-95979-1_5

with water management and contributes to asymmetric patterns of water access and distribution (Hurlimann et al. 2017). As such, the multiplicity of challenges creates a complex water management problem that has prompted the call for more holistic approaches and solutions to the complexity of water supply and access needs.

In Accra, Ghana, the urban water system is characterized by a conventional centralized water infrastructure that was constructed in the late nineteenth century. At the time, these consisted of large dams, treatment plants, pumping stations, a network of buried pipes, and large water storage and sewage tanks that provided sufficient water supply. Currently, these infrastructures have aged, and some components of the water system have exceeded their life cycle, as evident in the rupturing of the pipe networks, leakages, and nonrevenue water (NRW) losses (Koppelaar et al. 2018). The city's water system faces tremendous water demand pressure from the rapidly changing dynamics comprising the rising economic growth, rapid urban population, rapid urban sprawl, and industrial development. As a result, the water utility, Ghana Water Company Limited (GWCL), can only meet about 70% of the city's water demand (Asante-Wusu and Yeboah 2020). Compounding the challenge are climate change impacts such as long dry spells and flooding that threaten the quality, quantity, and regular flow of water supply. The problem of a recurring water supply deficit crisis in the city has evolved into a multifaceted challenge characterized by political contestation, economic hegemony, and social interactions that are deeply enmeshed in water management. Efforts toward sustainable water management call for transformative change and an integrated approach to explore the complexity of urban water problems. The notion of an integrated approach, in turn, requires a rethinking of the universality of water resources and systems and compels various water stakeholders to work together to transform the urban water sector into one that highlights resource perpetuity (Bahri et al. 2016).

This chapter employs a transdisciplinary research process to examine the challenges of urban water management in Africa under conditions of rapid urban growth, urban sprawl, and industrial development, compounded by the impact of climate change on water supply, pollution, and urban health. This involved bringing together a range of different stakeholders engaged in the urban water sector. Drawing on this unique process, the chapter aims to illustrate the complex dynamics of water flows, supply, and interactions among social, economic, political, and environmental realities that shape the water management decisions in Accra and offer solutions toward the transformation of the urban water sector and SDG implementation.

5.2 The City of Accra

Accra, the capital city of Ghana, covers a total land area of 185 km^2 and currently has a population of 2.5 million, which is estimated to increase to 3.5 million by 2030 (Government of Ghana 2019). Accra receives treated water from two main water production plants, the Kpong and Weija water treatment works. Treated water is supplied through a network of pipes that link households and commercial and industrial areas in the city. In several suburbs, pipe networks are absent, and individual households have constructed their boreholes to access groundwater or receive freshwater through rainfall harvesting.

The city's dynamic resource flow systems are based on water availability, urban population density dynamics (formal and informal settlements), and resource demand, which derive from its status as the largest city in Ghana and as an important economic hub in West Africa (Silver 2014). Over the past years, the city has experienced significant growth in three broad areas: infrastructural development, urban population, and commerce. With a steadily growing economy, residential expansion, and the rapid spread of informal settlements (slums) in the city, water consumption and demand have intensified and overwhelmed the existing water system's capacity to deliver efficient and effective water services.

Informal settlements represent approximately 15.7% of the total land area and are mainly located in the central and consolidated area of the metropolis (see Fig. 5.1) (Engstrom et al. 2019).

Around 38.4% of Accra's urban population lives in slums, with a population density of 607.8 inhabitants per hectare, which is more than twice the average population density of the city as a whole (250.7 persons per hectare) (ibid). Water supply to informal settlements can be difficult to quantify due to the poor pipe network, different modes of water supply, and the poor spatial arrangement and planning of housing in these areas. Furthermore, slums are often located in areas that are unevenly exposed to environmental risks such as polluted rivers, flooding, dumpsites, open sewers, and industrial waste (Engstrom et al. 2019). The current challenge of urban water supply, demand, and access by different users linked to economic development, societal water demands, and environmental impact in the city depicts the complexity of managing water sustainably. This also results in challenges for the monitoring and implementation of development goals such as the SDGs.

5.3 Transdisciplinary Research (TDR) Process

A transdisciplinary research (TDR) process was employed as a methodological framework for data collection. It aligns with the emerging mode of knowledge production that accentuates the idea of a complex system's reality, the critical role of participation, and the transformation of scientific practices. The TDR process combines academic and non-disciplinary knowledge, skills, and expertise of multiple actors to co-identify, co-formulate, and resolve complex societal problems (Reed and Abernethy 2018). Through the process, researchers work with society to identify practical solutions to complex sustainable development and sustainability problems while producing knowledge and fostering social learning (Siew et al. 2016).

The TDR process consisted of two main stages. At the first stage, a Learning Alliance

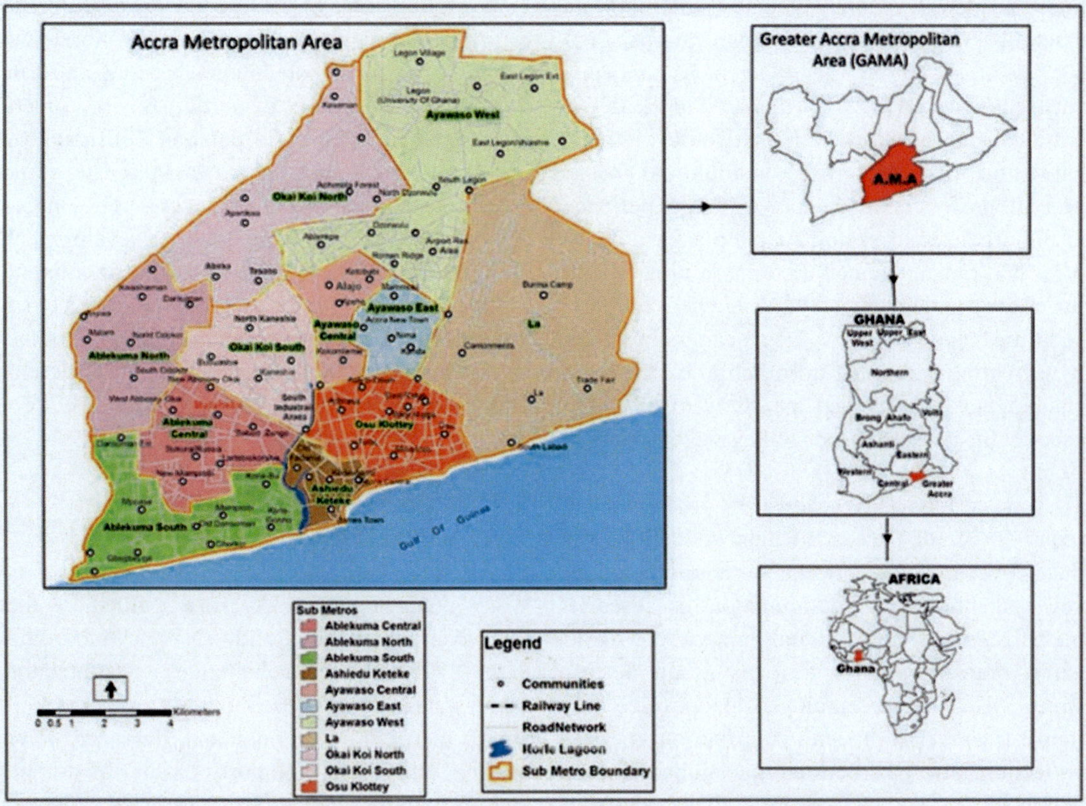

Fig. 5.1 Map of Accra (Amoako and Boamah 2015)

Platform (LAP) was established to bring different actors from government agencies, research institutions, private sectors, civil society, and nongovernmental organizations together for participatory inquiry and knowledge production on water management in Accra. The LAP is based on the assumption that different stakeholders in a platform can cooperate, learn, and share experiences in an action-based research process to develop innovative solutions that bring about lasting change. Such solutions are capable of addressing complex problems that have proven difficult for a single group of stakeholders to solve and for which alternative options are required to address risk and uncertainty (Moreno-Leguizamon et al. 2015).

A total of five platforms were organized over a period of 3 months, on the last Thursday of each month. Each consisted of 20 participants and lasted for 60 min. The type of participants varied per meeting. At the first platform meeting, a wide array of key stakeholders from the research institutes, academia, public agencies, metropolitan assembly, nongovernmental organizations, civil society organizations, donor agencies, and private sector stakeholders participated in the deliberations. Participants were invited to the platform based on their expertise, institutional background, professional orientation, and experience in the urban water sector. At the platform meetings, participants gained a common understanding of the problem of water management in Accra and thus helped co-define the research objectives by providing additional information on the local challenges. This allowed the study to take into account the research needs of key stakeholders in Accra.

At the second stage of the TDR process, a key requirement for successful data collection was the active involvement and participation of all key stakeholders. A combination of research methods, namely, individual interviews, world cafés, brainstorming, and focus group discussions (FGDs), were employed to optimize participant participation and contribution to data collection. An unstructured questionnaire was administered to collect specific information from individuals. These questionnaires proved useful in collecting vital information from participants who by nature would not express their views openly in FDGs (Siew et al. 2016; Ruiz et al. 2017). At the world café, a smaller group of five participants extensively discussed the issues related to the subject matter. This method allowed participants to switch groups and interact with more participants from different groups to generate more ideas and promote learning and knowledge sharing among actors. At the focus group discussions, a group of six participants deliberated on the issues in the water sector in Accra, including the current status and dynamics of the urban water management, water reform processes, urban system performance, and the potential for achieving sustainable water management using new and innovative approaches. At the end of the FDGs, each group presented the outcomes and recommendations of their discussion. This session narrowed and consolidated the views of participants, thus improving the outcome of the world café session.

Overall, a total of 100 people from different organizations and with different professional backgrounds participated in the learning platform meetings. The number of participants per meeting ranged from 18 to 24 persons, consisting of mid-level and senior-level officials who were research scientists, water engineers, policy makers, advocates, water regulators, development specialists, and academics. Applying a mix of research methods for data collection proved to be very useful for data triangulation, as well as the collection of comprehensive data about a complex issue (Abowitz and Toole 2009).

5.4 Current Water Management Practice

Water management in Accra is guided by the 2007 National Water Policy (NWP). According to the NWP, three main challenges confronted the urban water sector. These included a water deficit amidst increasing water demand, degraded water quality resulting in high cost of water treatment, and expanding and improving existing water sources. The policy adopts a holistic and integrated approach as a fundamental means to achieve sustainable water resource development, use, and management in Ghana (Government of Ghana 2007). Deeply rooted in the NWP are the equity and pro-poor principles of ensuring safe and adequate water supply to all people, particularly poor and vulnerable urban households. Putting these principles into action, the government implemented four key policy strategies, including decentralization, removal of water subsidies, Private Sector Participation (PSP), and stakeholders' participation in water governance.

The decentralization strategy aimed to differentiate and assign clear responsibilities and powers of implementation to various institutions within the urban water sector. For example, urban wastewater management was devolved from water supply, which was managed by the GWCL, to become the responsibility of the Waste Management Department (WMD) of the Accra Metropolitan Assembly (AMA). From the learning platform, it was revealed that the decentralization approach tended to be purely institutional rather than decentralizing the physical sewage system, which employed on-site and/or cluster treatment systems to collect, treat, and reuse wastewater from high-density low-income communities. Underlying this approach was the principle of recovering, recycling, and reusing (3R) wastewater to reduce environmental pollution impact, optimize recycled water use while reducing pressure on available water resource, and make treated water available for social well-being and the economy (Nansubuga et al. 2016).

Based on the discussion with stakeholders, it became clear that current institutional decentralization had not encouraged the integration and coordination of water and wastewater management activities in Accra. As in many other African cities, for a long time, water service delivery in Accra was heavily subsidized by the government with the intention to fulfill the fundamental rights of urban residents to safe and adequate water. In this position, the right to water access against the need to recover the cost of treated water has become contentious (Jaglin 2002). The need for cost recovery measures, prompted by the declining financial prospects of the water utilities and pressure from external donors, has led to water subsidy withdrawals in the water sector of many sub-Saharan African countries. Proponents of cost recovery measures argue that subsidies to the lowest water tariff block do not exclusively benefit the targeted low-income residents since low-income households are not even connected to the piped water network (Banerjee et al. 2017). Nonetheless, it appears that the impact of subsidy withdrawal on the improvement of water service delivery in many African cities, including Accra, has remained insignificant.

The PSP strategy introduced a private entity into urban water supply management to make the urban water sector viable and financially self-sufficient while improving accountability. This implementation of the cost recovery mechanisms under the PSP proved to be controversial, often facing fierce resistance from water consumers and civil society groups about perceived alienation of the poor urban residents (Hirvi 2012). Critics argued that the economically biased position to recover cost and make a profit through water tariff adjustments contravened the principle of water as a social good. They further argued that the PSP would rather exclude already disadvantaged communities, characterized by low-income dwellers, informal settlements, and poor infrastructure from the urban water system (Adu-Ampong 2013). In the end, the impact of the PSP strategy was considered minimal in achieving the goals of cost recovery, financial sustainability, and improved efficiency in water service delivery. Stakeholder engagement and participation in urban water governance were deemed to provide a valid platform for consensus building and ownership of decisions made. However, the reality was that such a platform was often prone to contestation among stakeholders and serving as a tool for elite capture.

Follow-up action to the NWP consisted of the development and implementation of the 2012 Water Sector Strategic Development Plan (WSSDP), which was expected to deliver sustainable access to clean water and basic sanitation for all citizens by 2025 (Government of Ghana 2014). Under the plan, Accra was expected to experience an increase in water coverage by

85% in 2015 and 100% in 2025. The plan was for Accra to further experience sustainable financing for investments, operation, and maintenance of water services and full recovery of rehabilitation, operation, and maintenance costs of urban water service by 2015. Additionally, water institutions with expertise, information, equipment, logistics, and funding are to perform their functions efficiently and effectively in a transparent and well-coordinated manner by 2025. Finally, the plan aims for the sustainable harnessing, utilization, and management of water resources by 2025. According to the Ghana Statistical Service (GSS), Ghana experienced an increase in the population percentage with improved drinking water services from 37.3% in 2013 to 44.3% in 2017 (Ghana Statistical Service 2017; Government of Ghana 2019).

Overall, the progress of implementation of the WSSDP was encouraging. In Accra, the facilities at the two waterworks serving the city improved their capacities significantly following the rehabilitation and expansion of the water treatment plants. Water supplied to the city has increased in volume according to data from GWCL. For example, the capacity of water supplied from Kpong waterworks increased from 220,454 m^3 per day in 2010 to 434,454 m^3 per day in 2015 (GWCL 2015, 2018). There was a consensus about the improvement in water governance at the national level, with the expectation that this would influence governance practice at the city level. Sewage and wastewater management was receiving some attention as the government pursued public-private partnerships to inject capital in the subsector. Through a public-private partnership mechanism, the sewage and wastewater subsector was witnessing some capital injection from private enterprises.

5.4.1 Dynamics of Water Supply

As of 2018, water produced and supplied daily to Accra and its environs was estimated to be approximately 679,000 m^3 (GWCL 2018). This amount of treated water is sufficient to meet the daily water requirements of the entire population of Accra. Despite the amount produced, the city continues to experience water shortages and unreliable supply to many areas due to the inadequate distribution of the pipe network and limited water infrastructure. Domestic water consumption accounts for more than 50% of the total water supplied to the city. The average water use per capita is directly correlated to the degree of water flow. For example, for areas with continuous water flow, the average water use is estimated at 100 liters per capita per day (l/c/d), 83 l/c/d for intermittent good flow, and 57 l/c/d for intermittent poor flow (Koppelaar et al. 2018). This is also linked to the socioeconomic status and geographical location of households. For example, households in well-planned residential areas with well-connected pipe networks have high per capita water consumption compared to households in informal settlements (slums) which accommodate the majority of the low-income residents, and water supply occurs mainly through water vendors such as water tankers and kiosks.

Water flows from water treatment, transmission, and supply to Accra are regarded as a highly energy-intensive process. Data from the GWCL suggests that 40–45% of the company's budget is expended on energy for processing raw water, treatment, transmission to booster stations, and distribution. The average water tariff per month is considered very high for the average household in the city. For example, domestic consumers pay US$ 0.53 per m^3 (GWCL 2018). However, officials at the GWCL argued that the current water prices did not reflect the true cost of water production and supply, even though this cost in Accra slightly exceeded the average cost of drinking water (US$ 1.16/m^3) from 44 cities in sub-Saharan Africa (SSA) in 2019 (Yates et al. 2020). With an average daily water consumption per capita in SSA of 0.05 m^3, the majority of low-income water consumers paid a high cost of approximately 25% of their daily income (US$ 2.40) on water service delivery (WaterAid 2016). Over the years, Accra has experienced a surge in water tariffs due to the rising production costs, including the increased costs of treatment, chemicals, and energy, and the high percentage of water losses in the supply and distribution chain. According to the Public Utilities Regulatory Commission (PURC), as of 2019, domestic water tariffs increased from US$ 0.53 per m^3 to US$ 0.57 per m^3, and commercial tariffs increased from US$ 1.21 per m^3 to US$ 1.61 per m^3 (GWCL 2020). While water tariff increases have not slowed water consumption or metabolic activities in the city, it has alienated many low-income urban households from accessing potable water (Dapaah and Harris 2017).

5.4.2 Urban Water Patterns and Access

From this analysis, the rapid urban transformation of Accra is concomitant with rising water requirements, thus compelling water managers to increase the amount of supplied water. In the case of Accra, the rapid urban expansion has kept pace with the development of utility systems and infrastructural expansion. According to the GSS, the population density of Accra has increased considerably, while land-use intensity has been mixed. Land use, mainly for construction, has increased through expansion rather than densification as reflected in the city's scattered structural urban arrangement and many vacant terrestrial spaces. The dispersed nature of the urban expansion in several parts of Accra reflects low urban compactness, which is a measure of the fragmentation of the landscape and infrastructure network. In these areas, water infrastructure expansion and connections, such as pipe network extensions, are easily hindered and thus restrict water flow throughout the city. Emerging socioeconomic growth trends suggest that future development and urban planning could lead to highly unplanned densification of the population and infrastructure that would complicate the structural landscape of Accra (Gaisie et al. 2019).

Different parts of the metropolis receive potable water through various modes of water supply. These arrangements suggest the existence of what Dinarès (2014) referred to as parallel metabolism for the same resource in Accra. For instance, water is supplied to urban elites in well-planned residential areas through networked pipe infrastructure, while water vendors supply to unplanned neighborhoods and slums with limited access to water supply infrastructure. While the presence of established infrastructure for water service delivery in certain parts of Accra allows for improvement in water supply efficiency, the continuous focus on renovating existing water infrastructure by the GWCL at the expense of developing new water infrastructure in new and unplanned areas tends to lead to a disproportionate flow of water delivery services.

5.4.3 Institutional and Political Arrangements for Water Management

Over the years, management reforms have contributed to well-established institutional arrangements for the management of the water system in Accra. As a result, several water-related institutions have emerged with clear roles and responsibilities. For example, the Ghana Water and Sewage Corporation (GWSC) was transformed into the GWCL in 1999, and the Water Resources Commission (WRC) and the Community Water and Sanitation Agency (CWSA) were newly established in 1996 and 1998, respectively. Stakeholders in the learning platform suggested that the effectiveness of these institutions in implementing interventions improved following the water sector institutional reforms. This raised optimism for an integrated approach to water management. However, the urban water sector is still confronted with sectoral fragmentation and weak capacity to coordinate different water-related policies, strategies, water uses, and users (Hirvi and Whitfield 2015).

Water management in Accra is inherently characterized by hydro-politics, dominated by asymmetrical power relationships that are deeply intertwined with the broader political context of national development. Several studies have highlighted the influence of power relationships and political status on the management of urban water (March 2015). The political interference in top executive and managerial appointments, contract awarding, and procurement processes based on networking and rent-seeking tend to create institutional crises, tensions among professional staff, and weakened capacity among the institutions to efficiently and effectively implement appropriate strategies for sustainable water management (Hirvi and Whitfield 2015). Such practices call for management reforms in the urban water sector. A major response to the institutional crisis as part of the reforms was the introduction of the public-private partnership (PPP) arrangements, which received mixed reviews from stakeholders in the water sector. For example, proponents of the PPP argued that its implementation was instrumental in improving the efficiency of water flow by removing institutional roadblocks, promoting autonomy, and streamlining revenue generation (Hirvi 2012; Adu-Ampong 2013). However, opponents argued that with the PPP based on the principle of water as an economic good tended to ostracize a policy goal of providing water as a fundamental human right, particularly for low-income urban residents.

5.4.4 Water Management Limitations

The NWP and WSSDP are national strategic frameworks that guide water management at the city level. According to stakeholders, water managers operating at the city level are confronted with the hurdle to adapt national-level policies and strategies to city-level context. Two probable explanations emerged from the discussion. The first was that national-level priorities were sometimes at variance with priorities at the metropolitan or municipal level. This was because some national preferences were sometimes derived from the ruling political party's manifestos, and these occasionally deviated from the national plan as well as from the priorities of the metropolis. Second, national average figures were often lower than figures for the Accra metropolis due to the difference between cities in the country. For example, Accra has a very high population compared to the rest of the country, and therefore, national figures can sometimes be misleading when used as a benchmark for planning or decision-making. Stakeholders argued that the development of a strategic plan that detailed specific interventions for Accra could help reduce the disparities that often arose when relying on city-level figures produced at the national level. In the absence of a specific water management strategy, Accra faces a dilemma of long-term planning for the sustainable management of water service delivery.

While the NWP and WSSDP strongly advocated for an integrated approach to water management through stakeholder participation, decentralization, and sectoral coordination, the current practice appears to be supply-oriented with limited emphasis on demand management options. Nonetheless, current practices fall short of achieving a holistic and integrated goal of improving water efficiency, improving water access and social equity, and maintaining ecological integrity. The reality is that current water management interventions in Accra are not holistically coordinated during planning and implementation. Instead, water problems in Accra were conceived as governance and economic issues. Strategies under implementation seemed politically and economically biased, with minimal regard for environmental sustainability and social equity. Such biases could be traced back to a neoliberal ideological influence, which highlights the economic value of water. Even so, economic goals such as improvement in efficiency, reducing nonrevenue water losses, and cost recovery of the urban water system have barely been achieved.

An aspect of water management practice often neglected in Accra was the management of the metabolic processes of water inflow and outflow and exchanges between the urban water system and the sinks. The urban water metabolic process involves the water flows and fluxes, socioeconomic dynamics, and driving mechanisms that facilitate urban water systems' delivery and wastewater discharge between cities and the hinterlands (Huang et al. 2013). As water is extracted from sources in the hinterland, treated, and then supplied to Accra, it is consumed to facilitate socioeconomic activities such as drinking, cooking, industrial use, irrigated agriculture, and sanitation. It then exits the system as untreated wastewater that pollutes the rivers, lagoons, and ocean. In addition to the limited wastewater treatment and recycling systems, the intensification of upstream activities such as agriculture and mining contributes to the high level of pollution of water bodies downstream which serve as water sources for Accra. Such impacts on the ecosystems have far-reaching consequences on the water supply to Accra, especially considering the declining raw water quality sources and the rising cost of treating highly polluted water (Oteng-Ababio et al. 2017). It is critical to take these water flow processes and interactions between the city and the hinterland into account, if there is to be an integrated approach to water management.

Accra is also characterized by a mixed dynamic of positive outcomes and limitations to sustainable water management. For example, regarding water access, data from the GWCL suggested that the city was on the verge of producing sufficient potable water to meet the daily water demand. However, leakages, limited pipe network coverage, and water theft explain 50% of water that is not accounted for. These thus derail efforts toward universal water access for all residents by 2025 or even 2030. A major limitation to water management in the city is the absence of demand management interventions, such as public education, economic incentives, and subsidies for the adoption of water-saving technologies, water use restrictions (quota system), bans on water use during emergencies, and penalties for water wastage, amongst others. These would limit water demand and create consciousness about water saving and wise water use.

One of the key principles of NWP and WSSDP was to ensure cross-sectoral coordination and collaboration. For example, the GWCL working with the GSS and the Accra Metropolitan Assembly could produce accurate population and household data in Accra to determine the current proportion of household with water access while working toward the target of universal coverage. Another instance could be a coordinated effort between the GWCL, WRC, and Environmental Protection Agency to manage water sources and their ecosystem to reduce water pollution and improve the surface water quality. However, cross-sectoral coordination and collaboration among water sector agencies remain a major hurdle, partly due to an "institutional ego" or silo mentality where water agencies prefer to work in silos without consultations or collaborations with other water-related institutions.

5.4.5 Urban Water Service Delivery Challenge to SDGs

African cities have implemented several interventions to improve water service delivery. Yet, these interventions have remained inadequate to address the crisis of total water availability and access. Insufficient access to clean and safe water will disproportionally affect low-income households living in the informal settlements. This contradicts the slogan of the United Nations 2030 Agenda for the SDGs to "leave no one behind" (Griggs et al. 2013; Stokstad 2015).

Urban water issues are directly connected to at least 11 out of the 17 SDGs. The lack of water could be the basic form of poverty given water's role in human's existence (SDG 1). Access to good quality water is essential for sustainable urban irrigated agriculture and food security (SDG 2). The provision of high-quality and safe drinking water could prove cost-effective in reducing illness and improve the well-being of all residents (SDG 3). Ready access to clean water and sanitation (SDG 6) means removing time-consuming long-distance water collection practices that restrict children's participation in school (SDG 4) and disproportionately burden young girls and women (SDG 5). Freeing children, young girls, and women from the struggles of accessing clean water and sanitation provision could unlock opportunities for decent work and economic growth (SDG 8). Sustainable water access and security reduce social inequality, particularly for poor urban inhabitants (SDG 10). An integrated approach to sustainable water management means promoting inclusive, resilient, and sustainable cities (SDG 11) while enforcing sustainable consumption norms (SDG 12) that accommodate long-term adaptation of water resources to climate change (SDG 13).

5.4.6 Toward an Integrated Approach for Sustainable Water Management in Accra

An integrated approach to sustainable water management demands pursuing an interconnected set of strategies to achieve multiple goals, including improving water efficiency, tackling social equity in water allocation, and maintaining the integrity of the environment. In the urban context, there is a particular need to achieve improved water efficiency in water supply, wastewater treatment, and stormwater management in a holistic manner (Bahri et al. 2016). Accra's water metabolic process can be described as a linear flow metabolism, given that water inflow into the system is discharged as wastewater without any form of recycling and reuse to close the urban water cycle loop. Such a linear metabolism—from water to waste—is no longer sustainable in an era where water is increasingly becoming scarce and where environmental sustainability has become a major concern in the management of the urban water cycle. The urban water cycle characterizes the holistic nature of the interconnected components of the water system that function dynamically to deliver multiple goals, including potable water production and supply, wastewater and stormwater management, and ecosystem protection (Moglia et al. 2018). A shift toward a circular metabolism, with emphasis on wastewater recycling and reuse, has the potential to increase water resource efficiency by augmenting water supplies with recycled water while drastically reducing environmental loads in Accra.

The status of Accra as a city in transition opens up opportunities for urban water managers to consider an integrated approach for sustainable water management. Within the framework of an integrated approach, three interconnected goals are fundamental to the sustainable management of the urban water system: ensuring the well-being of all water users through water provisions in an equitable and socially desirable manner; building a prosperous economy using water within sustainable thresholds (meaning it should be economically efficient); and maintaining the integrity of water sources to ensure long-term water access (Longo et al. 2016). These interconnected goals appeal to water managers and policy makers as they stimulate the quest for water resource perpetuity in the urban water cycle. The integrated approach recognizes and aligns with a range of formal (organizations, legislation, and policies) and informal (norms and conventions) institutions that govern water in cities (Grit et al. 2015). It involves managing urban water across different agencies and organizations by engaging all stakeholders and users to be part of the process. It takes into account the upstream and downstream issues and incorporates them into planning measures at the basin level (Closas et al. 2012). Such consideration is essential for sustainable access to water flow, considering that upstream activities (such as agriculture) and downstream activities (such as releasing untreated wastewater effluents) can cause stress to the water supply system by altering the local hydrological cycle and reducing environmental flows (ibid).

The outcomes of the WSSDP implementation in Accra suggest that interventions were not well packaged to address the multidimensional challenges that confronted the water system. Key challenges such as water system leakages, high volumes of NRW, a low percentage of treated wastewater, limited access to water data, weak institutional coordination, and weak understanding of the interactions between the water system and sources remained. The city continues to experience a substantial water deficit despite the increased volume of treated water produced. Fundamental to the pursuance of a sustainable water system in Accra is the basic principle that an integrated approach acknowledges that water supply, wastewater, and stormwater systems are purposely linked, and thus coordinated management is essential. It considers the entire urban water cycle with interconnected components that function dynamically to deliver multiple urban services, such as public health, ecosystem protection, water access, and well-being that are critical to the sustainability of African cities and the achievement of the SDGs. Critical for the integrated approach is the planning and coordination of individual processes to maximize efficiency and productivity of the collective system while minimizing the collective impact on the water system (Closas et al. 2012).

5.4.7 Enablers of an Integrated Approach to Sustainable Water Management

Efforts toward building a sustainable city may require a transition from supply-oriented interventions, toward demand management and conservation-based interventions. Given this, water managers and policy makers should define and implement a set of targeted strategies that could collectively help address the water management challenges identified in Accra. This chapter proposes five priority strategies to help transform the water sector in Accra.

Public Participation and Awareness Creation

An integrated approach to sustainable water management is labeled as a people-focused endeavor, rather than an approach that is technologically oriented (Sharp 2017). Therefore, public participation in the decision-making process is integral to developing strategies and solutions shaped by stakeholders' views and demands (Dapaah and Harris 2017). Institutions like the GWCL and PURC now acknowledge that water management decisions on issues such as cost recovery, tariffs' adjustment, and access for low-income residents should be informed by consumers' views and submissions and not only by external assumptions. Notwithstanding its positive influence, effective participation remains a difficult task in practice.

It is still an area that receives little investment and where even well-thought-out mechanisms could be easily altered by the stakeholders involved. Awareness creation is one of the often marginalized strategies that provide the opportunity for target beneficiaries to gain a clear understanding of the importance of interventions and how they could contribute toward their successful implementation. In the Accra metropolis, the adoption of rainwater harvesting techniques to conserve water was stalled by the lack of understanding and information on the techniques (ibid). However, studies have shown that regular media and communication campaigns on drought and the necessity for water conservation successfully lead to reductions in water consumption (Quesnel and Ajami 2017; Ziervogel 2019).

Strengthening the Capacity of Key Stakeholders

A major hindrance to sustainable water management in the Accra metropolis has been the display of limited understanding of the integrated approach to water management, as well as existing structures that promote or lead to a persistence of working in silos. In this regard, focusing on knowledge and skills development of water managers and policy makers is an essential prerequisite for designing an integrated process for the implementation of water management interventions.

Cost Recovery and Tariff Adjustment

The adjustment of water prices was a critical aspect of urban water demand management. Strong evidence exists to suggest that increasing water charges contribute significantly to the reduction in water consumption (Mylopoulos et al. 2017). Therefore, setting water tariffs at full cost could incentivize the reduction in water demand and promote efficient water use. However, in African cities like the Accra metropolis where the dominant population represents low-income urban residents, this may be complicated. Some scholars have argued that increasing water tariffs could deprive low-income households of the ability to pay for water services (Amankwaa et al. 2014). However, several studies have shown that in reality, low-income urban residents pay higher water charges than high-income residents as the former usually buy water from secondary service providers—water vendors and kiosks—who charge 10–20 times the rates charged by the GWCL (Oteng-Ababio et al. 2017). This indicates that there is a willingness and ability to pay, especially if a reliable and much more affordable service would be available from the GWCL.

Reducing Unaccounted for Water (UFW)

In developing countries, the Unaccounted for Water (UFW) accounts for 45 million m^3 per year, which is sufficient to serve the daily water needs of about 200 million people (Bhagat et al. 2019). In Accra, UFW is estimated to be 50% of the total volume of water produced. The GWCL estimates a cost of not less than US$ 200 million to restore the entire water transmission and distribution network to significantly reduce UFW. Seeking funding to cover this venture could prove to be daunting since annual investment contribution by the Ghanaian government to urban water supply in the entire country according to the GWCL is US$ 160.5 million, with a matching fund of US$ 189.5 million from development partners (GWCL 2015, 2018). Nonetheless, some cost-effective measure could be taken to reduce UFW, such as:

- Active leakage control: Several water experts consider this as a cost-effective action that would require fewer logistics and minimal effort to quickly detect, identify, and repair of leakages to reduce the volume of water loss and possibly third-party damages.
- Pipe replacement: Many pipelines installed several decades ago in Accra were weak and prone to leakages. Replacing them could guarantee high returns in terms of the initial reduction in NRW and UFW despite the high investment cost. A smart way to undertake the replacement of pipes is to conduct targeted monitoring to determine points or locations in the system prone to frequent leakages and bursts.

Financial Investment

Financial investment and finance mechanisms are essential ingredients for driving the application of water technologies and innovations to support sustainable urban water management. The lack of finance mechanisms may compel water managers to implement ad hoc stop-start water projects rather than developing long-term systematic programs in a coordinated way. The limitation of the centralized, formal water systems of infrastructure to deliver efficient water service to the entire urban population has prompted heterogeneous ways of water service delivery entailing informal systems (Cirolia 2020; Kimari and Ernstson 2020). In most African cities, informal systems such as water tankers, kiosks, and vendors serve as the major mode of water service delivery to areas not connected to the water network. Despite the importance of informal water service delivery in Africa, scholars have pointed to the limited attention given to the systems and stakeholders that govern these processes (Pieterse 2019; Cirolia 2020). Financing water service delivery in African cities means prioritizing the integration and regularization of informal water service delivery systems while incrementally investing in the extension of these systems in areas outside the reach of a formal pipe supply and distribution system (Cirolia 2020).

5.5 Conclusion

This chapter has highlighted water management limitations and challenges confronting the city of Accra, in Ghana. While considerable effort has been made to promote sustainable universal water service delivery in Accra, several bottlenecks remain, such as an inadequate infrastructure network, high water losses, water resource depletion, and pollution, among others. The politics and governance of urban water, along with attempts to promote participatory water management, tend to create inequality and wider social power relationships. Nonetheless, institutional reforms, the evolving policy environment, and structural changes in the water sector present an opportunity for practitioners to rethink water management in ways that highlight the critical role that water plays in addressing an interconnected set of SDGs.

An integrated approach to urban water management advocates for greater emphasis on investment in water infrastructure and appropriate technologies that increase efficiency, promote wastewater recycling and reuse, and encourage a circular water flow metabolism instead of linear flow to reduce water losses. At the same time, the approach promotes improved equity in water allocation and access, as well as improved urban water governance. Drawing on a unique TDR process, this chapter has highlighted the versatility in incorporating the views of different stakeholders and thus facilitating dialogue and participation that are essential elements of an integrated approach as a suitable tool for urban water management decision-making. Such an approach also stimulates thinking about how improved human well-being is connected to economic growth and environmental sustainability through equitable water access, consumption, and water resource perpetuity.

The transition toward sustainable water management requires the implementation of a management approach that takes into account water flows, processes, and available supply to meet the water needs and demands of both current and future generations in the city. This must be accomplished while maintaining water infrastructure systems and the integrity of water sources and sinks to regenerate and absorb return flows, respectively. The current water management approach in the Accra metropolis is yet to reflect this goal. Nonetheless, the current efforts regarding institutional arrangements, policy environment, and structural changes in the urban water system present an opportunity to effect a shift in the current management paradigm. The integrated approach seeks to promote a mix of strategies or solutions to achieve multiple goals: optimize efficiency in water production, supply, and usage; promote wastewater recycling and reuse; ensure equitable water allocation and distribution; and ensure the maintenance of the environmental sink function and water resource base.

References

Abowitz DA, Toole M (2009) Mixed method research: fundamental issues of design, validity, and reliability in construction research. Journal of Construction Engineering and Management 136:108–116

Adu-Ampong E (2013) Water privatisation policy in Ghana: stalled for good or a strategic pause. http://www.researchgate.net/publication/260417347_Water_Privatisation_Policy_in_Ghana_Stalled_for_Good_or_a_Strategic_Pause. Accessed 20 October 2021

Amankwaa EF, Owusu AB, Owusu G, Eshun F (2014) Accra's poverty trap: analysing water provision in urban Ghana. Journal of Social Science for Policy Implications 2:69–89

Amoako C, Boamah FE (2015) The three-dimensional causes of flooding in Accra, Ghana. International Journal of Urban Sustainable Development 7:109–129

Asante-Wusu I, Yeboah IEA (2020) Geography of water insecurity in three localities in the Accra-Tema city-region, Ghana. African Geographical Review 4:308–323

Bahri F, Brikké BF, Vairavamoorthy K (2016) Managing change to implement integrated urban water management in African cities. Aquatic Procedia 6:3-14

Banerjee S, Foster V, Ying Y, Skilling H, Wodon Q (2017) Cost recovery, equity, and efficiency in water tariffs: evidence from African utilities. Policy research working paper no. WPS 5384. World Bank, Washington, DC

Bhagat SK, Welde W, Tesfaye O, Tran MT, Al-Ansari N, Salih SQ, Yaseen ZM (2019) Evaluating physical and fiscal water leakage in water distribution system. Water 11:2091

Cirolia LR (2020) Fractured fiscal authority and fragmented infrastructures: financing sustainable urban development in sub-Saharan Africa. Habitat International 104:102233

Closas A, Schuring M, Rodriguez D (2012) Integrated urban water management—lessons and recommendations from regional experiences in Latin America, Central Asia, and Africa. Water Partnership Program. World Bank, Washington, DC

Dapaah EK, Harris LM (2017) Framing community entitlements to water in Accra, Ghana: a complex reality. Geoforum 82:26–39

Dinarès M (2014) Urban metabolism: a review of recent literature on the subject. Documents d'Anàlisi Geogràfica 60:551–571

Engstrom R, Pavelesku D, Tanaka T, Wambile A (2019) Mapping poverty and slums using multiple methodologies in Accra, Ghana. In: 2019 Joint Urban Remote Sensing Event (JURSE) 1–4. Institute of Electrical and Electronics Engineers (IEEE)

Gaisie E, Kim HM, Han SS (2019) Accra towards a city-region: devolution, spatial development and urban challenges. Cities 95:102398

Ghana Statistical Service (2017) Ghana living standards survey round 7. Ghana Statistical Service, Accra, Ghana

Ghana Water Company Limited (GWCL) (2015) Strategic investment plan (2007-2025). https://www.gwcl.com.gh/water_sector_strategic_development__plan.pdf. Accessed 2 August 2020

Ghana Water Company Limited (GWCL) (2018) Proposals for review of aggregate revenue requirement and tariff. http://www.gwcl.com.gh/tarrif_paper.pdf. Accessed 10 October 2020

Ghana Water Company Limited (GWCL) (2020) Public Utilities Regulatory Commission (PURC) publication of water tariffs. Ghana Water Company Limited, Accra, Ghana

Government of Ghana (2019) Ghana Voluntary National Review (VNR) report on the implementation of the 2030 Agenda for Sustainable Development. https://ghana.un.org/sites/default/files/2019-10/23420VNR_Report_Ghana_Final_print.pdf. Accessed 20 October 2021

Government of Ghana (2014) Water sector strategic plan (2012-2025): sustainable water and basic sanitation for all by 2025. Ministry of Water Resources, Works and Housing. https://www.gwcl.com.gh/water_sector_strategic_development__plan.pdf Accessed 20 October 2021

Government of Ghana (2007) National water policy. Ministry of Water Resources, Works and Housing https://www.gwcl.com.gh/national_water_policy.pdf. Accessed 10 October 2020

Griggs D, Stafford-Smith M, Gaffney O, Rockström J, Öhman MC, Shyamsundar P, Steffen W, Glaser G et al. (2013) Sustainable development goals for people and planet. Nature 495:305-307

Grit R, Jörg L, Steffen D, Gerel O (2015) Integrated urban water management: development of an adapted management approach. Environmental Earth Sciences 73:709–718

Head BW, Ross H, Bellamy J (2016) Managing wicked natural resource problems: the collaborative challenge at regional scales in Australia. Landscape and Urban Planning 154:81–92

Hirvi M (2012) Water privatization and social citizenship: the case of urban water sector in Ghana. Journal of Civil Society 8:351–368

Hirvi M, Whitfield L (2015) Public-service provision in clientelist political settlements: lessons from Ghana's urban water sector. Development Policy Review 33:135–158

Huang CL, Vause J, Ma HW, Yu CP (2013) Urban water metabolism efficiency assessment: integrated analysis of available and virtual water. Science of the Total Environment 452–453:19–27

Hurlimann A, Wilson E, Keele S (2017) Framing sustainable urban water management: a critical analysis of theory and practice. In: Urban water trajectories. Springer Publishing, Switzerland, pp 53–68

Jaglin S (2002) The right to water versus cost recovery: participation, urban water supply and the poor in sub-Saharan Africa. Environment and Urbanization 14:231–245

Kimari W, Ernstson H (2020) Imperial remains and imperial invitations: centering race within the contemporary large-scale infrastructures of East Africa. Antipode 52:825–846

Koppelaar RHEM, Sule MN, Kis Z, Mensah FK, Wang X, Triantafyllidis C, Van Dam KH, Shah N (2018) Framework for WASH sector data improvements in data-poor environments, applied to Accra, Ghana. Water 10:1278

Longo SB, Clark B, Shriver TE, Clausen R (2016) Sustainability and environmental sociology: putting the economy in its place and moving toward an integrative socio-ecology. Sustainability 8:437

March H (2015) Taming, controlling and metabolizing flows: water and the urbanization process of Barcelona and Madrid (1850–2012). European Urban and Regional Studies 22:350–367

Moglia M, Cook S, Tapsuwan S (2018) Promoting water conservation: where to from here? Water 10:1510

Moreno-Leguizamon C, Tovar-Restrepo M, Irazábal C, Locke C (2015) Learning alliance methodology: contributions and challenges for multicultural planning in health service provision: a case study in Kent, UK. Planning Theory & Practice 1:79–96

Mylopoulos N, Fafoutis C, Sfyris S, Alamanos A (2017) Impact of water pricing policy and climate change on future water demand in Volos, Greece. European Water Journal 58:473–479

Nansubuga I, Banadda N, Verstraete W, Rabaey K (2016) A review of sustainable sanitation systems in Africa. Reviews in Environmental Science and Bio/Technology 15:465–478

Oteng-Ababio M, Smout I, Yankson PWK (2017) Poverty politics and governance of potable water services: the core–periphery syntax in metropolitan Accra, Ghana. Urban Forum 28:185–203

Pieterse E (2019) The potential for sustainable urbanisation in Africa. Alfred Herrhausen Gesellschaft. https://www.africancentreforcities.net/wp-content/uploads/2019/04/Paper_Edgar-Pieterse_The_Potential_for_Sustainable_Urbanisation-in-Africa_2019.pdf. Accessed 20 August 2020

Quesnel KJ, Ajami NK (2017) Changes in water consumption linked to heavy news media coverage of extreme climatic events. Science Advances 3:e1700784

Reed MG, Abernethy P (2018) Facilitating co-production of transdisciplinary knowledge for sustainability: working with Canadian biosphere reserve practitioners. Society & Natural Resources 31:39–56

Ruiz AG, Dobbie M, Brown R (2017) Insights and future directions of transdisciplinary practice in the urban water sector. Journal of Environmental Studies and Sciences 7:251–263

Silver J (2014) Incremental infrastructures: material improvisation and social collaboration across post-colonial Accra. Urban Geography 35(6):788-804.

Sharp L (2017) Reconnecting people and water: public engagement and sustainable urban water management. Routledge, London

Siew TF, Aenis T, Spangenberg JH, Nauditt A, Döll P, Frank SK, Ribbe L, Rodriguez-Labajos B et al. (2016) Transdisciplinary research in support of land and water management in China and Southeast Asia: evaluation of four research projects. Sustainability Science 11:813–829

Stokstad E (2015) Sustainable goals from UN under fire. Science 347 (6223):702-703

WaterAid (2016) Water: at what cost? The state of the world's water 2016. https://washmatters.wateraid.org/sites/g/files/jkxoof256/files/Water%20%20At%20What%20Cost%20%20The%20State%20of%20the%20Worlds%20Water%202016.pdf. Accessed 10 October 2020

Yates S, Kruger A, Yaari E, Mpakama Z, Baloyi X, Menouer L, Earle A (2020) Water tariffing: what's at stake? Stockholm International Water Institute. https://www.siwi.org/wp-content/uploads/2020/03/AEWPP_-Water-Tariffing-2020_WEBB.pdf. Accessed 11 October 2020

Ziervogel G (2019) Unpacking the Cape Town drought: lessons learned. Report for Cities Support Programme undertaken by African Centre for Cities. https://www.africancentreforcities.net/wp-content/uploads/2019/02/Ziervogel-2019-Lessons-from-Cape-Town-Drought_A.pdf. Accessed 22 August 2020

Check for updates

6 Waste Recycling and Repurposing to Address SDG 11 in Burkina Faso: Do Multi-stakeholder Platforms Matter?

Safiétou Sanfo, Oblé Neya, Sylvestre Da, Seyni Salack, Joseph Amikuzuno, Bizoola Zinzoola Gandaa, Kwame Oppong Hackman, and Kehinde Olufunso Ogunjobi

Abstract

Poor waste management and uncontrolled waste disposal cause pollution and blocked drainage facilities, leading to flooding and water stagnation, increasing the risk of diseases, and eroding local progresses toward achieving the Sustainable Development Goals (SDGs). Recycling and repurposing liquid and solid waste for urban and periurban agriculture, green spaces, and green energy on the other hand benefit social and ecosystem resilience and can contribute to SDG11. This chapter considers the uneven progress in Burkina Faso where the transgressive behavior of garbage producers leads to illegal garbage dumping, equipment obsolescence, insufficient maintenance, and the lack of support from authorities. We show how the creation of a participatory Multi-Stakeholder Platform (MSP) can lead to better collection, recycling, and repurposing of wastes. The results showed that (1) the practice of liquid and solid waste management involves several interacting stakeholders, (2) these interactions generate complex problems hardly resolved by a single (few) stakeholder(s), and (3) an MSP is a good approach toward solving these complex problems.

Keywords

Urban waste management · Waste-water-energy nexus · Multi-stakeholder platform (MSP) · Burkina Faso

S. Sanfo (✉) · O. Neya · S. Da · S. Salack
K. O. Hackman · K. O. Ogunjobi
WASCAL, Competence Centre,
Ouagadougou, Burkina Faso
e-mail: sanfo.s@wascal.org; neya.o@wascal.org; da.s@wascal.org; salack.s@wascal.org; Hackman.k@wascal.org; ogunjobi.k@wascal.org

J. Amikuzuno · B. Z. Gandaa
University of Development Studies (UDS),
Tamale, Ghana
e-mail: amikj26@yahoo.com; naa.bizoola1970@gmail.com

6.1 Introduction

As a result of rapid urbanization, waste management, sanitation, and hygiene have become major urban concerns in West Africa. Inadequate disposal and household waste gathering in the streets block drainage systems, streams, and landfills and threaten health in residential areas in West

S. Croese, S. Parnell (eds.), *Localizing the SDGs in African Cities*, Sustainable Development Goals Series, https://doi.org/10.1007/978-3-030-95979-1_6

African cities (Onibokun 2002). This is not just a problem of poverty. In Burkina Faso, rapid economic and population growth are the result of population expansion and rising affluence. Waste generation in the capital city of Ouagadougou increases in direct correlation with income growth (UNEP 2011). With 600,000 metric tons of waste generated per year, an increase in population (from three million people in 2020 to an expected nearly six million in 2030, according to INSD (2019)), waste production will also increase exponentially. Out of the 600,000 metric tons of waste, only 300,000 metric tons are collected. The other 300,000 metric tons remain uncollected, resulting in illegal dumping of garbage which prevents rainwater from draining properly. Uncollected waste in the city of Ouagadougou favors flash flooding, causes water pollution, and increases the risks of cholera, diarrhea, and other waterborne diseases. The insufficient awareness of the population about the benefits of repurposing organic waste and domestic sewage combined with the inadequate waste management infrastructure further worsens the problem of sustainable waste disposal and recycling. Although research is increasingly showing the importance and the potential of the waste-water-energy nexus, this approach is only in its infancy in Burkina Faso. This is likely to delay progress on achieving the SDGs.

When it comes to urban and periurban agriculture (UPA), food ingredients (i.e., legumes, vegetables, and fruits) for the capital city of Ouagadougou are provided by gardeners in the center and suburbs of the city. Besides legumes, vegetable, and fruit gardening, UPA activities include plant nurseries for trees and flowers and dairy and poultry production for meat. UPA contributes to food security and promotes green urban spaces (GUS) as well as soil biodiversity (Hoornweg and Munro-Faure 2008; Bellfield 2015). However, this contribution is becoming increasingly unreliable due to negative effects of rapid demographic increase. Rural-urban migration is decreasing the availability of land in the capital city and increasing related land tenure insecurity and conflicts. Low investment in the recycling of liquid and solid wastes (e.g., food scraps, domestic sewage, livestock manure) to meet agricultural and green energy needs (compost, treated water, and biogas) has also led to heavy reliance on chemical fertilizers in UPA and dependence on wood and charcoal energy as household cooking fuels. Increase in wood and charcoal fuel consumption has adverse impacts on the urban environment, including outdoor and indoor air pollution, with associated health problems. Therefore, a good use of the integrated approach of waste, energy, and water could help to achieve SDG 11 and related urban goals.

In the past, urban waste, energy, and water management were generally considered and treated separately. Nowadays, we are witnessing a more integrated approach in the management of these key sectors and the ways in which nexus thinking can contribute to meeting global development goals (Boas et al. 2016; Liu et al. 2018). Understanding the synergies between waste, water, and energy, as well as related sectors such as food, enables trade-offs to be identified and addressed in city management and contributes to urban resilience and efficiency (Sperling and Berke 2017; Terrapon-Pfaff et al. 2018). For example, water is not only needed to support livelihoods, such as drinking water services, domestic and livestock purpose, aquaculture, irrigation, and food production, but also to produce energy, including hydroelectricity and biofuel production. Conversely, the production of drinking water and waste treatment requires energy, waste can be used to generate bioenergy, and treated wastewater can be used for crop irrigation. Reducing the quantity of waste in drainage canals of urban areas can prevent flash floods, and waste recycling also produces clean water and compost as soil nutrients. Therefore, a better collection, recycling, and repurposing of solid and liquid waste benefit both UPA and GUS.

Foregrounding the waste-water-energy nexus is key for any city or region wanting to make progress on the SDGs. Our argument is not just that the SDGs themselves are overlapping, are linked, and can be reinforcing of each other but that, in an urban context, it is not possible to intervene in one part of the urban system (like water or waste) without understanding the impact of those actions on other dynamics. This chapter builds on a study that investigated the potential of trans- and inter-disciplinary, participatory, and proactive research approaches to better address the waste-water-energy nexus in Burkina Faso. These approaches bring together different key stakeholders to manage the complexity of the framework. The chapter will answer the following questions: How can better liquid and solid waste management produce more effective utilities and products, including treated water, compost, and biogas? How can stakeholder collaboration through multi-stakeholder platforms (MSPs) help resolve the complex issues associated with the waste-water-energy nexus in an urban city like Ouagadougou? Across the chapter, we show how the waste-water-energy nexus contributes to the advancement of a wide range of SDG targets and paves the way for the implementation of SDG 11, to "make cities and human settlements inclusive, safe, resilient and sustainable," particularly target 11.6 which sets "to reduce by 2030 the negative environmental impact of cities per inhabitant, including by paying particular attention to air quality and management, particularly municipal, of waste." We will also demonstrate how it contributes to many other SDGs, including SDG 3 (Health), SDG 6 (Water and Sanitation), SDG 7 (Energy), SDG 8 (Work and Economic Growth), SDG 12 (Responsible Consumption), SDG 13 (Climate Action), and SDG 14 (Life Below Water) (Wang et al. 2018; Forti et al. 2020).

6.2 Waste-Water-Energy Nexus: Tracking the Links Toward SDG 11

An acknowledgement of the nexus between the waste, water, and energy sectors and their connection to other sectors such as food is central to achieving sustainable development in its environmental, economic, and social dimensions (Biggs et al. 2015; Stephan et al. 2018). However, despite the need for more integrated approaches to urban management in African cities, nexus approaches are only at their infancy stages. The roots of the waste-water-energy nexus lie in waste-to-energy processes, wastewater treatment, and anaerobic digestion processes. These have been well developed in the global North to maximize the efficient use of resources while reducing current environmental impacts (Kalair et al. 2019; Portney et al. 2017; Mancini et al. 2021). Waste recycling and repurposing are understood to mean all the provisions allowing for the collection, transport, and environmentally sound disposal of waste, taking into account health (public health), economic, social (attitudes of populations), and environmental considerations (Cheng and Urpelainen 2015; Cobbinah et al. 2017).

Efficient recycling and repurposing of liquid and solid organic wastes into compost and irrigation can supply energy for communities while improving sanitation conditions and fertilizing agricultural lands (Kaudal and Weatherley 2018). According to Gremillion and Avellan (2016), a nexus approach centered on wastewater treatment and recovery can link cycles of water, waste, energy, as well as food systems. Rich in carbon and nutrients, wastewater has the potential to supply greenhouse gas (GHG) neutral energy for households and irrigate and fertilize crops. Capturing the resources contained in wastewater can improve human and ecosystem health, reduce GHG emissions, and improve food and energy security. Access to energy in turn is essential to meeting development challenges, such as poverty, gender inequality, poor health and education services, and food insecurity (Bhatia and Angelou 2015; Masud et al. 2007; Spalding-Fecher et al. 2005). Waste-to-renewable energy technology, in particular anaerobic digestion and biogas technologies, is therefore well positioned to address global waste challenges and contribute to the achievement of some of the key SDGs (Kaudal and Weatherley 2018; Ghodsvali et al. 2019). Global progress is, however, dependent on implementation in every part of the world, including less well-researched places that face severe challenges of urban management, like the city of Ouagadougou in Burkina Faso.

6.3 Context of Burkina Faso

From Burkina Faso's independence (August 5, 1960), various waste management methods have been implemented in the city of Ouagadougou, but they have always been of limited effectiveness due to the limited importance given by the authorities to waste recycling and repurposing. The lack of human and financial resources and the poor hygiene practices of local inhabitants have also compounded the issue. The rapid increase in waste production due to a high demographic growth (4.4% per year according to INSD 2019) has resulted in the removal and treatment of solid waste becoming a major problem for the city of Ouagadougou. The modernization of Burkinabè society and especially the fast pace of urbanization of the city have made it urgent to (re)organize waste management more efficiently—a challenge which keeps evolving.

Five main periods can be distinguished in the history of waste management in Ouagadougou. During the period immediately after independence and until the 1970s, Ouagadougou implemented a waste management system inherited from colonial times, in which the municipal board was the main entity responsible for the task. In 1968, the municipality transferred responsibility to a private company called *Société Nakoulma*. Between 1979 and 1986, due to a lack of financial resources, the contract between the municipal authority Ouagadougou and *Société Nakoulma* was terminated, and the municipality took back the responsibility for waste management. In 1986, the *Direction Nationale des Services d'Entretien de Nettoyage et d'Embellissement* (DINASENE) was created and was devoted to the task throughout this period of the revolution, which was also marked by collective operations of public cleaning known as "Operations mana-mana." Financial and other difficulties arose again in the 1990s, and the National Office of Sanitation and Environment (*Office National de l'Assainissement et de l'Environnement,* ONASENE) established a public-private partnership (PPP) approach to handling waste management with waste collection fees varying between UD$ 0.50 and UD$ 2 per household, depending on the neighborhood (Bayili 2002). New waste management companies were created, but there was a lack of formal governing structure and constant failure to comply with environmental regulations. The collected wastes were often dumped at inappropriate places, sometimes only a few hundred meters away from settlements.

In 1993, poor management and the deterioration of the situation forced the municipal council to take over the management of the sector through the Department of Municipal Technical Services (*Direction des Services Techniques Municipaux*, DSTM). From 2000 on, the World Bank supported the municipality through its Urban Development Project (UDP). The main objective was to lay the foundations for the "polluter pays" principle through the partial recovery of operating costs. The polluter pays principle is one of the internationally recognized principles that influences the shaping of environmental policy at both the national and international level. By decentralizing and implementing a strategic plan for the sanitation of the city of Ouagadougou (French acronym PSAO), in 2001 a department responsible for the cleanliness and hygiene of the city was established. Private waste management companies were created, and opportunities for the improvement of local waste recycling capacity emerged. In 2004, a Master Scheme for Waste Management (*Schéma Directeur de Gestations des Déchets*, SDGD) was set up by the Sustainable Development Department (*Direction du Développement Durable*, French acronym DDD) under the framework of the Canadian DESSAU-SOPRIN study. The main objective was to improve the collection of household waste, in particular by setting up 35 collection centers and a technical center located on the northern outskirts of the capital, later called Center for Waste Processing and Recovery (*Centre de Traitement et de Valorisation des Déchets*, CTVD). To this end, the city was divided into 12 zones (*Arrondissements*), whose management was entrusted to private entrepreneurs (small- and medium-sized enterprises, SMEs). By establishing these centers and this new division, the city's Sustainable Development Department tried to reduce the number of uncontrolled landfills to cope with the threat of an unhealthy and degraded environment (Ouattara 2006).

From 2009 onward, the rapid growth of the population, uncontrolled urbanization (due to the rural exodus, among other reasons), and impoverishment have made it difficult to organize the sector (AAPUI Arcade 2009). As result of rural-urban migration, the city of Ouagadougou has been expanded by informal settlements with poor and marginalized population that are largely disconnected from the formal systems of city management (AAPUI Arcade 2009). In informal settlement areas, locally known as "non-lotis," settlement patterns are unplanned, with no specific access to roads. The situation is made worse by the lack of running drinking water and electricity. In these neighborhoods, unlike the formal areas, waste is dumped in the nearest public space or in the canals. The unplanned expansions of the city and these practices have put added pressure on the waste management system, which was already experiencing difficulties in the formal areas. While the actors active in the sector, including households, collectors, associations, SMEs, and local authorities, have accused one another of being responsible for the situation, some observers have argued that the issues faced by the waste management system were multidimensional: legislative and regulatory; financial; institutional and organizational; technical; political and strategic; anthropological; environmental; and scientific. The combination of these uncontrolled factors explains the current situation (Dos Santos 2015), which is characterized by unhealthy practices, clandestine collectors, and illegal dumping. Uncontrolled waste disposal causes pollution, inhibits storm water drainage, aggravates floods, and increases disease risk and therefore puts the achievement of the SDGs at risk. After describing the methods (see Sect. 6.4), the results are discussed in Sect. 6.5. The last section of this chapter concludes by discussing the study's implications for reporting progress on SDG 11.

6.4 Exploring Inter- and Trans-disciplinary Approaches to Facilitate SDG 11 Implementation

6.4.1 Study Area

The city of Ouagadougou (1° 31′ 05″ W, 12° 21′ 58″ N) is the capital of the Province of Kadiogo. Since 2009, the city has had 12 districts divided into 55 areas (see Fig. 6.1), with residential areas in both formal and informal settlements. In the informal settlements, the buildings have roofs essentially made of metal sheets without ceilings, which is not very favorable to health. The situation is made worse by the lack of drinking water and electricity. The walls are generally made of traditional and poor-quality construction material, locally referred to as *banco*. This is indicative of poverty levels among this population but also of the perceived temporary nature of informal settlements and the expectation of relocation by the government to formal settlements. In this area, settlement patterns are unplanned with no specific access to roads. In contrast, the buildings in formal settlements are covered with metal sheets and sometimes have a ceiling, or a concrete roof, thus improving local living conditions. However, one still encounters *banco* walls that contrast with concrete walls, indicating unequal incomes among the population. While the dwellings in formal settlements benefit from health and sanitation services such as household waste collection, informal settlements lack basic services and seem to be left behind (Ouattara 2006).

Ouagadougou is a rapidly growing city. With the current population estimated at 2.6 million inhabitants, the population of Ouagadougou could reach six million in 2030, or three times the current population (INSD 2019). This rapid urban population growth is a major challenge for the waste management sector. Using participatory, proactive, and MSP approaches, this chapter builds on a trans-disciplinary study that investigates how to better collect, manage, and repurpose organic liquid and solid waste to benefit UPA and GUS and generate biogas for urban and periurban sustainable development.

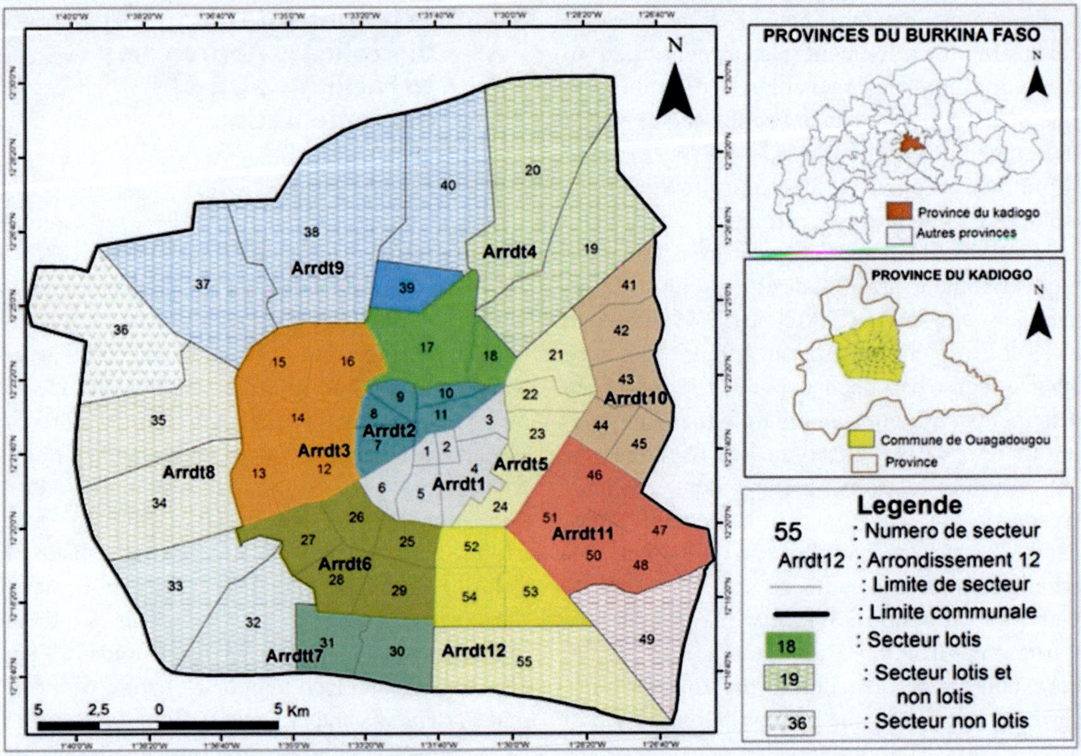

Fig. 6.1 Map of Ouagadougou. (Source: authors)

Trans-disciplinary research and the creation of MSPs require the collaboration between academics and non-academics from several disciplines to meet, interact, and deepen discussions for knowledge sharing (Pohl et al. 2017). Such an approach is therefore crucial for overcoming silo practices and thinking and for the implementation of nexus approaches to sustainable development (Ghodsvali et al. 2019; Sesan et al. 2021; Totin Vodounon et al. 2021). This study was a team effort with several partner institutions, including the municipal assemblies, private companies, and other relevant stakeholders. The research team from the West African Science Service Centre on Climate Change and Adapted Land Use (WASCAL) in Ouagadougou, Burkina Faso, and the University for Development Studies (UDS) in Tamale, Ghana, included agronomists, economists, climatologists, botanists, geographers, water engineers, anthropologists, and hydrologists. Researchers and development professionals worked together as a team, with each team member having specific roles and responsibilities in building the MSPs in order to advance the sharing of best practices and integrating and producing new knowledge to improve the waste management system for the benefit of UPA and GUS in the city of Ouagadougou.

6.5 Stakeholder Consultation and Dialogue Toward Multi-Stakeholder Platforms

Several meetings and workshops with relevant stakeholders were convened in order to build Multi-Stakeholder Platforms (MSPs) for knowledge creation and sharing toward proper waste disposal for clean cities and sustainable development. The various stakeholders organized themselves into working groups with specific roles for the smooth running of the MSPs. Throughout the city of Ouagadougou and beyond, our stakeholders were variously linked within a value network of waste recycling and recycled products. Key stakeholders included government agencies; cleaners, including septic tank cleaners; waste collectors and sorters; waste transporters; transformers, including compost producers, farmers, and livestock raisers; growers, including market gardeners, and nursery owners and assistants; and households and other waste producers. Stakeholder consultation and dialogue through workshops contributed to a better understanding of the places and different roles of the multiple actors within the waste value network and socio-ecological landscape. Stakeholders met each other, some for the first time, described the work that they do, and presented the various issues that concern them within their particular contexts (see Table 6.1).

During meetings, experts and water and sanitation engineers coordinated discussions on the following themes: centralized waste recycling facilities and their management systems; organic waste processing and the possibilities of enhancing those capacities; community engagement and participation in organic and plastic waste and domestic sewage collection and processing; and problems and challenges associated with organic and plastic waste recycling and lessons learnt.

Issues related to processing organic waste into green energy and nitrogen-rich fertilizer were led by waste-to-energy companies with support from livestock farmers and vegetables growers. All stakeholders contributed to the building and the running of the MSPs.

In order to better explore the links and relationships that surround the collection, recycling, and reuse of organic liquid and solid waste in Ouagadougou, we used the value network approach, superimposed on the landscape of the commune and its periphery (Block et al. 2008; Nederlof et al. 2011; Hein et al. 2017). Importantly, we brought some of the key stakeholders in this network together, whose roles in the provision and management of key urban services are not always well known or understood (Jimu 2008). Workshops used a landscape approach and stakeholder mapping exercises. While landscape approaches are used by many researchers and development practitioners in rural and forested areas, they are also applicable to cities because urban landscapes similarly represent socially produced spaces with natural resources, waterways, vegetation, soils, and climate (Hein et al. 2017). Stakeholder workshops

Table 6.1 List of engaged stakeholders and their respective roles (Source: authors)

Stakeholders	Role
Government offices and agencies • Department of Sanitation Exploitation (DEX-ASS, *Direction de l'Exploitation Assainissement*) of ONEA • Municipalities • DSPH (DSHP, *Direction de la Salubrité et l'Hygiène Publique Department*) • Municipal Department of Landscape Development and Park Management (DAPGP, *Direction des Aménagements Paysagers et de la Gestion des Parcs*),	ONEA, Burkina Faso's National Office for Water and Sanitation develops, manages, and protects drinking water distribution and manages household and business sewage collection in urban and periurban areas. It produces biogas, compost, and treated water for irrigation. ONEA also provided two well-trained and experienced experts on liquid waste repurposing who helped coordinate all activities and MSPs. The municipalities and DSPH oversee the collection, treatment, and recycling of solid waste, the waste collection center (*Centre de Collecte des Déchets*, CCD), and the Waste Treatment and Valorisation Centre (*Centre de Traitement et de Valorisation des Déchets*, CTVD). The Municipal Department of Landscape Development and Park Management has responsibility over public parks and the Green Belt of Ouagadougou.
Waste collectors and sorters "Economic Interest Groups" (*Groupements d'Intérêt Economique*, GIEs) • *Action Protection de l'Environnement of Tampouy*, Arrondissements 3 and 9, and Peng Wende of Bogodogo, Arrondissement 5	Waste collectors and sorters are women's associations. Women organize themselves into groups. Each group is responsible for an intervention area. Household waste collection is done from door to door. The working equipment consists of carts equipped with donkey traction systems, donkeys, and an old bag for waste collection. In the early 2000s, the municipal administration encouraged associations of mostly women who were already picking up garbage in their neighborhoods to group together in *arrondissement*-based Economic Interest Groups (GIEs, *Groupements d'Intérêt Economique*) so that they and some small- and medium-sized businesses (PMEs, *petites et moyennes entreprises*) could answer bids for tender from the municipality for trash collection.

(continued)

Table 6.1 (continued)

Stakeholders	Role
Cleaners • Mechanized septic tank cleaners: Septic Tank Cleaners Association of Faso (*Association des Vidangeurs du Faso*, AVIF) • Brigade Verte • Association that trims the trees	Several groups of people—associations and small and medium businesses—work to clean the city of Ouagadougou, and most are in some way connected with organic waste through the potential to produce compost, water, and gas. Ouagadougou has associations that clean the cemeteries, a sewer system that removes septic sludge from government buildings, large hotels, and other large buildings in the downtown area. Almost all the rest of the liquid waste from households, businesses, schools, hospitals, and other buildings flows into septic tanks or latrines. The sludge is removed from both by either mechanized or manual septic tank cleaners.
Waste transporters • EBTE • SONAF	EBTE and SONAF are contracted by the city of Ouagadougou to transport mixed waste bins from some of the CCDs to the CTVD. EBTE is the oldest, and SONAF, only a few years old, is an offshoot of EBTE. The two SMEs collaborate together like older and younger companies.
Waste transformers Solid and liquid waste transformers • Compost: Women Association Wend Bénédo • Irrigation water: Kossodo Step by ONEA • Biogas: PNB-BF	Transformers recycle solid and liquid wastes into products for use in other economic sectors. In the organic waste value network of Ouagadougou, some of the organic garbage and vegetation waste produced in the city is transformed into compost. Manure and septic sludge are processed through biodigesters into biogas and compost. In ONEA's anaerobic and aerobic basins, sewage is transformed into water for irrigating gardens. PNB-BF, the National Biodigester Program of Burkina Faso, supported by SNV, a Dutch development organization, focuses on promoting and installing household biodigesters for periurban cattle farmers.

(continued)

Table 6.1 (continued)

Stakeholders	Role
Growers Market gardeners • Association Koom Ramenyé in Kossodo, Arrondissement 4 • Cooperative of Boulmiougou, in Boulmiougou, Arrondissement 6 Nursery owners • Burkina Verdoyant, in Ouaga 2000, Arrondissement 12 • Teel Tissé, in Bissiguin, Arrondissement 8	Wherever there is water and space in the commune, and often with permission from the municipal administrations of the commune and arrondissements, people have planted market gardens and tree nurseries. The market gardeners and nursery owners are very small businesses, most likely informal, but they provide people and households with incomes and the population of the commune with vegetables, fruits, ornamental and medicinal plants, and trees. Many growers are grouped into site-based associations or cooperatives.
Farmers and livestock raisers	Farmers on the periphery of the commune and people who raise livestock in their households within the commune and in the periphery provide manure and sometimes compost to the growers within the value network. There are few dairy farmers within the 5-kilometer periurban limit. Their need for pasture has pushed them further away from the growing density of houses around the urban area.
Nongovernmental organizations (NGOs)	Different NGOs have intervened in the waste collection, sorting, and repurposing of association's life and work. For example, CREPA trained the women to create a high-quality compost, and others, such as Grand Lyon from France, have built infrastructure and brought equipment, such as a large sifter, and a mechanical chopper for tree limbs and other vegetation.
Households and other liquid and solid waste producers	Households, hotel, restaurants, and government offices produce waste. Households must pay an average fee of US$ 1.29 to collect their waste depending on their place of residence.

are the sort of collaborative knowledge sharing exercises that drive integrated solution design and politically negotiated settlements that lay the foundation for effective SDG implementation. In the different stakeholder workshops, we took a landscape perspective to envision how different stakeholders were linked together by streets and waterways and through diverse markets and green spaces across the commune (Figs. 6.2 and 6.3).

In the workshops, the landscape perspective allowed us to bring many things to the table, such as the complexity of different challenges and diverse possibilities. In the workshops, we could discuss the integration of concepts like the recuperation and reuse of organic waste, agriculture and horticulture, green urban spaces, and global development goals such as the SDGs. In addition, the landscape perspective allowed us to establish networks among stakeholders, including government partners, civic organizations, and the private sector.

6.6 Stakeholder Interactions in the Waste-Water-Energy Nexus to Address SDG 11 in Burkina Faso

Knowledge exchange through MSPs in the waste management sector played a key role in enabling interactions among relevant non-academic stakeholders, as well as between them and academic stakeholders. The waste value network's story takes place in a commune that is rapidly changing, demographically and politically in national and municipal administrations and policies. Stakeholder consultation and various stakeholder meetings through the MSPs showed that the liquid and solid waste value network in Burkina Faso is a complex system and a story of potentials in which several stakeholders interact. The investigations showed many common points among the stakeholders and the issues that connect them. While all stakeholders are concerned for their incomes and the financial well-being of their families, most stakeholders in nearly all stakeholder groups expressed personal interest in the cleanliness of the commune.

The results showed that the stakeholders could be classified into four categories: government agencies, cleaners, transformers, and horticulturists (gardeners and plant nursery owners) (see Table 6.1). One of the peripheral groups of stakeholders, namely, farmers and livestock raisers, was added to the transformer category because their manure and compost are sold to gardeners and plant nursery owners. At first sight, it would seem that all these different actors should be almost linearly connected: from waste producers through collectors, transformers, and horticulturists to consumers, but the results indicated that relationships are ramified and sometimes interwoven.

Fig. 6.2 Women waste collectors and sorters discussing and listing the problems they face at the MSP meeting in Ouagadougou, June 11–13, 2019. (Source: authors; photo credit: Qualifa Inoussa)

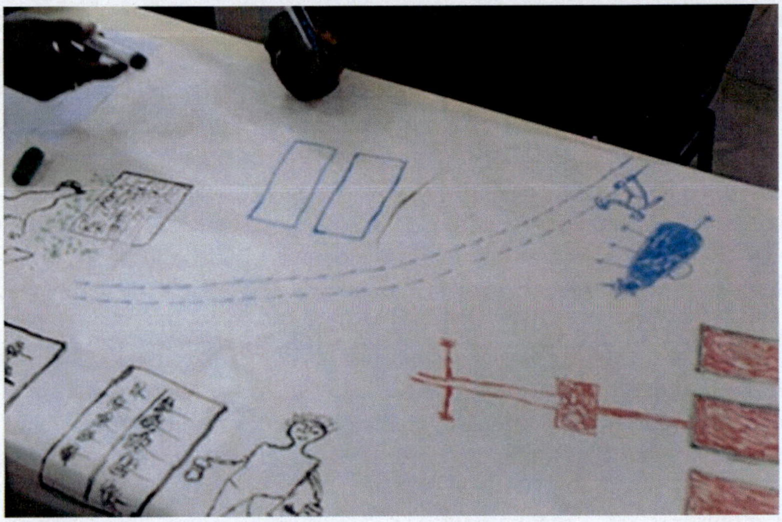

Fig. 6.3 Vegetable growers drawing a map of their landscape at the MSP meeting in Ouagadougou, August 8–10, 2018. (Source: authors; photo credit: Qualifa Inoussa)

Within the waste value network, stakeholders are informally linked through inputs and outputs (Table 6.2). For instance, the compost women depend on inputs of a variety of organic garbage and refuse by waste collectors and cleaners. Their inputs include the sorted waste from the waste collection center. They also receive other organic refuse, from associations that clean the cemeteries and others who trim vegetation along streets and elsewhere in the commune. The output, i.e., the compost from the waste composter, is sold to horticulturists, representing vegetables growers and plant nursery operators. Dialogues through MSPs revealed that there were conflicts between these different actors. For instance, the market gardeners were in conflict with ONEA, Burkina Faso's National Office for Water and Sanitation, whereas the plant nursery operators and the women who composted were in disagreement with the municipalities.

The conflict between the market gardeners and the ONEA is related to the use of the treated wastewater. The gardeners attributed the color and smell to "product," thinking some sort of chemicals had been added to the water. They claimed that the water burned the leaves of their plants such as spinach, okra, and indigenous herbs (*amaranth* and *malohiya*), used in sauces. However, the gardeners had never raised these suspicions. They reported being fearful that the land and treated water could be taken away from them if they were perceived as being hostile toward, or critical of, ONEA administrators. Most poor and marginalized women who garden are key agents of their families' well-being and economic development. Therefore, leaving ONEA's market gardening sites means no income for these women and their families. This could exacerbate the situation of food insecurity and the lack of schooling for children, especially the daughters of these women. Likewise, the plant nursery owners and the women recycling waste into compost had conflicts with the municipalities. The plant nursery operators have much better visibility for sales and a more constant source of water, but neighbors constructing multi-story buildings complain that the nurseries with their tall trees would block future businesses within these buildings. The municipality decided to expropriate the lands which had created frustrations among the two groups of stakeholders.

The conflict between the municipality and the Wend Bénédo—the association of women who manufacture compost—is mainly related to financial issues. The Association Wend Bénédo was created in 2005 to answer the needs of the commune, which wanted to both reduce the amount of garbage that goes into the landfill and also create employment for women. Though Wend Bénédo operates like a small cooperative business, the Public Health and Hygiene Department (DSPH) still pays a small salary to

Table 6.2 Stakeholder interactions (Source: authors)

Group of stakeholders	Inputs	Outputs	Associates
Government offices and agencies • Department of Sanitation Exploitation (DEX-ASS, *Direction de l'Exploitation Assainissement*) of ONEA • DSPH (*Direction de la Salubrité et l'Hygiène Publique Department*) • Municipal Department of Landscape Development • Municipalities	– Training, infrastructure, and equipment – International financing – Master plan for waste	Title (gray card), insurance, inspection, certificate and allow operations sewage and septic sludge system	Waste collectors and sorters, cleaners, waste transporters, growers, farmers and livestock raisers, nongovernmental organizations, households, and other liquid and solid waste producers
Waste collectors and sorters Economic Interest Groups (Groupements d'Intérêt Economique, GIEs) • *Action Protection de l'Environnement of Tampouy,* Arrondissements 3 and 9, and • *Peng Wende of Bogodogo,* Arrondissement 5	Trash, adapted uniforms, face masks and rubber gloves, shovels, brooms, sieves, and other tools	Mixed garbage in large bins, organic garbage in large bins, paper, cardboard, plastic, other reusable items, fine soil, sand, gravel, compost, and donkey manure	Government offices and agencies (DSPH, Ministry of Environment), waste transporters, waste transformers (Wend Bénédo), GIEs (garbage sorters), municipalities
Cleaners • Mechanized septic tank cleaners: Septic Tank Cleaners Association of Faso (*Association des Vidangeurs du Faso,* AVIF) • Brigade Verte • Association that trims the trees	Truck, working pump, hoses, boots, gloves, cloth, face masks	Septic sludge	Government agencies (Ministry of Environment, ONEA, municipalities), waste transporters, municipalities, waste transporters
Waste transformers Solid and liquid waste transformers • Compost: Women Association Wend Bénédo • Irrigation water: Kossodo Step by ONEA • Biogas: PNB-BF	Training, infrastructure, and equipment (concrete basins, large sifter, mechanical chopper, organic garbage, motorized tricycle, lands, rubber gloves and boots, shovels, rakes, pitchforks, small sifters, buckets, and large pans)	Compost	Growers, household, business owners, government agencies (municipalities, ONEA, DSPH)
Growers Market gardeners • Association Koom Ramenyé in Kossodo, Arrondissement 4 • Cooperative of Boulmiougou, in Boulmiougou, Arrondissement 6 Nursery owners • Burkina Verdoyant, in Ouaga 2000, Arrondissement 12 • Teeli Tissé, in Bissiguin, Arrondissement 8	Land, water from ONEA's wells or basins, buckets and watering cans, motorized pumps and fuel, pickaxe or small hoe, shovels, wheelbarrows, rakes, clippers, cotton waste, seeds, chemical fertilizers, pesticides and herbicides, black earth (black dirt), manure and compost, and information and knowledge from the Ministry of Agriculture	Vegetables and fruits, ornamental plants, medicinal plants, fruit trees, garden services, and various plants	Growers, vegetable and fruit vendors, households, and government offices (municipalities, Department of Landscape Development)

(continued)

Table 6.2 (continued)

Group of stakeholders	Inputs	Outputs	Associates
Farmers and livestock raisers	Land, barns, stables, milking sheds, borehole, motorized pump, fuel, storage tank, breeding stock, veterinary care and pharmaceuticals to maintain health, motorized tricycle, and biodigesters	Livestock for meat, milk, eggs, hides, compost, and biogas	Growers, government offices (municipalities, ONEA, DSPH (DSHP, *Direction de la Salubrité et l'Hygiène Publique*) Department, Department of Landscape Development), and households
Nongovernmental organizations		Training, construction of infrastructure, and supply of equipment	International association directors, project directors, agents
Households and other liquid and solid waste producers			Waste collectors, growers, government offices and agencies (municipalities, farmers, and livestock raisers)

the women (US$ 1.80 per day) and provides them with space and water at no cost. But the association buys phosphate and manure to add to their compost mixture, and has bought a motorized tricycle and pays for its fuel. It also pays two men (US$ 2.70 per day) to drive the tricycle and help them with heavy work. When they need to sort trash that comes from other sources than the waste collection centers, the association hires more women to help them. The DSPH would like them to become independent of the small salaries the commune pays them. The women in turn were very angry because they will face external challenges. The association has so far not been able to sell enough compost to expand. It is also possible that an association is not the correct structure to foster an independent and flourishing business. Moreover, the women's association blamed the municipality for not being active in training them on how to raise funds. These women, while they have many good ideas, lack some of the skills necessary to carry the ideas out, including understanding how to find technical and financial partners and how to apply for national or international funding.

Another key outcome from the discussions around the MSP exercises is the lack of communication and trust among stakeholders. For example, several people do not know that there are different types of compost for sale at the Waste Treatment and Valorisation Centre (CTVD). The treatment process for wastewater by ONEA was also clarified during the discussion of the gardeners' exercises. The roles and responsibilities of the different organizations were also not always clear among partners, such as the municipality and the associations that process waste.

These conflicts and miscommunication prevent a smooth integration and relationship among the different stakeholders for good practices of the waste-water-energy-food nexus (Daniels 2004; d'Alençon et al. 2018). A wide growing literature has shown that a smooth relationship between formal and informal actors plays a key role to the effective functioning of the waste-water-energy-food nexus (Sarkar and Pansera 2017; Smith et al. 2017). Smith et al. (2017) have reported that the responsiveness of some informal actors to the growing needs of many urban dwellers places them in the role of grassroots innovators, who build technologies, processes, and linkages from the bottom up and, in so doing, fill important service gaps despite their marginal status in society.

In addition, during the workshops, the stakeholders were able to list issues that prevent them from being effective in the waste-water-energy-food nexus and proposed new landscape features

(Table 6.2). Besides addressing their issues, a major outcome of the discussions enabled through the MSP was a common and increased understanding among network actors of the value-adding components of the waste value network and the interactions that enable them to co-found solutions for the many existing problems. Through the processes of knowledge co-production and exchange facilitated by the MSP, some of the long-standing barriers to relevant stakeholders' cooperation and to an effective and efficient functioning of the waste-water-energy-food nexus, in the study areas, were lowered. Some gains became immediately apparent, including the establishment of channels of communication between government agencies and horticulturists in Ouagadougou regarding the quality of treated water used in market gardens and the cancelling of plans by city authorities to evict plant nursery operators from the spaces they occupy in the city. The success of a multi-stakeholder platform in solving complex problem requires human (leadership) and financial resources but also effective collaboration, trust, time, and attention by all participating stakeholders.

6.7 Conclusion

Investing in the waste-water-energy-food nexus for the sustainable development of African cities is vital for the achievement of the SDGs, especially SDG 11 (Herslund and Mguni 2019; Bel and Warner 2008; Araral 2009; Tan 2012). The city of Ouagadougou has a rapidly growing population which has led to an increase in waste production and a growing demand for water, energy, electricity, and food to feed the young and expanding population (Dos Santos 2015). Moreover, the needs for water, sanitation, energy, and food of Ouagadougou city dwellers are already exposed to climate and environmental stressors (ibid.).

Recycling and repurposing liquid and solid waste from the city of Ouagadougou for UPA, GUS, and green energy enhance sanitation and contribute to food security, urban embellishment, and population well-being. Our study illustrates the possibility of promoting the effective functioning of the waste-water-energy-food nexus through consultations and collaboration between stakeholders in the liquid and solid waste value chain. Using a proactive, participatory, and transdisciplinary approach, the study illustrates the processes of engaging all actors through co-construction and knowledge sharing organized around a multi-stakeholder platform.

Stakeholder's dialogues revealed complex problems and conflicts among stakeholders active in the waste-water-energy-food nexus. Results indicated that many of the complex problems and difficulties that different groups of stakeholders face within the value network of solid and liquid wastes resulted from the fast demographic growth and the rapid urbanization that have not been followed by changes and improvements in infrastructure, the network of waste collection, recycling, and repurposing. A common concern was the uncontrolled garbage dumping everywhere in the city, as well as the garbage dumped in the open storm sewers.

Uncontrolled liquid and solid waste disposal is caused by unlawful behavior of the population, and government agencies have not been able to manage the consequent growth in waste production. The lack of equipment, infrastructure, and training, the outdated laws and regulations, a slow bureaucracy, and high turnover of public officials, from ministers to mayors, further contribute to the disorganization and slow development of the system. Old infrastructures must be upgraded, and new structures built in order to accommodate all of the waste that the city produces and to benefit the waste-water-energy nexus. The engaged consultation framework could be structured as a multi-stakeholder platform that holds general assemblies for representatives of all concerned organizations that clean the city of Ouagadougou and recycle and reuse its waste. Besides these assemblies, a multi-stakeholder platform also holds various ad hoc meetings that bring appropriate members together around specific issues. Human and financial resources and institutional reform are necessary to increase infrastructure capacity for managing liquid and solid waste, but consolidated MSPs are needed to accommodate all the necessary changes, interactions, and transformations.

References

AAPUI Arcade (2009) Schéma Directeur d'Aménagement du Grand Ouaga, Horizon 2025 (SDAGO). Rapport final, Ministère de l'Habitat et de l'Urbanisme, Direction générale de l'urbanisme et des travaux fonciers, Ouagadougou. Volume 1, Le portrait du Grand Ouaga, p 200; volume 2, La stratégie d'aménagement, p100

Araral E (2009) The failure of water utilities privatization: synthesis of evidence, analysis and implications. Policy and Society 27(3):221-228

Bayili PP (2002) Journée d'information et de sensibilisation du personnel de la mairie de Pô sur le thème 'organisation et gestion des déchets solides au Burkina Faso, Aspects législatif, réglementaire, institutionnel, organisationnel et technique'. Rapport de synthèse

Bel G, Warner M (2008) Does privatization of solid waste and water services reduce costs? A review of empirical studies. Resources, Conservation and Recycling 52(12):1337-1348

Bellfield H (2015) Water, Energy and Food Security Nexus in Latin America and the Caribbean. Global Canopy Programme. https://cdkn.org/wp-content/uploads/2015/04/The-Water-Energy-Food-Nexus-in-LAC-April-2015.pdf Accessed 20 October 2021

Bhatia M, Angelou N (2015) Beyond connections: energy access redefined. ESMAP technical report 008/15. World Bank, Washington, DC

Biggs EM, Bruce E, Boruff B, Duncan JMA, Horsley J, Pauli N, McNeill K, Neef A et al. (2015) Sustainable development and the water–energy–food nexus: a perspective on livelihoods. Environmental Science & Policy 54:389-397

Block DR, Thompson M, Euken J, Liquori T, Fear F, Baldwin S (2008) Engagement for transformation: value webs for local food system development. Agricultural Human Values 25:379-388

Boas I, Biermann F, Kanie N (2016) Cross-sectoral strategies in global sustainability governance: towards a nexus approach. International Environmental Agreements: Politics, Law and Economics 16:449–464

Cheng C, Urpelainen J (2015) Who should take the garbage out? Public opinion on waste management in Dar es Salaam, Tanzania. Habitat International 46:111-118

Cobbinah PB, Addaney M, Agyeman KO (2017) Locating the role of urbanites in solid waste management in Ghana. Environmental Development 24:9-21

d'Alençon AP, Smith H, de Andrés EA, Cabrera C, Fokdal J, Lombard M, Mazzolini A, Michelutti E et al. (2018) Interrogating informality: conceptualizations, practices and policies in the light of the New Urban Agenda. Habitat International 75:59-66

Daniels PW (2004) Urban challenges: the formal and informal economies in mega-cities. Cities 21(6):501-511

Dos Santos S (2015) Accès à l'eau et enjeux socio-sanitaires à Ouagadougou. Espace, Populations, Sociétés 2006(2-3):271-285

Forti V, Balde CP, Kuehr R, Bel G (2020) The global e-waste monitor 2020: quantities, flows and the circular economy potential. United Nations University (UNU), United Nations Institute for Training and Research (UNITAR), International Telecommunication Union (ITU), International Solid Waste Association (ISWA); Bonn, Geneva, Rotterdam

Ghodsvali M, Krishnamurthy S, de Vries B (2019) Review of transdisciplinary approaches to food-water-energy nexus: a guide towards sustainable development. Environmental Science and Policy 101:266–278

Gremillion P, Avellan T (2016) Wastewater as a resource: the water-waste-energy nexus in sub-Saharan Africa. Policy brief no. 01/2016. United Nations University Institute for Integrated Management of Material Fluxes and of Resources (UNU-FLORES)

Herslund L, Mguni P (2019) Examining urban water management practices—challenges and possibilities for transitions to sustainable urban water management in sub-Saharan cities. Sustainable Cities and Society 48:101573

Hein A, Jankovic M, Feng W, Farel R, Yune J, Yannou B (2017) Stakeholder power in industrial symbioses: a stakeholder value network approach. Journal of Cleaner Production 148:923-933

Hoornweg D, Munro-Faure P (2008) Urban agriculture for sustainable poverty alleviation and food security. Food and Agriculture Organization (FAO) Position Paper on Urban Agriculture, New York

Institut National de la Statistique et de la Démographie (INSD) (2019) Annuaire statistique 2016. Institut National de la Statistique et de la Démographie, Ouagadougou, Burkina Faso

Jimu IM (2008) The role of stakeholders in the provision and management of water kiosks in Nkolokoti, Blantyre (Malawi). Physics and Chemistry of the Earth, parts A/B/C 33(8–13):833-840

Kalair AR, Abas N, Ul Hasan Q, Kalair E, Kalair A, Khan N (2019) Water, energy and food nexus of Indus Water Treaty: water governance. Water-Energy Nexus 2:10–24

Kaudal BB, Weatherley AJ (2018) Agronomic effectiveness of urban biochar aged through co-composting with food waste. Waste Management 77:87–97

Liu J, Hull V, Godfray HCJ, Tilman D, Gleick P, Hoff H, Pahl-Wostl C, Xu Z et al. (2018) Nexus approaches to global sustainable development. Nature Sustainability 1:466–476

Mancini G, Luciano A, Bolzonella D, Fatone F, Viotti P, Fino D (2021) A water-waste-energy nexus approach to bridge the sustainability gap in landfill-based waste management regions. Renewable and Sustainable Energy Reviews 137:110441

Masud J, Sharan D, Lohani BN (2007) Energy for all: addressing the energy, environment, and poverty nexus in Asia. Asian Development Bank, Mandaluyong, the Philippines

Nederlof S, Wongtschowski M, van der Lee F (eds) (2011) Putting heads together: agricultural innovation platforms in practice. KIT Publishers, Amsterdam

Onibokun AG (ed) (2002) La gestion des déchets urbains: des solutions pour l'Afrique. Éditions Karthala, CRDI, Paris

Ouattara A (2006) Les processus d'urbanisation et l'aménagement urbain à Ouagadougou. In: Hien PC, Compaoré M (eds) Histoire de Ouagadougou des origines à nos jours, chapter 11. DIST-CNRST, Ouagadougou, pp 283-313

Pohl C, Truffer B, Hirsch Hadorn G (2017) Addressing wicked problems through transdisciplinary research. In: Frodeman R, Thompson Klein J, Pacheco RCS (eds) The Oxford Handbook of Interdisciplinarity. Oxford University Press, Oxford, pp 319-331

Portney K, Hannibal B, Goldsmith CL, McGee P, Liu X, Vedlitz A (2017) Awareness of the food–energy–water nexus and public policy support in the United States: public attitudes among the American people. Environment and Behavior 50(4):001391651770653

Sarkar S, Pansera M (2017) Sustainability-driven innovation at the bottom: insights from grassroots ecopreneurs. Technological Forecasting and Social Change 114:327-338

Sesan T, Sanfo S, Sikhwivhilu K, Dakyaga F, Aziz F, Yirenya-Tawiah D, Badu M, Derbile E et al. (2021) Mediating knowledge co-production for inclusive governance and delivery of food, water and energy services in African cities. Urban Forum. https://doi.org/https://doi.org/10.1007/s12132-021-09440-w

Smith A, Fressoli M, Abrol D, Around E, Ely A (2017) Grassroots innovation movements. Routledge, London and New York

Spalding-Fecher R, Winkler H, Mwakasonda S (2005) Energy and the World Summit on Sustainable Development: what next? Energy Policy 33:99–112

Sperling JB, Berke PR (2017) Urban nexus science for future cities: focus on the energy-water-food-x nexus. Current Sustainable Renewable Energy Reports 4:173–179

Stephan RM, Mohtar RH, Daher B, Irujo AE, Hillers A, Ganter JC, Karlberg L, Martin L et al. (2018) Water–energy–food nexus: a platform for implementing the Sustainable Development Goals. Water International 43(3):472-479

Tan J (2012) The pitfalls of water privatization: failure and reform in Malaysia. World Development 40(12):2552-2563

Terrapon-Pfaff J, Ortiz W, Dienst C, Gröne M-C (2018) Energising the WEF nexus to enhance sustainable development at local level. Journal of Environmental Management 223:409-416

Totin Vodounon HS, Houédakor KZ, Amoussou E, Azalou Tingbé EM, Nantob M, Lambert Ayitchéhou K, Nabola-Bounou Enoumodji MK (2021) Contributing to the achievement of sustainable development goals: knowledge on water, sanitation and health risk in Cotonou and Lomé cities. International Journal of Sustainable Development & World Ecology 1-12

United Nations Environment Programme (UNEP) (2011) Towards a green economy: pathways to sustainable development and poverty eradication. https://www.unep.org/resources/report/towards-green-economy-pathways-sustainable-development-and-poverty-eradication-10. Accessed 25 October 2021

Wang X, Guo M, Koppelaar RHEM, Van Dam KH, Triantafyllidis CP, Shah N (2018) A nexus approach for sustainable urban energy-water-waste systems planning and operation. Environmental Science & Technology 52(5):3257-3266

7 Global Surgery and Healthy Cities: Adopting a Global Surgery Perspective to Harmonize SDGs and Enhance Urban Health in Africa

Salome Maswime and Ché L. Reddy

Abstract

While the Sustainable Development Goals (SDGs) may present a holistic and universal framework that acknowledges the varied and interlinking dimensions of development, practical implementation strategies to achieve the goals are often sub-optimally aligned with contextual realities and particularities. Global Surgery is an emerging discipline, a mode of inquiry, an applied science, and a highly participatory global social movement which sits at the interface of health and sustainable development. As such, this chapter argues that Global Surgery represents a crucial perspective to illustrate the cross-cutting nature of the SDGs and foreground the importance of comprehensive urban health interventions for vulnerable and marginalized populations in low- and middle-income countries (LMICs), particularly in the African context. We draw on key trends in surgical systems and empirical knowledge about healthy cities and urban slums in Africa to provide a practical framework to rethink and improve surgical healthcare, build healthier cities, and advance the SDGs in African cities.

Keywords

Global Surgery · Urban health · Surgical healthcare · African cities · Sustainable cities

7.1 Overview of Global Surgery, Urban Health Systems, and Sustainable Development

The SDGs aim to create a better and sustainable future for all people by addressing global inequities and interdependent challenges. Substantial emphasis is placed on improving population health outcomes by addressing the social determinants of health (UN 2015). It is widely accepted that health is the absence of disease and the product of dynamic interactions between the human being and their environment. The state of health is therefore multifactorial and influenced by myriad of factors: genetic, behavioral, and contextual factors, including political, social, ecological, and economic ones all playing a crucial role (Sartorius 2006). As an emerging global health subdiscipline and mode of critical inquiry, Global Surgery as a discipline provides a unique opportunity to examine sustainable urban development priorities by emphasizing different links

S. Maswime (✉) · C. L. Reddy
Global Surgery Division, Department of Surgery, Faculty of Health Sciences, University of Cape Town, Cape Town, South Africa
e-mail: salome.maswime@uct.ac.za; CheLen_Reddy@hms.harvard.edu

S. Croese, S. Parnell (eds.), *Localizing the SDGs in African Cities*, Sustainable Development Goals Series, https://doi.org/10.1007/978-3-030-95979-1_7

and interplays between the patient, urban environment, health system, and broader context. Such a perspective necessitates a multi-sectoral and inclusive approach to surgical healthcare, a cross-cutting theme of the SDGs.

Modern shifts in healthcare, particularly observed in Low- and Middle-Income Countries (LMICs) and urban settings, where the rate of change is rapid, the demand for surgical healthcare far outstrips its supply. This trend is expected to continue with significant implications for promoting sustainable development. The World Health Organization estimated in 2017 that half of the world lacked access to essential health services (WHO 2017). The finding that "five billion people do not have access to safe, affordable surgical and anesthesia care when needed" by the Lancet Commission on Global Surgery (LCoGS) was critical for defining the magnitude and scale of the surgical deficit globally (Fig. 7.1). Historically, global health (with its emphasis on public health, preventative care, and infectious disease in LMICs) has tended to neglect surgical healthcare, which has a critical role to play in preserving and improving health and well-being throughout the lifespan in all societies (Meara et al. 2015). Surgical healthcare, which refers to all surgical subdisciplines, perioperative care, and the entire surgical ecosystem, is especially needed in LMICs: "of the 313 million procedures undertaken worldwide each year, only 6% occur in the poorest countries, where over a third of the population lives," reported by the LCoGS. By expanding access to effective surgical healthcare services, substantial death and disability from injuries, noncommunicable diseases, and complications related to pregnancy could be averted. It is estimated that LMICs which fail to invest in and expand access to effective surgical healthcare could suffer cumulative economic losses of

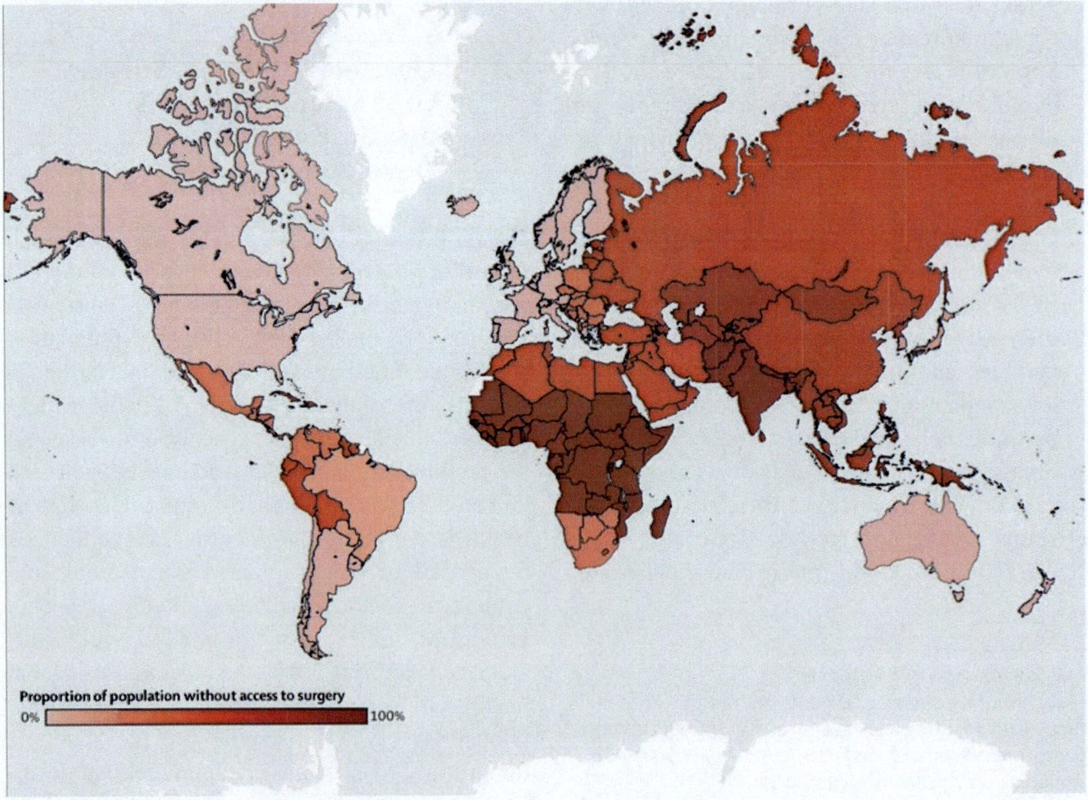

Fig. 7.1 Proportion of population without access to surgery, highlighting the surgery deficit in Africa (Meara et al. 2015)

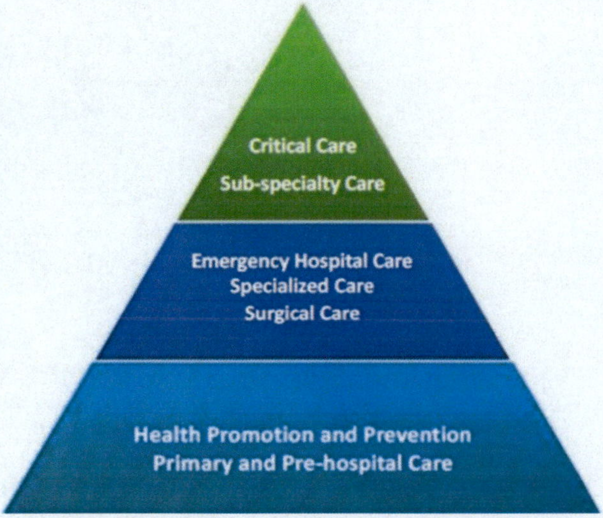

Fig. 7.2 Levels of healthcare including surgical care. (Source: authors)

up to US$ 12.3 trillion (2010 US$, purchasing power parity) from 2015 to 2030 (Meara et al. 2015).

Surgery is not a luxury. It is an essential component of comprehensive healthcare delivery of health systems and a foundational element for building and upholding a healthy urban environment. While the surgical burden of disease is important to measure population health, it is also crucial to assess the unmet need for surgical healthcare to capture the nature, extent, and scope of surgical disease. People may simply live with a surgical disease because they have no access to treatment. This empirical reality has been observed in global health through medical mission trips, where specialists travel into a city to perform surgeries. Surgical missions have provided a significant volume of surgical healthcare in many LMICs and are typically organized through universities, nongovernmental organizations, and religious organizations. Common surgical conditions treated during mission trips include cleft lip or palate, cataracts, complex fractures, and removal of benign head and neck solid tumors (Ginwala and Rickard 2015). Though well intentioned, recently the approach has gained attention for not adequately strengthening country health systems and building sustainable capabilities to deliver surgical healthcare, which has led to more concrete efforts to translate such vertical models into sustainable systems.

The health system is the principal structure involved in the delivery of healthcare within a country and is consequently the major mediator of population health. However, there are significant differences in health system performance between nations, leading to varying levels of population health between nations. People in many African nations have among the lowest life expectancy compared to people living in other continents, often suffer catastrophic and impoverishing health expenditures, and experience significant challenges accessing healthcare services. The life expectancy in sub-Saharan Africa in 2019 was 62 years, compared to the United States (78 years), Canada (82 years), or the United Kingdom (81 years) (World Bank n.d.). There are myriad reasons to explain causal mechanisms of health systems' underperformance, including insufficient financing, suboptimal governance organization, and inadequate resources such as workforce, hospitals, intelligence, and supplies. Surgical healthcare services are provided at all levels of the health systems and range from primary and prehospital healthcare to highly specialized and advanced surgical services and critical care at the tertiary level (Fig. 7.2). Due to the cross-cutting nature of surgical healthcare, strengthening surgical healthcare provides a unique opportunity to strengthen health systems in delivering effective surgical services and enhance demand-side factors, such as the social

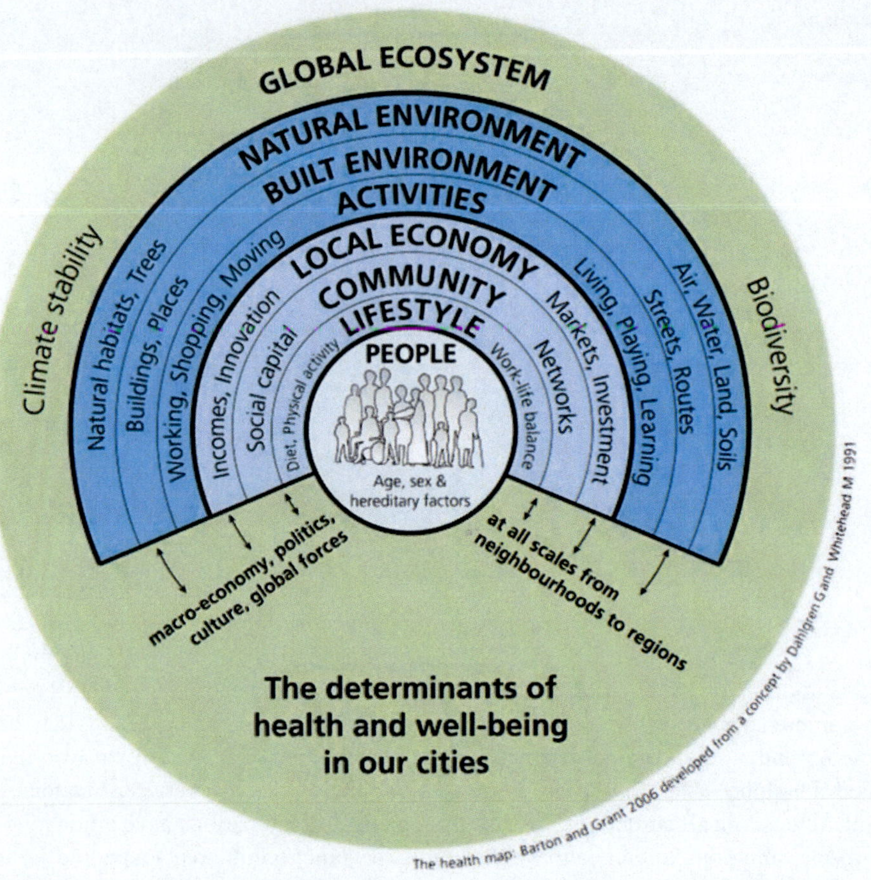

Fig. 7.3 The determinants of health and well-being in our cities (Barton and Grant 2013)

determinants of health in urban settings (Fig. 7.3). The rapid shifts in life and space that characterize the urban African landscape in epidemiology and demography make interdisciplinary efforts to bolster the health system a crucial facet for achieving the SDGs.

7.2 Interlinkages Between the Sustainable Development Goals: Opportunities and Synergies

The overarching aim of the SDGs is to achieve peace and prosperity for people on the planet now and in the future, with the recognition that strategies, which aim, for example, to improve health and education must align with strategies to eradicate poverty and global inequality (UN 2015). An approach that focuses on SDG 3 (health and well-being) in isolation is restrictive and has several limitations, especially in complex environments such as the African city. A comprehensive approach is crucial and entails exploring the interlinkages between all the SDGs as part of an intentional and holistic strategy to achieve specific sustainability objectives (Griggs et al. 2017). Cogent leadership, state capability, and civil assertiveness are prerequisites for any sustained plan to develop, implement, and scale an ambitious development agenda. Eradicating poverty and hunger is undoubtedly critical for good health and well-being but is not sufficient. For all people to flourish and participate meaningfully in society, quality education, gender equality, clean water and sanitation, affordable and clean energy, and purposeful work, driven by a growing and inclusive economy and participatory political environment, are all crucial ingredients.

These need to be addressed as part of a unified strategy for sustainable development.

Recognition of the links between health and environment has opened up spaces of dialogue, knowledge, and translation for the medical professions to contribute to public health and sustainable development policy and practice. Surgery, typically seen as a pure biomedical domain of medical practice, is one such area. Global Surgery also breaks new ground as it looks at issues that lie beyond the traditional remit of other medical fields that have contributed to public health such as primary healthcare. Global Surgery which is founded on principles of equity and social justice, with the aim of ensuring that people from vulnerable and underserved populations are able to receive safe and timely surgery when needed, aligns naturally with the basic principles and tenets of SDGs. The integrated nature of the SDGs and the empirical reality that some SDGs depend on access to surgical healthcare highlight the importance of reinforcing the value of joined up local and global action across silos.

The cross-cutting nature of surgical disease and surgical healthcare creates new synergies and opportunities between the SDGs. Achieving collective targets between SDG 3 (health and well-being), SDG 11 (sustainable cities and communities), and Global Surgery will require specific interventions that take a holistic and system approach to making measurable progress in both realms. For example, when electricity or oxygen supplies are not reliably provided, surgical operations are disrupted. Collaboration with urban planners and engineers to design and introduce interventions that make provisions for sustainable energy and the provision of oxygen with rigorous supply chain management to target hospital sites could make measurable progress in sustaining well-functioning operating theaters (Roa et al. 2019).

7.3 Global Surgery: An Evolving Priority in Global Health

Global Surgery may be described as an area of study, research, practice, and advocacy that seeks to improve health outcomes and achieve health equity for all people who need surgical, obstetrics, and anesthesia healthcare. Though Global Surgery has a special focus on LMICs and underserved populations or populations in crisis (Meara et al. 2015), marginalized groups in high-income countries (HICs) are also a major focus. It is perhaps this all-encompassing and inclusive spirit of Global Surgery, together with the fascination that clinical surgical discipline brings, that has ignited the highly participatory and vibrant Global Surgery movement over the years. Though the surgery discipline has evolved dramatically over the past century in terms of the level of sophistication in applying modern technologies to deliver more effective surgical healthcare, a significant surgical access gulf exists between high- and low-income countries, rural and urban areas, and racial and socio-economic groups, with some of the worst health outcomes occurring on the African continent (Table 7.1). Despite some of the greatest advances in surgery stemming from LMICs or LMIC surgeons that practice in HICs, such as the first heart transplant in South Africa, for instance, people in HICs benefit from these advances more than people in LMICs, where these diseases are often leading causes of death (Brink and Hassoulas 2009). Global Surgery uses partnerships with stakeholders and a multi-disciplinary approach to address healthcare challenges.

When Global Surgery first emerged in the past decade, it was described as the "neglected step-child of Global Health" (Farmer and Kim 2008), emphasizing the need to strengthen surgical healthcare.

While Global Surgery as a process has gained substantial support since then, within the academic surgical community in HICs, and more recently in LMICs, it remains low to moderate priority within global health. However, tremendous progress has been made in terms of the governance and organization of the Global Surgery community, dispelling myths about surgical cost and complexity, and aligning global surgical objectives with those of sustainable development and Universal Health Coverage (UHC). While surgery is rarely discussed explicitly in the emerging urban health agenda (Harpham 2009;

Table 7.1 Core indicators for monitoring of universal access to safe, affordable surgical and anesthesia when needed (Adapted from Meara et al. 2015)

	Indicator	Definition	Target
1	Access to timely essential surgery	Proportion of the population that can access, within 2 h, a facility that can do cesarean delivery, laparotomy, and treatment of open fracture (the Bellwether procedures)	A minimum of 80% coverage of essential surgical and anesthesia services per country by 2030
2	Specialist surgical workforce density	Number of specialist surgical, anesthetic, and obstetric physicians who are working, per 100,000 population	100% of countries with at least 20 surgical, anesthetic, and obstetric physicians per 100,000 population by 2030
3	Surgical volume	Minimum procedures per 100,000 population	80% of countries by 2020 and 100% of countries by 2030 tracking surgical volume; minimum of 5000 procedures per 100,000 population by 2030
4	Perioperative mortality	All-cause death rate before discharge in patients who have undergone a procedure in an operating theatre divided by the total number of procedures, presented as a percentage	80% of countries by 2020 and 100% of countries by 2030 tracking perioperative mortality; in 2020, assess global data and set national targets for 2030
5	Protection against impoverishing expenditure	Proportion of households protected against impoverishment from direct out-of-pocket payments for surgical and anesthesia care	100% protection against impoverishment from out-of-pocket payments for surgical and anesthesia care by 2030
6	Protection against catastrophic expenditure	Proportion of households protected against catastrophic expenditure from direct out-of-pocket payments for surgical and anesthesia care	100% protection against catastrophic expenditure from out-of-pocket payments for surgical and anesthesia care by 2030

Galea and Vlahov 2006; Vlahov et al. 2007), a Global Surgery approach makes clear why this omission is problematic in the battle for improved health and well-being in the city context.

Despite the huge unmet need for surgery, health systems do not adequately provide surgical healthcare services and do not address the underlying determinants of surgical disease at a population level. Recent data from the African Surgical Outcomes Study (Biccard et al. 2018), which affirmed the unmet need for surgical healthcare in African cities, also highlighted the poor quality of care provided. Patients in Africa, the study showed, were twice more likely to die after a surgical operation than patients in high-income countries, and African women were 50 times more likely to die after a cesarean section than women in high-income countries (Bishop et al. 2019). Over 90% of deaths occurred within the first 24 h from anesthetic complications, bleeding, and failure to rescue. The avertible mortality due to unsafe cesarean sections in African countries is thought to be associated with a shortage of blood banks, critical care units, ventilators, essential drugs, and skilled anesthesiologists and surgeons (Bishop et al. 2019; Maswime and Buchmann 2017a).

In our research on bleeding during and after cesarean sections in Johannesburg, South Africa, one of the most populous urban environments in Africa, we found that women were more likely to survive in a functional health system with a multidisciplinary team (Maswime and Buchmann 2017a, 2017b) One of the study objectives was to identify feasible and evidenced-based interventions that could reduce bleeding in the surgical theatre. However, despite the implementation and strengthening of such interventions to reduce bleeding in the surgical theatre, the geographical location where the cesarean section took place was a far more significant explanatory factor in determining which patients would survive or die from bleeding complications during or post-cesarean section. We concluded that bleeding during a cesarean section occurs in all parts of the world; however women in particular regions of the world and certain areas within thecity are less likely to survive bleeding complications during cesarean section. This is mainly because of deficiencies in their health systems, notably prehospital care, prolonged and ineffective referrals, and system-related theatre inefficiencies (Maswime and Buchmann 2017b). Bleeding women, even under the care of the best doctors, may not have access to blood products, an intensive care bed, oxygen, or even a ventilator. In well-resourced settings, a patient who has lost small amounts of blood may have access to an intensive care unit and receive the superior care that is available when needed. This case, though specific to obstetrics, occurs to varying levels in other areas of priority in global health, including oncology, trauma, and congenital anomalies, where weak health systems compromise the delivery of surgical healthcare. More recently, studies have demonstrated the importance of a strong surgical ecosystem in responding to COVID-19. Surgical resources—workforce, infrastructure, and supplies—all proved critical assets to managing COVID-19 globally (Chu et al. 2020a; Ma et al. 2020).

7.4 Healthy Cities and Global Surgery

A healthy city is defined as a city that is continually creating and improving physical and social environments and expanding community resources which enable people to mutually support each other in performing all functions of life while developing to their maximum potential (Duhl and Sanchez 1999). Healthy cities are indispensable assets to ensure that people live healthy lives. "A state of complete physical, mental and social well-being" was how the World Health Organization (WHO) defined health in 1948 and "not merely the absence of disease or infirmity." The Ottawa Charter in 1986, a landmark public health document, in turn highlighted the critical importance of creating "supportive environments" to enable the flourishing of human health (WHO 1987). Two fundamental concepts in the pursuit of healthy cities include (1) social justice and (2) spatial justice, which interact and align with Global Surgery goals, aspirations, and activities. Social justice lies at the heart of Global Surgery and is central to its mission. Social justice affects the way people live, chances of illness, and the risk of premature death. The inequities in health arise because of the circumstances in which people live, grow, work, and age and are also shaped by economic, political, and social forces. The life chances of children depend on where they are born (WHO 2008). Spatial justice is used often in urban studies discourse and critical geography and is based on the notion of the interrelationship between social inequalities and geographic space, which creates social advantages for some and disadvantages for others. What may be missing from this mode of analysis is a critical examination of how power, ideology, and the creation of subaltern subject identity establish and nourish those social and institutional formations to produce and perpetuate pervasive inequalities in the first place. South Africa is a good example of such a system, where inequalities have risen in the post-democratic dispensation. Such empirical realities highlight the need to critically examine the fundamental drivers of an unfair social order.

The idea of a healthy city aims to protect and advance human health (Duhl and Sanchez 1999; Barton and Grant 2013). Healthy cities promote healthy living and create spaces that enable healthy choices, promote healthcare of vulnerable groups (providing dignified housing, education, job opportunities), and address environmental hazards. The healthy cities framework intersects with Global Surgery along two dimensions: (1) reducing the need for surgery (spatial justice) and (2) improving access to surgery (social justice). There is growing evidence that there are patterns of health associated with inequalities. Higher socio-economic status and street connectivity, for example, have been associated with lower risk for disability and improved mental health (Greenfield 2018). Creating a healthy city has as much to do with city planning and infrastructure, as it does with providing access to hospitals and healthcare facilities. Specifically, it also has to do with how well city planning and health planning interact and intersect to achieve collective objectives. In this regard, local governments are a crucial stakeholder to improve urban healthcare, but many health service providers do not adequately engage urban stakeholders, targeting national governments and ministries of health instead.

7.4.1 Spatial Justice: Addressing the Social Determinants and Drivers of Diseases and Injuries

The unequal conditions we live in explain why certain people develop diseases and injuries that necessitate surgery or suffer the complications related to surgery while others don't. The social determinants of health are based on the principle that health starts in our homes, schools, workplaces, neighborhoods, and communities—and the political and economic structures that underpin these social institutions (Table 7.2). This implies that cities and communities should aim to create social and physical environments that promote health for all, including the access to essential resources we need to function in political and economic life, such as clean air and water, nutritious food, and even high-speed Internet (Artiga and Hinton 2018). Global Surgery addresses surgery not through biomedical but through a public health lens to address health system failures, human behavior, infrastructure, finance, town planning, and the built environment, as possible enablers and barriers to surgical healthcare. This approach has special traction in poor African settlements.

African urban residents require surgery at least as much as any other urban resident. How the context is understood and how delivery of these health services is conceived will determine health outcomes. Africa is one of the least urbanized continents, with less than 40% of the urban population living in urban areas, though the rate of urbanization is rapidly growing (Simos et al. 2017). Globally, there is rapid migration to urban areas. The majority of the worldwide population (4.2 billion) already lives in urban areas, which is expected to expand considerably to 6.5 billion people by 2050. In 2018, the World Bank estimated that 29% of the urban population lived in urban slums, but this number may be as high as 54% in sub-Saharan Africa. There are wide disparities between countries ranging from 0% of the urban population living in slums in Sweden to 88% in Sudan (World Bank 2018). There are striking similarities across slums in different countries with certain key features and characteristics that need urgent rectification, including overcrowding, limited access to basic services (water, sanitation, electricity, transportation), material instability (physical, legal, economic, political), and unhealthy living conditions. As a corollary, people living in urban slums across different countries often have more in common with each other in terms of their living material conditions than they have with fellow citizens who live in wealthier neighborhoods in the same country. These characteristics are important for population health because they create unsafe living conditions, which impact on population-level health, independent of health system capabilities. It is estimated that by 2050 more than 70% of the population in Africa will live in an urban area, the majority of which is likely to live in urban slums characterized by these conditions (Simos et al. 2017).

Table 7.2 The social determinants of health and underlying key factors

Economy	Environment	Security	Policy	People
Employment	Electricity	Food	Regulations	Crime
Poverty	Housing	Health	Laws	Violence
Income	Air quality	Education	Health plans	Culture
Health cover	Infrastructure	Water	Equity	Discrimination

Using a healthy cities framework for a global surgery approach compels us to look beyond the healthcare sector to the broader environment to address the key drivers of surgical disease in the first place. By improving transportation and road infrastructure, fewer motor vehicle accidents and deaths due to trauma could follow. Creating purposeful work and critically examining the drivers of alcohol abuse could help to reduce violence, a major distal cause of the surgical trauma burden. Ensuring that people have access to nutritious food and information available on the Internet could empower people to make behavioral choices that diminish noncommunicable disease risk, including cancer and cardiovascular diseases that require surgery. In a city with appropriate health infrastructure, there is timely access to healthcare and services required such as surgery. Municipalities can become drivers and champions of healthcare by creating an environment that promotes healthy lives.

7.4.2 Social Justice: The Imperative to Improve Surgical Systems

Weak and unstable health systems, which lead to poor service coverage and utilization of surgical healthcare, constitute a major driver of suboptimal surgical outcomes. The quality, or rather the effectiveness of surgical health services, is increasingly becoming an important theme in national policy efforts to improve surgery. The three-delay model is often used to classify preventable causes of maternal deaths and is increasingly being used in Global Surgery to identify health system failures. The model describes three barriers to receiving timely and appropriate healthcare (Thaddeus and Maine 1994).

- Delay in decision to seek care (the patient)
- Delay in arrival at the healthcare facility (the transport system)
- Delay in the provision of adequate care (the healthcare facility)

The first delay relates to whether a patient decides to access care or not. Such a decision is not always straightforward and depends on numerous factors, including socio-cultural elements and beliefs about health and well-being, levels of health education and awareness, and concerns about the ability to pay or receive quality care. Once the decision to seek care has been made, paramedic and transport systems are needed to assess, acutely manage, and transport patients in a timely manner to the appropriate level of care. Ambulances are required for patients with severe injuries, and helicopters are often necessary to airlift patients who require urgent transfers or over large distances. In some countries, animal carts or wheelbarrows are still being used to transfer patients with life-threatening complications. The third delay refers to the delay in accessing appropriate care. Receiving appropriate care requires well-trained healthcare workers, adequately resourced facilities, and information systems to capture important data that will inform clinical management. Finally, patients should receive care that is affordable and which does not require substantial out-of-pocket payment, which could further cause catastrophic or impoverishing expenses.

The real-life impact of these delays can be illustrated through hypothetical scenarios that show the outcomes of unequal levels of access to timely and appropriate healthcare (Table 7.3).

Finally, an insufficient surgical workforce represents a significant barrier to safe surgical care, and Africa has the lowest number of surgeons per capita, and critical care specialists, globally. According to Holmer et al. (2015), while low- and lower-middle income countries represent 48% of the global population, they only possess 20% of the surgical workforce (Holmer et al. 2015). The COVID-19 pandemic has further disrupted surgical healthcare and is likely to increase the unmet need for surgery in Africa. During the lockdown in South Africa, deaths in maternal health increased by 30%, because women were not able to access care (Pattinson et al. 2020). As health facilities adapted to provide care to the large number of patients with COVID-19, elective surgeries were cancelled in numerous African countries (Chu et al. 2020b; COVIDSurg Collaborative 2020). The pandemic has made the

Table 7.3 Hypothetical scenarios depicting the interrelationship between spatial and social justice in Global Surgery

Scenario 1: A pregnant women in a remote area (which may be in a part of the city that is cut off from health and transport service access) is not able to access antenatal care; she delivers a baby with a birth defect that requires surgical correction at birth. The condition is missed throughout the pregnancy because she does have an ultrasound done. The baby is born with the large intestines protruding through the baby's belly. It takes 6 h after a home delivery for her to get to a hospital that offers pediatric surgery. The baby might survive but might miss the opportunity to have corrective surgery within the first few hours of life, which will impact their quality of life. As a result, the child might not reach their full potential.
Scenario 2: A road traffic accident occurs between two cities. One person has health insurance and the other does not. The person with health insurance has severe injuries, but a private ambulance arrives within 15 min and transports them to the nearest hospital where they are immediately assessed and taken to the operating room for surgery. They are admitted to the high care unit for observation, and they recover well. They spend the next 6 months consulting with both the physiotherapist and an occupational therapist and return to work fully recovered and rehabilitated. The person without health insurance has less severe injuries but waits longer for the ambulance to get them to the nearest public hospital. They have sustained internal injuries, but this is missed by the junior doctor who is running a busy casualty. They are admitted to the inpatient's ward for observation. After a few hours, their blood pressure becomes unstable, and they are immediately wheeled into an ambulance and transferred to a hospital with specialists for further management. They die on the way to the hospital. On postmortem, it is discovered that their abdomen is filled with blood because of sustaining an injury that could have been repaired had they been sent immediately for a laparotomy. Both examples illustrate the failure to access timely, affordable, and appropriate care. In the first example, the pregnant woman did not attend an antenatal clinic, even though she may have known that pregnant women should seek routine care. There may have been many reasons for this. Perhaps they were financial, resulting from a lack of education and insight, or perhaps even fear or anxiety. What she did not know is that her baby would be born with a congenital defect that could have been detected during pregnancy. As a result, her baby was born at home instead of in a health facility. On discovering that her baby has a defect, she is rushed to the nearest hospital, but the local district hospital does not have pediatric surgeons. In this case, the time it will take for her to access appropriate care might result in the condition becoming inoperable, further compromising the chances of survival in the baby.
In the second example, comparing two individuals who are involved in the same road accident, at the same time, with perhaps similar injuries. The difference between them is that one has medical insurance, and one doesn't. This results in one arriving sooner at a healthcare facility and immediately receiving appropriate healthcare. The person who is relying on public healthcare does not get timely prehospital care and does not get appropriate care, resulting in a death that is related to poor management. On postmortem, it is clear that the patient needed an exploratory laparotomy, which they didn't have either because they were poorly managed or because they didn't have access to emergency surgical care.

need for global surgery more important than ever. This is especially true in Africa and even more so in African cities that are dense in areas where there is a concentration of poorer households and where communities are under-serviced and poorly planned. Efforts to improve surgical care will need to accelerate.

7.5 Global Surgery: An Interface Between Surgical Care, Public Health, and Public Policy

Since the launch of the LCoGS, several countries in Africa, Asia, and Latin America, including the small island states of the Pacific, are engaging in national planning processes to improve surgical healthcare. The LCoGS provided a flexible framework to catalyze such national efforts, with Zambia and Tanzania among the first nations to develop National Surgical, Obstetric and Anesthesia Plans (NSOAPs). In 2020, the United Nations Institute for Training and Research published the NSOAP Planning manual as a toolkit to further assist countries develop and implement NSOAPs within broader national health strategies. At the global level, the World Health Assembly passed Resolution 68.15, calling for emergency and essential surgery inclusion in UHC (UNITAR 2020), and more recently at a regional level, the Southern African Development Community passed an intergovernmental resolution to pri-

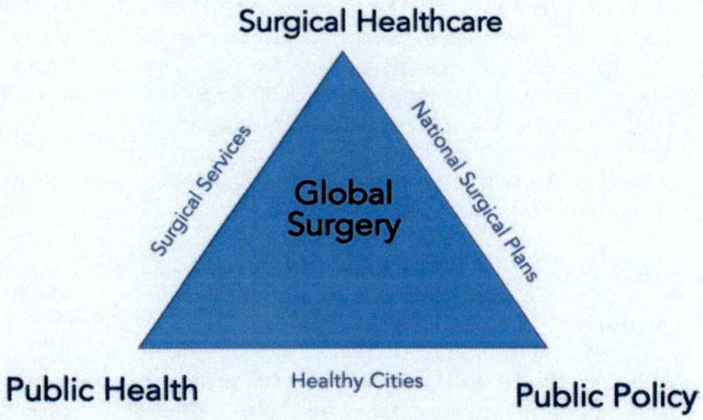

Fig. 7.4 A proposed framework for Global Surgery as the interface between surgical healthcare, public health, and public policy. (Source: authors)

oritize surgical healthcare within broader efforts to achieve UHC, regional health security, and economic development.

Efforts to improve surgical healthcare in African cities need to be driven collaboratively by the Global Surgery community, public health community, and policy makers, with the aim of improving service delivery and healthier cities (Fig. 7.4). The Global Surgery process allows for the creation of partnerships and for engaging with key stakeholders to improve surgical healthcare. The following recommendations were made by the Lancet Commission on Global Surgery to achieve universal coverage:

- Governments in LMICs to strengthen surgical services and the national health systems
- Global health and development organizations to include indicators of surgical care within existing health goals and monitoring systems
- Funding agencies to invest in surgical care as a tool for poverty alleviation and general welfare gains
- International partners to support local leaders in their efforts to provide equitable surgical care
- The general public to lobby for access to surgical services for all

7.6 Conclusion

The SDGs set out a vision of universal affordable healthcare and cities that work for all.

This creates a unique opportunity for health promoting policies across different sectors and at different scales—including, crucially, cities where the majority of the global poor now live. Healthy cities are also the foundation for effective primary health and successful global surgery delivery but will require enhanced dialogue between stakeholders in global surgery, public health, and public policy. We need to address poverty, unemployment, lack of education, and inequality while also striving to create healthy cities that advance health. Achieving the SDGs for health in African cities will become possible when governments, both local and national, recognize the thread between all the SDGs and the varied environments that people live in.

References

Artiga S, Hinton E (2018) Beyond health care: the role of social determinants in promoting health and health equity. Issue brief. Henry J. Kaiser Family Foundation. https://files.kff.org/attachment/issue-brief-beyond-health-care. Accessed 22 July 2021

Barton H, Grant M (2013) Urban planning for healthy cities. Journal Urban Health 90(1):129-141

Biccard B, Madiba T, Kluyts H, Munlemvo D, Madzimbamuto F, Basenero A, Gordon CS, Youssouf C et al. (2018) Perioperative patient outcomes in the African Surgical Outcomes study: a 7-day prospective observational cohort study. The Lancet 391(10130):1589-1598

Bishop D, Dyer R, Maswime S, Rodseth R, van Dyk D, Kluyts H, Tumukunde J, Madzimbamuto F et al. (2019) Maternal and neonatal outcomes after caesarean delivery in the African Surgical Outcomes study: a 7-day prospective observational cohort study. Lancet Global Health 7(4):e513-e522

Brink J, Hassoulas J (2009) The first human transplant and further advances in cardiac transplantation at Groote Schuur Hospital and the University of Cape Town. Cardiovascular Journal of Africa 20(1):31-35

Chu K, Smith M, Steyn E, Goldberg P, Bougard H, Buccimazza I (2020a) Changes in surgical practice in 85 South African hospitals during Covid-19 hard lockdown. South African Medical Journal 110(9):916-919

Chu K, Reddy CL, Makasa E (2020b) The collateral damage of the COVID-19 pandemic on surgical health care in sub-Saharan Africa. AfroSurg Collaborative. Journal of Global Health 10(2):020347

CovidSurg Collaborative (2020) Elective surgery cancellations due to Covid-19 pandemic: global predictive modeling to inform surgical recovery plans. British Journal of Surgery 107:1440-1449

Duhl L, Sanchez A (1999) Healthy cities and the city planning process: a background document on links between health and urban planning. World Health Organization Regional Office for Europe, Copenhagen

Farmer P, Kim J (2008) Surgery and global health: a view from beyond the OR. World Journal of Surgery 32(4):533-536

Galea S, Vlahov D (eds) (2006) Handbook of urban health: populations, methods, and practice. Springer Publishing, New York

Ginwala R, Rickard J (2015) Surgical missions: the view from the other side. JAMA Surgery 150(4):289-290

Greenfield E (2018) Age-friendly initiatives, social inequalities, and spatial justice. Hastings Center Report 48(S3):S41-S45

Griggs D, Nilsson M, Stevance A, McCollum D (eds) (2017) A guide to SDG interactions: from science to implementation. International Council for Science, Paris

Harpham, T (2009) Urban health in developing countries: what do we know and where do we go? Health Place 15(1):107-116

Holmer H, Lantz A, Kunjumen T, Finlayson S, Hoyler M, Siyam A, Montenegro H, Kelley ET et al. (2015) Global distribution of surgeons, anaesthetists, and obstetricians. Lancet Global Health 3(S):S9-S11

Ma X, Vervoort D, Reddy CL, Park KB, Makasa E (2020) Emergency and essential surgical healthcare services during COVID-19 in low- and middle-income countries: A perspective. International Journal of Surgery 79:43-4

Maswime S, Buchmann E (2017a) Why women bleed and how they are saved: a cross sectional study of caesarean section near-miss morbidity. BMC Pregnancy Childbirth 17:15

Maswime S, Buchmann E (2017b) Near-miss maternal morbidity from severe haemorrhage at caesarean section: a process and structure audit of system deficiencies in South Africa. South African Medical Journal 107(11):1005-1009

Meara J, Leather A, Hagander L, Alkire C, Alonso N, Ameh E, Bickler SW, Conteh L et al. (2015) Global surgery 2030: evidence and solutions for achieving health, welfare, and economic development. The Lancet 386:569-624

Pattinson R, Fawcus S, Gebhardt S, Nilt R, Soma-Pillay P, Moodley J (2020) The effect of the first wave of Covid-19 on use of maternal and reproductive health services and maternal deaths in South Africa. Obstetrics Gynaecology Forum 30(4):36-44

Roa L, Jumbam D, Makasa E, Meara J (2019) Global surgery and the Sustainable Development Goals. British Journal of Surgery 106(2):e44-e52

Sartorius N (2006) The meanings of health and its promotion. Croatian Medical Journal 47(4):662-664

Simos J, Naissem FB, Naissem J, Sokona FM, de Dieu Konongo J, Sani A, Corburn J, Karanja I et al. (2017) Healthy cities in Africa: a continent of difference. In: de Leeuw E, Simos J (eds) Healthy cities: the theory, policy and practice of value-based urban planning. Springer Publishing, New York, pp 89-132

Thaddeus S, Maine D (1994) Too far to walk: maternal mortality in context. Social Science Medicine 38(8):1091-1110

United Nations (UN) (2015) Transforming our world: the 2030 Agenda for Sustainable Development (70/1). United Nations General Assembly, New York. https://sustainabledevelopment.un.org/post2015/transformingourworld. Accessed on 19 July 2021

United Nations Institute for Training and Research (UNITAR) (2020) National surgical, obstetric and anaesthesia planning manual, 2020 edition. United Nations Institute for Training and Research, Geneva

Vlahov D, Freudenberg N, Proietti F, Ompad D, Quinn A, Nandi V, Galea S (2007) Urban as a determinant of health. Journal of Urban Health 84(1):16-26

World Bank (2018) Urban population living in urban slums. https://data.worldbank.org/indicator/EN.POP.SLUM.UR.ZS. Accessed 22 July 2021

World Bank (n.d.) Life expectancy at birth, total (years)—sub-Saharan Africa. https://data.worldbank.org/indicator/SP.DYN.LE00.IN?locations=ZG&view=map. Accessed 20 July 2021

World Health Organization (WHO) (2017) World Bank and WHO: half the world lacks access to essential services, 100 million still pushed into extreme poverty because of health expenses, 13 December. https://www.who.int/news/item/13-12-2017-world-bank-and-who-half-the-world-lacks-access-to-essential-health-services-100-million-still-pushed-into-extreme-poverty-because-of-health-expenses. Accessed 20 July 2021

World Health Organization (WHO) (2008) Closing the gap in a generation: health equity through action on social determinants of health. World Health Organization, Geneva

World Health Organization (WHO) (1987) Ottawa charter for health promotion. https://www.who.int/publications/i/item/ottawa-charter-for-health-promotion. Accessed 15 July 2021

Part III

Barriers or Opportunities: Data, Finance, Corruption

Check for updates

8 Data and the Localization of Sustainable Development Goals in Africa: The Case of SDG 11 in Lagos and Accra

Peter Elias and João Porto de Albuquerque

Abstract

This chapter examines the critical role of data for the localization of the Sustainable Development Goals (SDGs) in Africa, with a specific focus on SDG 11 (Sustainable Cities and Communities) in Lagos and Accra. The chapter argues that while the importance of digital technologies and data for the implementation of the SDGs has been acknowledged, data inequality remains a fundamental challenge. This inequality reflects existing global socio-spatial inequalities. If not carefully considered, these can be perpetuated even further. The chapter builds on results of the ongoing research project, SCiLeD: Standardizing City-Level Data-Gathering for Achieving SDG 11 in Africa, to examine the concrete processes, challenges, and opportunities available for African cities to implement data innovations aimed at localizing SDG 11. It draws on lessons learned and provides future directions based on how the SCiLeD project applied a transdisciplinary research approach for co-designing and co-producing knowledge. This approach allowed for the generation of data that is disaggregated to the neighborhood level, involving community mappers and profilers to reflect and respond to local needs and realities, thereby illustrating its transformative influence and potential toward achieving SDG 11 in African cities by 2030.

Keywords

SDG localization · SDG 11 · City-level data · Lagos · Accra · Transdisciplinary research (TDR)

8.1 Introduction

In the past few years, the importance of digital technologies and data innovations for the implementation of the United Nations Agenda 2030 has been increasingly acknowledged. Early calls for a "data revolution for sustainable development" (UN 2014) have been recently complemented by an emphasis on the crucial role of a "digital revolution" to support transformations to sustainability (Corbett and Mellouli 2017; Pappas et al. 2018; Sachs et al. 2019).

More recently, the crisis resulting from the COVID-19 pandemic has brought about an accel-

P. Elias (✉)
Department of Geography, University of Lagos, Lagos, Nigeria
e-mail: pelias@unilag.edu.ng

J. P. de Albuquerque
School of Social and Political Sciences, University of Glasgow, Glasgow, UK
e-mail: Joao.Porto@glasgow.ac.uk

S. Croese, S. Parnell (eds.), *Localizing the SDGs in African Cities*, Sustainable Development Goals Series, https://doi.org/10.1007/978-3-030-95979-1_8

erated adoption of digital technologies in many parts of the world, which have enabled many people to carry on social, economic, and education activities amid restrictions of physical contact. However, the pandemic crisis has also clearly shown how the most marginalized and impoverished people globally were the ones to feel its most severe impacts: they were less able to benefit from the use of digital tools and less represented in the data used for decision-making. As underlined by the Sustainable Development Goal (SDG) Report 2020, "huge data gaps still exist in terms of geographic coverage, timeliness and the level of disaggregation required" for assessing country-level progress toward the SDG targets and indicators (UN 2020a, p. 4). The report also states that investments in data and innovation are key to responding to the crisis and supporting policy making and enabling progress toward the SDGs. Similarly, the UN Research Roadmap for the COVID-19 Recovery (UN 2020b) calls for investments in data systems and infrastructure, considering them the linchpin to the implementation of impactful research efforts to support just and effective recovery. To this we would add that investments into data innovations for localizing the SDGs today not only provide an essential instrument to assess achievements of countries in relation to the SDGs but have also the potential to create a solid evidence base and a digital infrastructure which will be essential for a post-2030 Sustainable Development Agenda.

However, despite the acknowledged importance of data innovations and digital infrastructures, we would like to argue in this chapter that a fundamental problem for SDG data innovations needs to be explicitly addressed: data inequalities reflect socio-spatial inequalities, which, if not carefully considered, can imply that SDG data will enshrine and perpetuate those inequalities (Ulbrich et al. 2019). This fact is behind an apparent paradox related to data, which can be well exemplified in the context of cities, and thus directly relevant to SDG 11 (Sustainable Cities and Communities). On the one hand, the recent surge in the use of digital technologies and emerging sources of data (e.g., low-cost sensors, high-resolution satellite imagery, data generated by citizens using mobile phones) has given rise to ideals of "smart cities" (Przeybilovicz et al. 2018; Townsend 2013), which are built upon powerful artificial intelligence algorithms to make sense of the resulting "big data." This would enable fine-grained and sophisticated understandings of urban dynamics to support policy and decision-making based on the unprecedented availability of urban data (Kitchin 2014), thus enabling a "new science of cities" (Batty 2013). However, in an apparent paradox observed in the context of crisis management (Restrepo-Estrada et al. 2018), a related and frequently overlooked phenomenon happens simultaneously to the perceived overabundance of data, namely, an information dearth. Existing data about cities frequently lacks geographical and temporal coverage, and the putative abundance of data is concentrated in some well-known urban areas (frequently wealthy areas of Western cities), whereas for many other areas and decision-making tasks, there is a real lack of actionable information that is able to support decision-making.

A practical consequence of this Janus-faced problem of urban data can be found in the very first target of SDG 11: "by 2030, ensure access for all to adequate, safe and affordable housing and basic services and upgrade slums." Notwithstanding the need to track progress in SDG 11.1 with data on the number of people currently living in deprived urban neighborhoods, we presently lack an internationally agreed definition of "slum area" or standardized data for reporting on the physical locations and characteristics of these areas (Lilford et al. 2019), with current methods for mapping them being siloed and disconnected (Thomson et al. 2020). Not surprisingly, analyses of existing official datasets in large cities of the global South have observed that the data available is often concentrated in wealthy areas, with poor and deprived neighborhoods having much less digital footprint (Macaya et al. 2020). To the welcome emphasis on the importance of data innovations and data infrastructures of current policy discourses mentioned above, we thus would like to emphasize the need for data innovations which not only produce high-quality, timely, and comprehensive data for policy making, but which address existing inequalities by also enabling currently marginalized voices to use data generation as an opportunity for shared learning and for empowering local transformations (de Albuquerque and de Almeida 2020).

Against this backdrop, in the remainder of this chapter, we investigate the following questions: what are the key challenges and requirements for data innovations that support the SDGs to be truly equitable, inclusive, and empowering? How do these challenges and requirements manifest in African countries? In order to investigate these questions, we present results of an ongoing research project SCiLeD: Standardizing City-Level Data-Gathering for Achieving SDG 11 in Africa, which enables us to discuss the concrete challenges and opportunities of the African cities of Accra (Ghana) and Lagos (Nigeria) to implement data innovations aimed at localizing SDG 11. The chapter concludes with a summary of key lessons learned and future directions for research and practice.

8.2 Background: Sustainable Development Agendas and African Countries

African countries have been active players and partners to several global aspirations such as 2030 Agenda for Sustainable Development, the Paris Agreement on climate change, the Addis Ababa Action Agenda (AAAA) on financing for development, the Sendai Framework for Disaster Risk Reduction, and the New Urban Agenda. This is motivated by the aspiration of the continent to achieve sustainable development as demonstrated in the African Union Agenda 2063: The Africa We Want. Governments of several countries in Africa have, however, demonstrated a lack of national and local institutional capacities in the implementation of these aspirations amid growing inequality and poverty. Many cities in Africa are confronted with evidence of inequality and poverty including water shortages, poor sanitation, inadequate housing, and increasing rates of air pollution, traffic congestion, and health challenges. Rapid urban population growth and urbanization continue to diminish the capability of city governments to tackle urban development challenges, which are particularly exacerbated by inequality, poverty, and uncontrolled development of urban slums. In West Africa, rapid growth of urban population and urbanization are creating large urban centers which pose a big challenge for city governments lacking the required resources for sustainable development. These have contributed to the deterioration of quality of life, high level of poverty, inequality, and the proliferation of slums.

This case study focuses on two West African cities—Accra and Lagos—the largest cities in Ghana and Nigeria, respectively. They also share common history, having started as small coastal fishing settlements between the fifteenth and sixteenth centuries, to later become prominent colonial administrative centers and more recently serving as national capitals. They also have common sociopolitical antecedents in their formation as colonial urban centers, which became the basis for the rapid urbanization and population growth experienced by them (Akinyele 2014; Songsore and Stephens 2008). Although these two African cities have common chronicles, they also differ significantly in terms of population size, demography, and culture which have distinct social, economic, and environmental dimensions. Slum/informal settlement development is one of the biggest urbanization challenges confronting Accra and Lagos. Yet, the two cities account for a huge percentage of economic growth, industrial activities, and contributions to gross domestic product (GDP) of their respective countries (Filani 2012).

8.3 Case Study: Agenda 2030 in Accra and Lagos

Accra and Lagos are both being transformed in ways that allow them to become more sustainable and competitive and better able to tackle urban challenges (Filani 2012). Unfortunately, they both have huge data deficits despite several agencies involved in the collection of economic and environmental data. Furthermore, typical economic and environmental data in these cities are collected in silos and in different time periods and frequencies, stored in different formats, and used for varied purposes, thus failing to ensure safe, inclusive, resilient, and sustainable cities and communities. Thus, it is obvious that social and spatial inequalities are strongly connected with data inequalities especially in slums and informal settlements in Accra and Lagos. This may make it difficult for these two big West African cities to measure, track, and monitor progress toward Agenda 2030.

Agenda 2030 was launched in 2015 to tackle various social, economic, and environmental dimensions of development challenges including inequality, poverty, climate stresses and shocks, as well as the proliferation of slums. Within this context, our work in Accra and Lagos focuses on the urban-specific SDG 11 and its aim to "make cities and human settlements inclusive, safe, resilient and sustainable." Transforming urban slums through efficient land use planning, improved housing quality, reliable transport system, adequate public space, affordable waste management, and others is highly central to SDG 11. They are also interconnected with other goals in Agenda 2030 including SDG 3, 6, 8, 9, and 13. Reliable and accurate data on the underlying distributions, patterns, trends, or disparities inherent in cities are critical for urban planners, researchers, and governments to effectively measure, track, and monitor implementation and performance of Agenda 2030.

8.4 The Standardizing City-Level Data-Gathering for Achieving SDG 11 (SCiLeD) Project

Motivated by the goal of addressing the urban data gap in Africa, the Standardizing City-Level Data-Gathering for Achieving SDG 11 in Africa (SCiLeD) project aims to generate city-level data with respect to the proportion of urban population living in slums, in informal settlements, or with inadequate housing in the West African cities of Lagos and Accra. The project was inspired because of the recognition of the missing gaps in the administrative records, official statistics, census, and surveys collected by ministries, departments, and agencies in many African countries. Most available data do not contain information about fringe urban dwellers and communities including slums and informal settlements; hence they are often masked in highly aggregated national, regional, municipal, or district data. This often makes decision-making, planning, and policies difficult and unable to address the needs of the more than 65% urban population living in slums and informal settlements. The lack of evidence in terms of accurate, relevant, and suitable data hampers city governance and intensifies poverty, inequality, and the proliferation of slums and informal settlements in Africa.

The approach adopted is the transdisciplinary research (TDR) which connects governments, academia, businesses, civil society organizations, and communities. The aim is to produce, share, and use data as well as incorporate multiple perspectives and sources, shared or social learning, reliability, and empowerment (Moser 2016; Osborne 2015; Lang et al. 2012). This is to safeguard the participatory process of integrating urban data and address the twin problems of information overload and information dearth. Information overload refers to high-volume data streams and unstructured data, among others, while information dearth refers to a lack of spatial and temporal coverage of datasets and low integration into decision-making. The TDR approach for participatory urban data collection and governance in Accra and Lagos followed a series of interrelated activities. First, in each city, there was a focal point bringing together key experts from governments, academia, businesses, civil society organizations, and local communities which became the Local Think Tank Group (LTTG) of the project. This group started by reviewing the SDG 11 targets and indicators and aligning them with the aspirations and priorities of the participating cities, namely, Lagos and Accra.

Based on this exercise, the LTTG selected five targets out of the total ten targets of the SDG 11, namely, housing and basic services with slum upgrading (SDG 11.1), sustainable transportation (SDG 11.2), participatory urbanization and planning (SDG 11.3), air quality and waste management (SDG 11.5), and disaster risk reduction (SDG 11.6). Relevant stakeholders within these sectors were identified, mapped, and selected for engagement (see Table 8.1). Some of the criteria for the selection included interest, influence, organizational mandate and/or activities, as well as those affected by these issues.

Table 8.1 SDG 11 stakeholders and their relevance

Stakeholders	Relevance
City government officers	City government officers are key stakeholders because they make, approve, and implement urban policies; deliver and manage local services; wield enormous power and influence on the nature and quality of urban development; track progress; and monitor performance of progress toward achieving SDG 11
Academic urban researchers	Urban researchers in academia are significant stakeholders because they are the custodians of knowledge; control processes and technical skills associated with the co-design and co-production of knowledge; and disseminate knowledge
Civil society organizations (CSOs)	CSOs are relevant stakeholders: through advocacy activities, they may demand or are invited to participate in the process of co-design and co-production; they influence behavior and actions through grassroots mobilization
Local communities	Local communities organized through organizations such as the Nigeria Slum/Informal Settlement Federation and the People's Dialogue on Human Settlements, Ghana should be the principal beneficiaries of SDG implementation; they are service user and best informed about local needs and assets
Private businesses	Private businesses are key stakeholders involved with agenda setting because they are interested and informed about how things ought to be done to improve the quality of life among the poor vulnerable groups and communities
Local media	Local media are relevant stakeholders because they are responsible for informing the public; publishing performance indicators; and dissemination of information

The third step in the process involved convening an inception workshop which brought together government ministries, departments and agencies, academia, civil society, international organizations, the media, and the Slum Dwellers International (SDI) Nigeria Slum/Informal Settlement Federation and Ghana Federation of the Urban Poor. These actors were arranged according to the various sectors to form thematic clusters. Five thematic clusters were formed—housing and basic services, sustainable transportation, participatory planning, air quality and waste management, and disaster risk reduction. During the inception workshop, an institutional survey was concurrently conducted by administering structured questionnaires to the participants to examine the status of urban data in the context of the SDG 11 and the selected targets and indicators.

After the inception workshop, the survey continued by administering the same set of questionnaires on an additional number of organizations which were identified as relevant in each cluster. The specific objectives of the institutional survey were to (1) identify and characterize major institutions involved in urban data governance; (2) assess the state of urban data governance practices and constraints by institutions; (3) examine the performance of the selected institutions with respect to data collection, storage, sharing, and integration; and (4) determine which of the FAIR (findable, accessible, interoperable, reusable) data principles were used in urban data governance. These principles represent key elements of good data governance which enhance data discovery and innovation and data integration and reuse by the community of knowledge producers and users (Wilkinson et al. 2016).

By leveraging their grassroots presence and experience in profiling and mapping activities, the Slum Dwellers International (SDI) Nigeria Slum/Informal Settlement Federation led the household survey and slum mapping in selected slum communities in the two cities (see Fig. 8.1).

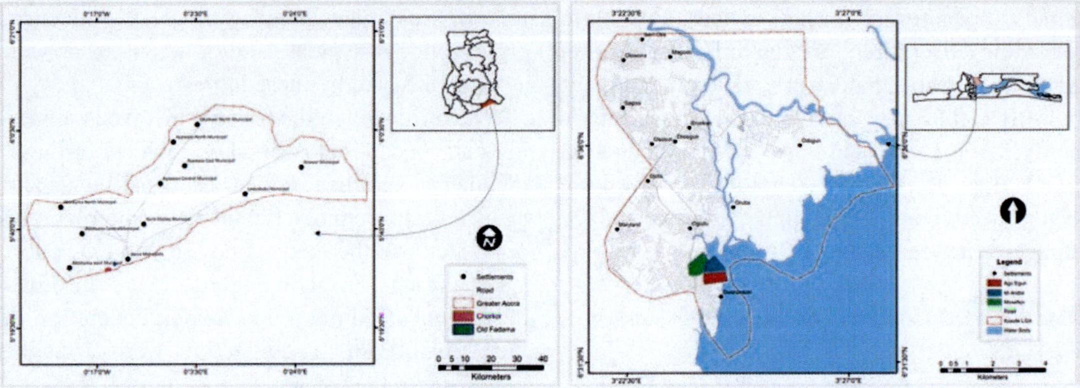

Fig. 8.1 Selected slum communities in Accra (left) and Lagos (right)

In Lagos, three waterfront communities were purposively selected because they were prioritized by the city government, civil society organizations, and the local community stakeholders as of primary interest to ongoing and future urban development projects. These communities included Mosajejo, Idi Araba, and Ago-Egun in Oworonshoki, Kosofe Local Government Area. Kosofe is the third largest Local Government Area with a population of 682,772 (Nigerian Population Commission 2006), while the population of the other selected communities is estimated between 250 and 4000 persons. These communities are full residential areas with high population densities and inadequate access to urban basic services. They are located close to the Lagos Lagoon where the land is freehold and ownership is by inheritance. The housing condition is generally poor owing to construction materials and space allocation.

Two medium and large slum communities were selected in Accra, namely, Old Fadama and Chokor. The enumerators and mappers were selected from the communities, trained, and equipped with tablets loaded with questionnaires on open-source software (KoBoCollect) by the supervising SDI-approved CSOs, namely, People's Dialogue on Human Settlements, Ghana, and Justice and Empowerment Initiative, Lagos. A research team from the Universities of Ghana and Lagos led the data cleaning, analysis (quantitative and qualitative), and presentation using graphical charts and photographs. After this, small area mapping was led by the urban modelers and students of the two universities in the team with contributions from community profilers/mappers to map slum boundaries and facilities and assets in the selected slum communities.

The last step involved two-phased post-fieldwork engagement strategies. The first phase of this engagement consisted of giving data back to the communities, and the second was the technical validation workshop. Giving data back to communities involved dissemination and sharing of the outputs of data analysis in the form of graphical charts and maps. The graphical and pictorial communication gave insight into the conditions of slum communities and households. It further created awareness about their needs and empowered residents to use the data to engage local authorities to negotiate and influence their envisaged futures. The technical validation workshop in turn enabled stakeholders from both scientific and non-scientific communities to interrogate, verify, and validate the outputs of the study.

8.4.1 Key Outputs of SCiLeD

Some of the outputs of the transdisciplinary research process show the benefits of the participatory process of collaborative knowledge co-design and co-production. It also highlights the pathway for departing from the traditional siloed approach in urban data governance. Based on consensus agreements by the stakeholders on cities' priorities and preferences in Accra and Lagos, five SDG 11 targets with their respective indicators were selected. These formed the core themes for the project, which resonated well with stakeholders and propelled their buy-in throughout the lifecycle of the project. The process of stakeholder identification, mapping, and engagement was used to create a community of data producers and users. These form the basis for understanding the urban data ecosystem on the one hand and the urban data value chain on other hand.

The process of community-led slum mapping and profiling in this project increased participatory data collection which includes and empowers marginalized communities to obtain evidence which they can use to negotiate with city authorities for improved service delivery and well-being. In Lagos, for instance, the city government uses data from community-led participatory slum profiling and mapping to make decisions about revitalization. The priority community and issues in the report guided the city government in making informed decisions and design sustainable solutions. The disaggregated data provided insights into the intensity of the urban challenges and needs, which were previously masked in the aggregated data. The local communities and the civil societies also used the disaggregated data to pinpoint areas requiring urgent interventions and guidance on the nature of the interventions.

The assessment of the institutional capacity for SDG tracking and monitoring in turn provided important insight into practices such as data collection mandate, methods, collaboration, analysis, storage, retrieval, sharing, and constraints including ownership, stewardship, and accessibility, which define action plans for improved urban data governance. Table 8.2 is the result of an institutional survey on the status of urban data in Accra and Lagos. It describes institutional data management practices in the two cities across different sectors according to selected SDG 11 targets, namely: SDG 11.1 (Housing and Basic Services), focusing on five institutions in Lagos and four institutions in Accra; SDG 11.2 (Sustainable Transportation), focusing on four institutions in Lagos and three institutions in Accra; SDG 11.3 (Inclusive and Sustainable Urbanization), focusing on five institutions in Lagos and two institutions in Accra; SDG 11.5 (Disaster Risk Reduction), which examines four institutions in Lagos and six institutions in Accra; and SDG 11.6 (WASH), where three select institutions were examined in Lagos and two select institutions examined in Accra. The five targets were selected because they were the top five ranked sectors of Accra and Lagos by the TTLG. The assessment was conducted with respect to how the institutions in the selected sectors manage their data in terms of data processes, the application of FAIR (Findable, Accessible, Interoperable, and Reusable) principles, and the levels of data collection.

The results of the survey showed that there are variations in data management practice within and across institutions in the two cities, represented in Table 8.2 with the symbol (√) indicating presence of a practice, while symbol (O) indicates an absence. For instance, data processes, the application of FAIR principles, and the levels of data collection in the housing sector were not uniform within and across both cities. The same applied to the sustainable transportation sector, the inclusive and sustainable urbanization sector, disaster risk reduction, and the WASH sector. Overall, all selected institutions in Accra corroborated the application of FAIR data principles, while institutions in Lagos did not conform to the use of FAIR data principles. This implies that just as there are variations in institutions in both cities, variations also exist in data processes, the application of FAIR data principles, and levels of data collection within and across both cities. It is therefore evident that there is a lack of harmonization in urban data management within and across the cities, which represents a major challenge to the capacity of city government institutions to implement SDG 11. The creation of small area maps through

Table 8.2 Institutional data management practices in Accra and Lagos (source: authors)

	Sectors	Institutions Lagos	Data management practices													Institutions	Data management practices											
			Data processes							Level of data collection							Data processes							Level of data collection				
Targets			Disaggregation	Aggregation	Conversion	Validation	Sorting	Summarization	FAIR principles	National	Regional/state	City level	Neighborhood level	Neighborhood level	Slum/informal communities	Accra	Disaggregation	Aggregation	Conversion	Validation	Sorting	Summarization	FAIR principles	National	Regional/state	City level	Neighborhood level	Neighborhood level
SDG 11.1	Housing	Lagos State Building Control Agency (LASBCA)	O	O	O	O	√	√	√	O	O	O	O	O	O	Ghana Standards Authority	O	O	O	O	O	√	O	O	O	O	O	O
		Lagos State Physical Planning Permit Authority (LASPPPA)	√	√	√	√	√	√	√	O	√	√	√	√	√	Ghana Statistical Services	√	√	√	√	√	√	√	√	√	√	O	√
		LAGOSHOMS	O	O	O	O	O	O	O	O	O	O	O	O	O	Land Use and Spatial Planning Authority	O	√	√	√	√	√	√	√	√	√	√	√
		Ministry of Physical Planning and Urban Development (MPPUD)	O	O	√	√	√	O	√	O	√	√	√	√	√	Ministry of Works and Housing	O	O	O	O	O	√	√	√	O	√	O	√
		Ministry of Housing	√	√	O	√	O	√	√	√	√	√	O	O	√													

SDG 11.2	Sustainable Transport	Federal Road Safety Commission (FRSC)	O	√	O	√	√	√	O	O	O	√	O	O	O	Greater Accra Passenger Transport Executive (GAPTE)	√	√	√	√	√	√	√	O	O	√	O	O
		Lagos Metropolitan Area Transport Authority (LAMATA)	O	√	O	√	√	√	√	√	√	√	√	√	√	Department of Urban Roads	√	O	√	√	O	√	√	√	√	√	O	O
		Lagos State Waterways Authority (LASWA)	O	O	O	O	O	O	√	O	O	√	O	√	O	Electoral Commission	O	√	O	O	O	O	O	O	√	O	O	O
		Ministry of Transportation (MOT)	√	√	√	√	√	√	√	√	√	√	√	√	O													
SDG 11.3	Participation Planning	Lagos State Office for Disability Affairs (LASODA)	O	O	√	√	O	√	O	√	O	O	√	O	O	Ministry of Employment and Labour Relations	√	√	O	√	√	√	√	√	√	O	O	O
		Ministry of Justice	√	√	√	√	√	√	√	O	O	√	√	√	√	Ministry of Gender, Children and Social Protection	√	√	O	√	√	√	√	√	O	O	O	O
		Ministry of Wealth Creation and Employment	O	O	O	√	O	√	√	O	O	O	O	O	O													
		Office of the Public Defender (OPD)	O	O	√	√	O	√	√	O	√	O	O	O	O													
		Ministry of Women affairs and Poverty Alleviation (WAPA)	O	O	O	O	O	O	√	√	√	√	O	O	O													

(continued)

Table 8.2 (continued)

SDG 11.5	Disaster Risk Reduction	Lagos State Fire Service	O	O	O	√	√	√	√	√	√	√	√	√	O	Ministry of Environment, Science and Technology	√	√	O	√	√	√	√	√	√	O	O	O
		Lagos State Emergency Management Agency (LASEMA)	O	√	O	√	√	√	√	√	√	√	√	√	√	Ambulance service	O	O	√	O	O	O	√	O	O	O	O	O
		Lagos State Environmental Protection Agency (LASEPA)	O	O	O	√	O	O	O	O	O	O	O	O	√	Amnesty International	√	O	O	√	√	√	√	√	√	√	√	√
																Ghana Red Cross	O	O	√	O	O	O	O	O	O	O	O	O
		Ministry of Environment	O	O	O	O	O	O	O	O	O	√	O	√	O	Ghana Police Service	O	O	O	O	O	O	O	√	√	O	O	√
																Ghana National Fire Service	O	O	√	√	O	O	√	O	O	O	O	O
SDG 11.6	WASH	Lagos State Water Corporation (LSWC)	O	√	O	√	√	√	√	O	√	√	√	√	O	Water Research Institute	O	O	√	O	O	√	√	√	√	√	O	O
		Lagos State Urban Renewal Agency (LASURA)	√	√	O	√	O	O	O	√	√	O	O	O	√	WaterAid	√	√	√	√	√	√	√	√	√	√	√	O

the SCiLeD project in turn highlighted the need for disaggregated data on economic and environmental deprivations and needs for sustainable urban development.

8.5 Important Outcomes of SCiLeD

Given the outputs stated above, the following are the related short-term and long-term outcomes of the SCiLeD processes.

8.5.1 Data for Action

The short-term outcomes of using data for action include awareness, cooperation, improved capacity, knowledge exchange, and social/shared learning about local needs and assets for sustainable development. The selected SDG 11 targets (11.1, 11.2, 11.3, and 11.4) sit well with the stakeholders as they align with the local priorities in the two cities and selected communities. Meanwhile, existing data are not disaggregated by local communities and their households and gender, which makes it difficult to use them for the SDG 11 indicators. There is increasing recognition of the need for community-level data collection and disaggregation which were previously unavailable for selected slums and informal settlements. There is also increasing interest by city government to ask relevant questions in the survey questionnaire used for slum profiling so as to capture indicators which address issues that affect slum/informal settlements. This enables place-specific prioritizations and guidance on interventions, as well as tracking and monitoring of performance. Co-design and co-production of knowledge also enhanced trust in the quality and accuracy of the data leading to its uptake for decision-making by city governments. The long-term outcomes of using data for action increased participation in decision-making, empowerment for improved well-being, and advocacy for change in slum communities.

8.5.2 Knowledge-Driven Solutions

The short-term outcomes of knowledge-driven solutions for building and managing marginalized people and places include linking data to decision, awareness of alternative solutions, and translating knowledge to actions. For instance, it resulted in a plan for the long-term translation of knowledge to actions including teaching behavioral change in the context of water, sanitation, and hygiene (WASH) which aligns with SDG 11.1. Another plan consists of overturning the scarcity of clean water in slum neighborhoods in Lagos by optimizing the use of ultraviolet circular surface sanitization equipment for a plug-and-play system to achieve sanitization within 60 s in the context of COVID-19. This entails wastewater disinfecting using an advanced oxidation process assisted by ultraviolet lamps, followed by biological treatment and/or chlorine disinfection to make water suitable for handwashing. Knowledge of this solution has resulted in a partnership for innovation involving existing stakeholders (civil society, academia, and local communities) in Lagos and new partners from the United Kingdom (academia and private sector). This has led to an opportunity to evaluate the project's impact and willingness to pay for services in the slum neighborhoods.

8.5.3 Networking for Upscaling

The short-term outcomes of building networks and promoting integrated solutions as exemplified in the SCiLeD project have led to collaborations, cross-learnings, and partnerships at all levels. By leveraging a transdisciplinary approach, we now have a growing community of slum mappers from local communities, local experts, tertiary institutions, research organizations, businesses, and funding agencies connected together to redefine our slum data observatory and an urban data ecosystem in general toward achieving the SDGs. New paradigms, tools, technologies, and platforms are beginning to integrate local, regional, and global resources, including technical, human, and financial, for slum mapping. This could redefine the future of SDG monitoring and implementation. The long-term outcomes of networking for upscaling will increase funding for knowledge production, global awareness about slums, and best practices in slum data observatory, among others.

8.5.4 Science for Society

Transdisciplinary research aims to identify and implement solutions for difficult sustainability issues and to foster social learning. The SCiLeD project brought together a multidisciplinary team which comprised natural and social scientists with specific knowledge and skills in social and environmental issues from two different countries and multi-sectoral stakeholders with diverse socio-cultural backgrounds. The short-term outcomes of science for society include an increased role for communities, the design of local solutions, improved usability of local knowledge, and evidence for the utility of citizen science for SDGs. This leads to the long-term outcomes of evolving a fair and equitable society and of doing research for societal impacts.

8.5.5 Science for Policy

Transdisciplinary research aims to bridge the recognized gap between scientists and policy actors through the involvement of non-scientific partners in research projects. The SCiLeD project team comprised experts and practitioners from the government organizations in Accra and Lagos who participated in four interrelated activities. These include (1) initiating, framing, and designing of the transdisciplinary process such as forming the team, framing the research problem, and co-designing the methods; (2) identification, mapping, and engagement with relevant stakeholders; and (3) selection of suitable tools and procedures for collection and uptake of knowledge. The short-term outcomes include effective decision-making and timely response and action, and the long-term outcomes include equitable allocation of resources, effective tools for slum revitalization, and data-driven solutions.

8.6 Key Lessons Learned

The transdisciplinary research approach used in the SCiLeD project instilled a culture of collaboration, openness, and trust among participants. It emphasized the importance of data about people and place while empowering the community to tell their own story. The project demonstrated that the coming together of governments and slum communities to create solutions is a possibility. On the flip side, bringing together diverse communities of data producers and users is not without its challenges. These include challenges around harmonizing different perspectives, priorities, mandates, and methods, which led to delays in decision-making and project time overrun.

Clearly, everyone must act to improve the living conditions and quality of life in slums and informal communities. The importance of inclusive and participatory approaches to finding solutions cannot be overemphasized. The experiences in SCiLeD underscore the challenge of multiple partners and their divergent perspectives, highlighting the difficulty in synthesizing the various perspectives and priorities which emerge when bringing together different stakeholders. It also creates a platform for productive debate on urban issues, generates ideas on alternative approaches to urban data governance, and strengthens science for policy and science for society research collaborations. Furthermore, it opens a new horizon for inclusive discussion and reimagining of urban data governance as key components of urban governance especially in ensuring safe, resilient, and sustainable cities and communities in Africa.

8.7 Agenda 2030 in Africa: Challenges and Opportunities for Research and Practice

The case studies of localization of SDG 11 in Lagos and Accra clearly imply important challenges and opportunities for research and practice.

8.7.1 Connecting Stakeholders and SDG Data Systems

One of the main challenges observed in the case study above, and in practical SDG data initiatives, is the challenge of connecting different stakeholders, their perspectives, and various data generation processes and systems. In response to these challenges, the IDEAMAPS (Integrated Deprived Area Mapping System) network was launched in 2020 with funding from a UK Research and Innovation (UKRI) grant. This was a departure from the siloed approaches to SDG data generation which include aggregated surveys, field-based mapping, manually digitized imagery, and machine-learning computer modeling. The network is piloting in three African cities: Accra, Lagos, and Nairobi. Accra and Lagos were included in the network so as to build on the gains from the SCiLeD project, including established institutional collaborations (between the Universities of Lagos and Ghana), trained slum mappers, established presence in local communities, proven skills of technology-driven slum profiling, cordial relationships with city government officials, and an active group of student volunteers. Nairobi was selected primarily as part of the process of building collaboration among African cities in the UK Research and Innovation (UKRI) grant. We partnered with the African Population and Health Research Center (APHRC) in Nairobi, where the principal investigator of the IDEAMAPS project is based, and local slum profiling and community mappers in Kibera, Nairobi, one of the largest slums in Africa, thereby leveraging their vast knowledge and experience.

The rationale for the network derives from the need to further harmonize several official statistics for planning and decision-making, which are usually aggregated at the city, district, or national level, whereas data from censuses, household surveys, vital registration, and health records are available at the individual or household level. Data from Earth observation, Geographic Information System (GIS), field observation of small areas, and big data may also be found at the neighborhood level. Despite the usefulness of these different sources of data for specific purposes, they are not well integrated.

The novel approach which is being developed by IDEAMAPS harnesses the various sources of data by bringing together neighborhood, city/regional, and national stakeholders to collaboratively collect data at the neighborhood level on economic and environmental data, using an interactive interface to depict the degree of deprivation and to support advocacy, planning and investment, and monitoring of SDGs (Thomson et al. 2020). As such, the IDEAMAPS project aims to produce integrated data for community advocacy, urban planning, and SDG monitoring which can assist various stakeholders to do their work more efficiently. Through the project, community stakeholders should be able to access timely, relevant information about their neighborhood(s) that would otherwise be inaccessible so that they can achieve tenure security, improve connectivity to core urban centers, negotiate with city government about their needs and aspirations, and initiate self-improvement projects. The IDEAMAPS data can also support national governments to effectively set priorities and accurately track and monitor SDG 11. Similarly, the data can help local governments understand the spatial extents of slums, informal settlements, and other deprived areas so that the whole city is integrated, well planned, and served. Other stakeholders including scientists, businesses, and civil society groups can make use of the IDEAMAPS data to monitor the outcome of development policy and programs on different aspects of societal challenges including health, education, and housing.

8.7.2 Citizen-Generated Data for SDG Monitoring and Empowerment

There is an increasing recognition of the critical contribution of citizen science data or simply citizen-generated data to researchers and, policy actors, and for SDGs (de Sherbinin et al. 2021; Fraisl et al. 2020; Fritz et al. 2019). Using the traditional data sources to measure the 244 SDG indicators may create barriers for developing countries to monitor and track progress toward the SDGs due to the low data availability and high costs of traditional data production (Fraisl et al. 2020; Bowser et al. 2020). According to the Global Partnership for Sustainable Development Data (GPSDD 2016), the cost for 144 developing countries to generate data for SDG indicators (in the so-called Tiers I and II) is estimated at US$ 283 billion per year up to 2030. Citizen-generated data have the potential of being available at higher frequencies and spatial disaggregation at a lower cost and with additional benefits. Citizen-generated data have been described as public participation in scientific research, since it can involve five dimensions of participation: contract, contribution, collaboration, co-creation, and collegiality (Shirk et al. 2012). Some of the common approaches to citizen-generated data in the literature include community-based participation (Asaba and Suarez-Balcazar 2018); community and participatory mapping (de Albuquerque et al. 2019); participatory sensing (Coulson et al. 2018); citizen sensing (de Albuquerque and de Almeida 2020); crowd sourcing (Howe 2006; Nov et al. 2010); community-based monitoring (Conrad and Hilchey 2011); volunteered geographic information (Sieber and Haklay 2015); and citizen-generated data (Datashift 2017). Although the value of citizen-generated data has been recognized, there has not been enough empirical validation of citizen contributions to SDG monitoring at an indicator level.

As we have clearly observed in our case studies in Lagos and Accra, citizen-generated data and citizen science projects are important initiatives helping to encourage community-based participation in SDG data collection, stewardship, and sharing. Community volunteers are getting more excited about the role of data in addressing community challenges by negotiating with city authorities using local evidence as well as for implementing and monitoring SDGs. Community-based participation in data generation has the potential not only to democratize data production but also to enhance informed decision-making. Community-led initiatives can thereby provide local evidence about their needs and aspirations which can be used to assist governments to make decisions. In this manner, community-led data generation would be able not only to give visibility to the local situation and priorities of communities but also to align this with SDG indicators which are disaggregated at the neighborhood level, thus helping to implement the principle to "leave no one behind."

For this vision to work though, it will be important to consider carefully how communities are engaged in SDG data generation and usage (de Albuquerque and de Almeida 2020). Citizen data generation can be as much an opportunity for closing existing data gaps, as well as for empowering communities to redefine which aspects of their socio-ecological environments should count as data for SDG monitoring. A frequently forgotten issue when discussing the localization of SDGs is that the definition of what counts as data for a given SDG indicator may depend on the context and is frequently open for varying interpretations (Ulbrich et al. 2019). In this manner, in addition to the important aspect of strengthening the institutional capacity of statistical offices in African countries to generate and use SDG data, emphasized by actors such as the Global Partnership for Sustainable Development Data, we contend that it will be imperative to think of capacity building for SDG data in broader terms.

Local capacities in citizen-generated data initiatives should be recognized, as well as institutionalized and integrated into government statistics and records. This can only be accomplished by creating enabling political and policy environments for community-led initiatives which integrate them into the co-production of processes of SDG localization meaningfully (Coaffee et al. 2021; Croese et al. 2021). This implies the need to look for new approaches to dialogue with communities and recognize their particular interests and worldviews instead of instrumentalizing them as "blind" sensors that generate data that is only relevant and intelligible for others (De Albuquerque and de Almeida 2020). Last but not the least, to truly democratize SDG data, we must also decolonize funding for research to ensure that sustainable development data definitions and processes are generated in the global South.

8.8 Conclusion

Agenda 2030 represents a set of ambitious goals which define our collective development aspirations. The aspiration of SDG 11—to ensure safe, inclusive, resilient, and sustainable cities and communities—sits well with the priorities and aspirations of African cities, where rapid urban population growth and urbanization are stretching governance capacity. As such, cities remain the epicenter for the intensification and management of global challenges which are threatening the achievement of the SDGs in Africa.

With less than 10 years left to tackle poverty and inequality amid the escalating impact of the COVID-19 pandemic, climate change, and economic recession, it is imperative to align with the philosophy and principles of SDG 11. The UN Decade for Action hopes to mobilize everyone, everywhere; demand urgency and ambition; and supercharge ideas to find solutions. To achieve this, there is a need to mobilize and partner with governments, businesses, civil society, and communities in knowledge co-design and co-production. Given the fact that existing data in Africa are highly aggregated, largely unreliable, mostly inaccessible, and generally nonspatial, we must move on to a new urban data governance paradigm which makes data Findable, Accessible, Interoperable, and Reusable (FAIR).

The transdisciplinary approach, and the activities used in the SCiLeD project and by extension the IDEAMAPS project, has demonstrated the possibility of identification, mapping, mobilization, and collaboration of various stakeholders in diverse sectors, disciplines, and cities in Africa. The project has also harnessed the strengths of different stakeholders to overcome their weaknesses toward standardizing city-level data-gathering for achieving SDG 11 in Africa. The project further resolved the lack of trust and transparency which have characterized urban data governance in Africa by emphasizing how to simplify existing data ecosystems and value chains among diverse sectors, disciplines, policy actors, civil societies, and the slum communities. The transdisciplinary approach to data collection and governance as exemplified in the SCiLeD project has further generated data that is disaggregated to the neighborhood level by involving community mappers and profilers. The data therefore reflects and responds to local needs and realities and, because it is accessible, has facilitated the use of data for empowering neighborhoods and places to contribute toward specific interventions that are inclusive, innovative, and sustainable. In this way, data is not only useful for tracking and monitoring progress but also for exhibiting the transformative influence and potential toward achieving SDG 11 in African cities by 2030.

Acknowledgments The research was supported by the LIRA 2030 Africa Programme, implemented by the International Science Council (ISC) in partnership with the Network of African Science Academies (NASAC) with support from the Swedish International Development Cooperation Agency (SIDA). The authors also acknowledge support from UK Research and Innovation GCRF Digital Innovation for Development in Africa for Network Fund. The views expressed herein do not necessarily represent those of the funders.

References

Akinyele RT (2014) Nigeria: contesting for space, identity and security. Connell and Rex Charles Publications, Ibadan

Asaba E, Suarez-Balcazar Y (2018) Participatory research: a promising approach to promote meaningful engagement. Scandinavian Journal of Occupational Therapy 25(5):309-312

Batty M (2013) Building a science of cities. In: The new science of cities. MIT Press, Cambridge, Massachusetts, pp 13-45

Bowser A, Cooper C, de Sherbinin A, Wiggins A, Brenton P, Chuang TR, Faustman E, Haklay M et al. (2020) Still in need of norms: the state of the data in citizen science. Citizen Science: Theory and Practice 5 (1):303

Coaffee J, de Albuquerque JP, Pitidis V (2021) Risk and resilience management in co-production. In: Loeffler E, Bovaird T (eds) The Palgrave handbook of co-production of public services and outcomes. Palgrave Macmillan, Cham, pp 541-558

Conrad CC, Hilchey KG (2011) A review of citizen science and community-based environmental monitoring: issues and opportunities. Environmental Monitoring and Assessment 176:273–291

Corbett J, Mellouli S (2017) Winning the SDG battle in cities: how an integrated information ecosystem can contribute to the achievement of the 2030 Sustainable Development Goals. Information Systems Journal 27(4):427–461

Coulson S, Woods M, Scott M, Hemment D, Balestrini M (2018) Stop the noise! Enhancing meaningfulness in participatory sensing with community level indicators. In: Proceedings of the 2018 Designing Interactive Systems Conference. Association for Computing Machinery (ACM), pp 1183-1192

Croese S, Massamba D, Raimundo IM (2021) Co-producing urban knowledge in Angola and Maputo towards meeting SDG 11. npj Urban Sustainability. doi:https://doi.org/10.1038/s42949-020-00006-6

Datashift (2017) Using citizen-generated data to monitor the SDGs. A tool for the GPSDD data revolution roadmaps toolkit. https://www.data 4sdgs.org/resources/making-use-citizen-generated-data. Accessed 27 August 2021

de Albuquerque JP, de Almeida AA (2020) Modes of engagement: reframing "sensing" and data generation in citizen science for empowering relationships. In: Davies T, Mah A (eds) Toxic truths: environmental justice and citizen science in a post-truth age. Manchester University Press, Manchester, pp 267–281

de Albuquerque JP, Yeboah G, Ulbrich P, Pitidis V (2019) Towards a participatory methodology for community data generation to analyse urban health inequalities: a multi-country case study. Proceedings of the 52nd Hawaiian Conference on System Sciences, 2019

de Sherbinin A, Bowser A, Chuang T-R, Cooper C, Danielsen F, Edmunds R, Elias P, Faustman E et al. (2021) The critical importance of citizen science data. Frontiers in Climate 3:650760

Filani MO (2012) The changing face of Lagos: from vision to reform and transformation. https://www.citiesalliance.org/sites/default/files/Lagos-reform-report-lowres.pdf. Accessed 15 October 2021

Fraisl D, Campbell J, See L, When U, Wardlaw J, Gold M, Moorthy I, Arias R et al. (2020) Mapping citizen science contributions to the UN Sustainable Development Goals. Sustainability Science 15:1735–1751

Fritz S, See L, Carlson T, Haklay M, Oliver JL, Fraisl D, Mondardini R, Brocklehurst M, Shanley LA, Schade S et al. (2019) Citizen science and the United Nations Sustainable Development Goals. Nature Sustainability 2(10):922-930

Global Partnership for Sustainable Development Data (GPSDD) (2016) The State of development data funding 2016 Report. https://www.data4sdgs.org/resources/state-development-data-funding. Accessed 1 March 2021

Howe J (2006) The rise of crowdsourcing. Wired Magazine 14(6). http://www.wired.com/wired/archive/14.06/crowds_pr.html. Accessed 27 August 2021

Kitchin R (2014) Conceptualising data. In: The data revolution: big data, open data, data infrastructures and their consequences. Sage, London, pp 1–26

Lang DJ, Wiek A, Bergmann M, Stauffacher M, Martens P, Moll P, Swilling M, Thomas CJ (2012) Transdisciplinary research in sustainability science: practice, principles, and challenges. Sustainability Science 7:25-43

Lilford R, Kyobutungi C, Ndugwa R, Sartori J, Watson SI, Sliuzas R, Kuffer M, Hofer T et al. (2019) Because space matters: conceptual framework to help distinguish slum from non-slum urban areas. British Medical Journal Global Health 4(2):e001267

Macaya JFM, Alves AF, de Albuquerque JP, Cunha MA (2020) Digital-by-default: exclusion through digital public service channels. Proceedings of Conf-IRM 2020

Moser SC (2016) Can science on transformation transform science? Lessons from co-design. Current Opinion in Environmental Sustainability 20:106–115

Nigerian Population Commission (2006) Nigeria Population and Housing Census. https://nationalpopulation.gov.ng. Accessed 15 October 2021

Nov O, Arazy O, Anderson D (2010) Crowdsourcing for science: understanding and enhancing SciSourcing contribution. Position paper: ACM CSCW 2010 workshop on the changing dynamics of scientific collaborations. https://www.sci.utah.edu/images/Workshops/cscw2010/nov_arazy_anderson.pdf. Accessed 27 August 2021

Osborne P (2015) Problematizing disciplinarity, transdisciplinary problematics. Theory, Culture & Society 32(5-6):3-35

Pappas IO, Mikalef P, Giannakos MN, Krogstie J, Lekakos G (2018) Big data and business analytics ecosystems: paving the way towards digital transformation and sustainable societies. Information Systems and E-Business Management 16(3):479–491

Przeybilovicz E, Cunha MA, Macaya JFM, de Albuquerque JP (2018) A tale of two "smart cities": investigating the echoes of new public management and governance discourses in smart city projects in Brazil. Proceedings of the 51st Hawaiian Conference on System Sciences, pp 1–10

Restrepo-Estrada C, De Andrade SC, Abe N, Fava MC, Mendiondo EM, de Albuquerque JP (2018) Geo-social media as a proxy for hydrometeorological data for streamflow estimation and to improve flood monitoring. Computers & Geosciences 111:148–158

Sachs JD, Schmidt-Traub G, Mazzucato M, Messner D, Nakicenovic N, Rockström J (2019) Six transformations to achieve the Sustainable Development Goals. Nature Sustainability 2(9):805–814

Shirk JL, Ballard HL, Wilderman CC, Phillips T, Wiggins A, Jordan R, McCallie E, Minarchek M et al. (2012) Public participation in scientific research: a framework for deliberate design. Ecology and Society 17 (2):29

Sieber RE, Haklay M (2015) The epistemology(s) of volunteered geographic information: a critique. Geo Geography and Environment 2(2):122-136

Songsore J, Stephens C (2008) The Accra waste project: from urban poverty and health to aid and trade. Ghana Geographical Association, Accra

Thomson DR, Kuffer M, Boo G, Hati B, Grippa T, Elsey H, Linard C, Mahabir R (2020) Need for an integrated deprived area "slum" mapping system (IDEAMAPS) in low-and middle-income countries (LMICS). Social Sciences 9(5):80

Townsend AM (2013) Smart cities: big data, civic hackers, and the quest for a new utopia. WW Norton & Company, New York

Ulbrich P, de Albuquerque JP, Coaffee J (2019) The impact of urban inequalities on monitoring progress towards the Sustainable Development Goals: methodological considerations. ISPRS International Journal of Geo-Information 8(1):6

United Nations (UN) (2020a) The Sustainable Development Goals report 2020. https://unstats.un.org/sdgs/report/2020/The-Sustainable-Development-Goals-Report-2020.pdf. Accessed 15 October 2021

United Nations (UN) (2020b) UN research roadmap for the COVID-19 recovery. Leveraging the power of science for a more equitable, resilient and sustainable future. https://www.un.org/en/pdfs/UNCOVID19ResearchRoadmap.pdf. Accessed 15 October 2021

United Nations (UN) and Independent Expert Advisory Group on a Data Revolution for Sustainable Development (2014) A world that counts: mobilising the data revolution for Sustainable Development. http://www.undatarevolution.org/report. Accessed 15 October 2021

Wilkinson MD, Dumontier M, Aalbersberg IJ, Appleton G, Axton M, Baak A, Blomberg N, Boiten J-W et al. (2016) The FAIR guiding principles for scientific data management and stewardship. Scientific Data 3:160018

9 Fiscal Decentralization, Comparative Data, and Sustainable Development: What Do We (Need To) Know About Financing Subnational Governments in Africa?

Liza Rose Cirolia, Jennifer Van Geesbergen, and Sylvia Croese

Abstract

Central to the successful implementation of the Sustainable Development Goals (SDGs) in urban Africa will be the establishment of robust multilevel governance and public finance systems that can be responsive to local context. There have been numerous decentralization reforms across Africa over the past 30 years. However, in many countries, the fiscal architecture of multilevel government and the fiscal empowerment of subnational governments remain uneven and contested. Drawing on subnational fiscal data on 21 African countries, this chapter explores subnational fiscal decentralization and empowerment in Africa. This chapter makes three arguments. First, due to the incredible local diversity across the continent, there is limited comparative data on subnational finance in Africa. Second, the available data foreground several challenges, including limited fiscal transparency and different fiscal accounting and reporting styles, meaning that existing fiscal data sets need to be considered within their individual contexts. Finally, there is considerable conflation between "local government" and "city government" with serious implications for fiscal decentralization that is tailored to the specific needs of rural, urban, and metropolitan areas. Building on these findings, the chapter argues that the prospects for financing sustainable development in Africa will depend on a better understanding of the complexities of the local fiscal space.

Keywords

Public finance · Subnational fiscal data · Fiscal decentralization · Multilevel governance

9.1 Introduction

The Sustainable Development Goals (SDGs) require significant levels of finance to be mobilized to meet development objectives. The

L. R. Cirolia (✉)
African Centre for Cities, University of Cape Town, Cape Town, South Africa
e-mail: liza.cirolia@uct.ac.za

J. Van Geesbergen
Independent Researcher/Consultant, Paris, France
e-mail: jennifer.van.geesbergen@gmail.com

S. Croese
University of the Witwatersrand, Johannesburg, South Africa

University of Cape Town, Cape Town, South Africa
e-mail: sylvia.croese@wits.ac.za

S. Croese, S. Parnell (eds.), *Localizing the SDGs in African Cities*, Sustainable Development Goals Series, https://doi.org/10.1007/978-3-030-95979-1_9

United Nations Conference on Trade and Development (UNCTAD) puts the global investment needs to meet the SDGs at between US$ 5 trillion and US$ 7 trillion per year. For developing countries alone, the investment needs range from US$ 3.3 trillion to US$ 4.5 trillion. This includes finance for basic infrastructure such as roads, rail and ports, power stations, and water and sanitation. It further extends to investment required to address food security such as agriculture and rural development, climate change mitigation and adaptation, health, and education (UNCTAD 2014). The economic and financial shocks associated with COVID-19 have further increased the levels of finance required for the achievement of the SDGs by 2030, especially at the local government level (UN 2020; OECD 2020).

Most suggested measures to address the SDG finance gap focus on ways to mobilize or increase sources of funding, such as Official Development Assistance (ODA), Foreign Direct Investment (FDI), and other private funding and domestic revenues (Move Humanity 2018), even if these have been declining (UN 2020). However, governments have the mandate and provide the institutional apparatus for the SDGs to be implemented. Public finance should therefore be central to debates on SDG implementation. Moreover, considering the importance of local governments in achieving the SDGs in urban contexts, such debates should consider the importance of subnational government finance.

While global agendas recognize the importance of local governments, most overlook the issue of local finance. For example, Agenda 2030 only makes reference to finance as part of SDG 17.1, which focuses on the need to strengthen domestic resource mobilization, including through international support to developing countries, to improve domestic capacity for tax and other revenue collection. Meanwhile, SDG 17.2 and 17.3 link to ODS and FDI levels. Similarly, the Addis Ababa Action Agenda (AAAA), the most important international framework dedicated to development finance,[1] makes no mention of cities at all and only has one mention of subnational governments.

The New Urban Agenda (NUA) is the only global agenda that provides suggestions for urban local government financing; however, these provisions are very general and fail to provide robust guidance (UN 2017).[2] Recent UN reports have started to include more attention to subnational financing, especially when it comes to infrastructure finance in contexts of rapid urban growth (UN 2018). However, action areas that are identified for subnational financing are generic and do not take the specificity of existing fiscal systems into account.

In most African countries, the public finance apparatus—through which funding for the SDGs is meant to flow—remains a work in progress. Reforms implemented through the decolonization period, and later structural adjustment, have laid a patchy groundwork for fiscal governance. In many countries, reforms are still underway, with powers and functions shifting, at times with and at times without the resources needed. These reforms are overlaid on dynamic processes of urbanization, wherein metropolitan regions, cities, and towns are growing and placing new pressures on local states. In order to assess the prospects for SDG financing in African cities, we must have a better understanding of the workings and dynamics of the local fiscal space.

In this chapter, we therefore draw attention to the importance of local government finance and raise questions about the ways in which finance flows, who controls it, and how much of it there actually is, in order to highlight the importance of

[1] The AAAA is the outcome document of the Third International Conference on Financing for Development which builds on two previous conferences that took place in 2002 in Monterrey and in 2008 in Doha; see Engberg-Pedersen (2016).

[2] They include attracting private sector and commercial investments, creating supportive frameworks for subnational borrowing, making use of land-based financing, and fiscal decentralization (see, e.g., article 15 (a) and (c) (iv)).

fiscal decentralization for SDG implementation in African cities.

9.2 Why Decentralization Requires Effective Multilevel Governance

Global development agreements generally represent the ambit of national governments, as the main signatories and entities responsible for implementation. However, many of the areas that development agendas address—such as water or health—have been devolved to subnational levels of government. Subnational government refers to all of the levels or tiers of government below the national. In many contexts, this includes regional governments, such as states or provinces, and local municipal governments. As such, subnational levels of government, despite their often-limited role in global forums, are central to the effective implementation of global agendas. An important part of localizing the SDGs, therefore, is establishing robust multilevel governance systems that can allow for different levels of government to play their role.

A key part of multilevel governance systems is determining what different levels of government can and cannot do, what they are and are not responsible for, and how the relationships between them will operate. As countries are internally differentiated, another important part of designing these systems is addressing the diverse needs and capabilities of different types of areas and by extension those authorities responsible for them. There are several interrelated concepts which together have produced a strong discourse around the most appropriate assignment of functions, political powers, and resources to subnational governments (Smoke 2001; Bird and Vaillancourt 2008). The first significant concept is the principle of subsidiarity (Moeti et al. 2007). Subsidiarity argues that public goods or service provisions should be assigned to the smallest or lowest level of government, granted that this is practical and that economies of scale are still possible (Boadway and Shah 2007). Subsidiarity is closely aligned with the neo-classical "decentralization theorem" which argues that assignment should be made to the lowest level at which provision is efficient and public good maximized (Oates 2008; Buchanan 1989). The classical economic case for fiscal decentralization is predicated on the assumption that welfare gains are maximized when decision-making is brought closer to constituents (Oates 2008).

The decentralization of functions, political power, and fiscal resources to subnational levels of government is key to this agenda. Decentralization refers to the process of shifting powers from central governments to subnational levels of governments. The decentralization of functions, political power, and resources is widely assumed to be a precondition for achieving good governance in developing countries (Faguet 2014). In theory, decentralization can achieve more accountable and efficient allocation of resources that respond to local needs and priorities, combatting top-down decision-making and corruption. As a development mantra, it has been pushed by multilateral institutions, think-tanks, and policy-oriented scholars (UN-Habitat 2009; Lincoln Institute and World Bank 2016). It features in key United Nations agendas including the AAAA, Agenda 2030, and NUA, all of which articulate the importance of empowering subnational governments to enable them to play more central roles in development processes (UN 2015, 2017).

9.3 Urban Africa's Multilevel Government Experience

There is much to debate regarding the applicability of these neo-classical arguments and their applicability to the African context. As the wider debates on decentralization have progressed, many questions have been raised regarding the assumptions that underpin these theories. Regardless, decentralization has been aggressively pursued in Africa since the 1980s, including in both political and fiscal decentralization reforms (Ribot 2002; Smoke 2003; Tanzi 2016; Farvacque-Vitkovic and Godin 1998).

Dovetailing with the Structural Adjustment of African economies, decentralization formed part of a suite of reforms that focused on "institution-building" and "good governance" (Becker et al. 1994; Clarke Annez et al. 2008). In some African countries, reforms were supported by central states in order to access debt relief. In other cases, decentralization allowed states to "neutralize regional ethnic tensions," offering fiscal, administrative, or political power to regional elites and curbing opposition (e.g., in South Africa, Ethiopia, and more recently Kenya) (Smoke 2003, p. 12).

Decentralization reforms have reshaped African countries, shifting powers, functions, and resources away from central governments. However, these reforms, both political and fiscal, have been hotly contested. In many cases, reforms have been resisted by national governments who fear the loss of both fiscal and political control (Wunsch 2001). Africa's capital cities have been a particularly important site for this resistance, as has been apparent in cities in metropolitan areas such as Gauteng and Dakar (Bekker and Therborn 2012). Private sector actors, such as construction companies and infrastructure lenders, have also resisted decentralization, preferring to do business with centralized technical agencies, rather than democratized local authorities. This is particularly true when it comes to large infrastructure projects that require coordination between several local governments and in cases where local governments are perceived to be heavily influenced by party politics. While great strides have been made in terms of democratic decentralization, the question of how money flows, and who controls and shapes budgets, is a vital part of understanding decentralization in practice. In other words, political decentralization and fiscal decentralization have not always gone hand in hand. Understanding the nature of the fiscal requires understanding more than just the legal and regulatory frameworks which underpin fiscal processes—it requires understanding how these operate in practice.

9.4 Method

Unpacking fiscal dynamics in Africa requires both understanding aggregates and comparisons across the continent, as well as the specificity of particular places. Both require contextualization of data, either among a family of cases or of the particular histories, arrangements, and development trajectories of countries and cities. For this purpose, this chapter consolidates and synthesizes material from several studies of subnational finance in Africa. The bulk of the insights are drawn from a study on subnational government in Africa, which collected data on 21 African countries.

The 21 countries, listed in Table 9.1, fall into different income groups and population sizes and were selected due to their availability of fiscal data (see Table 9.1, columns 2 and 3). The fiscal data was collected for 2016 as this was the year with the most complete data set in the year of collection. The authors for this chapter were integrally involved with the data collection process. The raw data was collected from enacted government budgets, central banks documents, public finance transparency forums, and organizations such as the International Budget Partnership (IBP), the Commonwealth Local Government Forum (CLGF), and United Cities and Local Governments of Africa (UCLGA) and using International Monetary Fund (IMF), UNICEF, and World Bank reports (when official government budgets were unavailable) (see Table 9.1, column 5). The collected data was standardized through a framework developed as part of a multi-year fiscal data observatory. For several countries, including South Africa, Nigeria, Ethiopia, and Namibia, more detailed fiscal breakdowns were available and have been used to provide nuance. The raw data is available online as part of the United Cities and Local Governments (UCLG) and Organization for Economic Cooperation and Development (OECD) as part of the OECD-UCLG World Observatory on Subnational Government Finance and Investment (OECD and UCLG 2019a).

Table 9.1 Overview of fiscal data for 21 African countries (Compiled by authors. Income categorization based on the World Bank Atlas method, published by the World Bank, using 2018 data. Population is based on the United Nations Department of Economic and Social Affairs (UNDESA), World Population Prospects, 2016 data. Country data based on available Fiscal Year (FY) and exchange rates in Purchasing Power Parity (PPP))

Country name	Income group (2018)	Population (2016)	Exchange rate in US$ PPP (2016)	Key sources of data used
Angola (FY 2017)	Lower middle income	29,816,748	84.07	Fiscal data: Ministério das Finanças (2018) Orçamento Geral do Estado//OECD/UCLG (2016) Subnational Governments Around the World Structure and Finance: A First Contribution to the Global Observatory on Local Finances//UNICEF (2018) National Budget Report. Other sources of information: Cities Alliance and UCLGA (2018) Assessing the institutional environment of local governments in Africa//SAPO 24 (2018) Angola Begins This Year to Decentralize Skills and Resources to Municipalities//Aalen, L. and Muriaas, R. L. (2017) Power calculations and political decentralisation in African post-conflict states//UNDP in Angola (2008) Decentralization and Local Governance//Election Calendar (2018) Electoral Institute for Sustainable Democracy in Africa//National Democratic Institute for International Affairs (1996)
Benin (FY 2016)	Low income	11,175,204	0.36	Fiscal data: Local Finance Department of the Ministry of Economy and Finance//National Commission on Local Finance (2018) Les finances locales du Bénin 2016//Ministry of Economy and Finance (2019) Projet de loi de finances 2019—document de programmation budgétaire et économique pluriannuelle 2019–2021. Other sources of information: Annual reports of the sectoral reviews 'Decentralization, devolution and spatial planning'//OECD (2017) African Economic Outlook 2017//UCLGA and Cities Alliance (2018) Assessing the Institutional Environment of Local Governments in Africa
Botswana (FY 2016)	Upper middle income	2,205,128	4.47	Fiscal data: Statistics Botswana (2017) National Accounts Statistics Report 2015//CLGF—Botswana country profile 2017/18//IMF (2017) Report on public investment. Other sources of information: UCLGA and Cities Alliance (2018) Assessing the Institutional Environment of Local Governments in Africa//Ministry of Finance and Development Planning of Botswana—web portal// Ministry of Local Government of Botswana—web portal//Kampamba, J., Leima, S. and Svensson, A. (2016) A comparative analysis of residential property tax assessment in Botswana and Sweden
Burundi (FY 2016)	Low income	10,827,024	627.57	Fiscal data: World Bank//Burundian Institute of Statistics and Economic studies—ISTEEBU (2017) Annuaire statistique du Burundi 2016//Bank of the Republic of Burundi (2016) Rapport annuel de la Banque de la République du Burundi. Other sources of information: World Bank (2014) Burundi Fiscal decentralization and local governance: managing trade-offs to promote sustainable reforms//Nicaise, Guillaume (2015) Évaluation de la performance des collectivités décentralisées: une étude comparée entre le Rwanda et le Burundi//UCLGA and Cities Alliance (2018) Assessing the Institutional Environment of Local Governments in Africa//Weneger, Alexander (2018) Les finances locales dans l'Afrique francophone subsaharienne: une étude comparative sur dix-huit pays et leurs systèmes des finances locales

(continued)

Table 9.1 (continued)

Country name	Income group (2018)	Population (2016)	Exchange rate in US$ PPP (2016)	Key sources of data used
Cabo Verde (FY 2016)	Lower middle income	537,497	46.89	Fiscal data: World Bank. Ministry of Finance Conta geral do Estado 2016. Other sources of information: African Development Bank (2018) Programme d'amélioration de la compétitivité du secteur privé et de développement de L'economie locale (PSC-LED) Phase I//African Development Bank (2018) Note pays Cabo Verde 2018//UCLGA and Cities Alliance (2018) Assessing the institutional environment of local governments in Africa//Judite Nascimento and Rui Jacinto (2015)//Villes et conditions de vie urbaine au Cap Vert
Eswatini (FY 2014 for expenditure and FY 2016 for revenue)	Lower middle income	1,124,753	4.89	Fiscal data: Ministry of Finance (2015) Swaziland Budget Speech//OECD (2016) Revenue Statistics//Ministry of Finance Open Data. Other sources of information: UCLGA and Cities Alliance (2018) Assessing the institutional environment of local governments in Africa//Local Authorities Managers Association of Swaziland—LAMAS//Ministry of Housing and Urban Development//Ministry of Tinkhundla Administration and Development//CLGF (2015) The Local Government System in Swaziland: Country Profile 2017–18
Ethiopia (FY 2015 and 2016)	Low income	106,400,024	8.68	Fiscal data: World Bank (2016) Ethiopia Public Expenditure Review//UNICEF (2017) Budget Brief, Ethiopia Other sources of information: UCLGA and Cities Alliance (2018) Assessing the Institutional Environment of Local Governments in Africa//UNDESA (2017) Ethiopia Voluntary National Review//Tilahun, M. (2014) Local Government in Ethiopia: Practices and Challenges. Journal of Management Science and Practice//Ayele, Z. A. (2014) The politics of sub-national constitutions and local government in Ethiopia. Perspectives on Federalism//Gebre-Egziabher, T., and Dickovick, J. T. (2010) Comparative assessment of decentralization in Africa: Ethiopian desk study. USAID//Yesegat, W. A., and Krever, R. (2018) Subnational Value Added Tax in Ethiopia and Implications for States' Fiscal Capacity. Budget Brief, Ethiopia
Kenya (FY 2016)	Lower middle income	50,221,473	47.03	Fiscal data: IMF-GFS Data//The Office of the Controller of Budget of Kenya. Other sources of information: UCLGA and Cities Alliance (2018) Assessing the Institutional Environment of Local Governments in Africa//George Githinji (2019) A look at presidential system of government in Kenya//Commonwealth Local Government Forum (2017) The Local Government System in Kenya. Country Profile//UN Sustainable Development Knowledge Platform (2017) Kenyan Voluntary National Review 2017//Ministry of devolution and planning, Implementation of the Agenda 2030 for Sustainable Development in Kenya//UNDP (2016) National Urbanization Strategies//International Journal for Innovation Education and Research (2018) Effects on fiscal decentralization on poverty reduction in Kenya

Malawi (FY 2015/16)	Low income	17,670,260	184.48	Fiscal data: IMF Government Finance Statistics—Malawi//Ministry of Finance, Economic Planning and Development. Department of Economic Planning and Development. Annual Economic Report 2016/17//UCLGA (2016) The State of Local Economic Development in Malawi//World Bank (2016) Malawi Economic Monitor Harnessing the Urban Economy. Other sources of information: UCLGA and Cities Alliance (2018) Assessing the institutional environment of local governments in Africa//United Nations Malawi (2018) Sustainable Development Goals: Localisation, Opportunities and Challenges for Malawi//Malawi Electoral Commission//CLGF. The Local Government System in Malawi. Country Profile 2017//Yasin Maoni (2013) Decentralization and Local Development In Malawi: Demerits of The Public-Private Partnership Act, 2011//Dr. Asiyati Chiweza (2010) A Review of the Malawi Decentralization Process: Lessons from Selected Districts: A joint study of the ministry of local government and rural development and concern universal
Mauritius (FY 2016)	Upper middle income	1,264,613	16.31	Fiscal data: IMF Government Finance Statistics//Ministry of Finance and Economic Development. Other sources of information: CLGF (2017) Country Profile Mauritius//Ministry of Finance and Economic Development (2018) Public Sector Investment Programme 2016/17–2020/21
Morocco (FY 2016)	Lower middle income	35,581,294	3.61	Fiscal data: OECD Revenue Statistics//MEF, Trésorerie Générale du Royaume du Maroc (Décembre 2016): Bulletin Mensuel de Statistiques des Finances Publiques and Bulletin mensuel de statistiques des finances locales//MEF- Direction des Études et des Prévisions Financières (Juin 2017): Tableau de Bord des Finances Publiques. Other sources of information: OCDE (2018), Dialogue Maroc-OCDE sur les politiques de développement territorial: Enjeuxet Recommandations pour une action publique coordonnée//M. Elkhdari (2018) Deconcentration, political and fiscal decentralization, in Morocco//Fonds d'Équipement Communal FEC (2016): Rapport d'Activité 2016//A. Houdret and A. Harnisch (2017) Decentralisation in Morocco: The Current Reform and Its Possible Contribution to Political Liberalisation//OECD and UCLG (2016): Subnational Governments Around the World: Structure and Finance//Cour des Comptes du Maroc (Mai 2015): La Fiscalité Locale—Synthèse
Mozambique (FY 2016)	Low income	28,649,007	0.31	Fiscal data: MEF (2018) Relatório de Execução do Orçamento do Estado, 2017//MEF/TA (2017) Conta Geral do Estado, Ano 2016. Other sources of information: UCLGA and Cities Alliance (2018) Assessing the institutional environment of local governments in Africa//MAP Consultoria (for MEF) (2017) O Sistema de Transferências Fiscais Intergovernamentais em Moçambique//CLGF (2017) Mozambique Country Profile//USAID/ARD (2010) Comparative Assessment of Decentralization in Africa: Mozambique Country Assessment Report//Y-A Fauré and C. Udelsmann Rodrigues (2011) Descentralização e desenvolvimento local em Agola e Moçambique: Processos, terrenos e atores

(continued)

Table 9.1 (continued)

Country name	Income group (2018)	Population (2016)	Exchange rate in US$ PPP (2016)	Key sources of data used
Namibia (FY 2015/16)	Upper middle income	2,402,603	6.33	Fiscal data: Auditor General Audit report on the service delivery of regional councils in the National Assembly//Namibia Ministry of Finance (2017) Budget Statement: 2017/18//Namibia Statistics Agency. Annual National Accounts 2016. Other sources of information: CLGF. The Local Government System in Namibia. Country Profile 2017/18//UCLGA and Cities Alliance (2018) Assessing the Institutional Environment of Local Governments in Africa
Nigeria (FY 2016)	Lower middle income	190,873,311	94.09	Fiscal data: BudgIT civic organization (2017) State of State 2017 Report//Budget Office of the Federation//Revenue Mobilization Allocation and Fiscal Commission//Central Bank of Nigeria (2017) Statistical Bulletin: Public Finance Statistics. Other sources of information: Akpan H. Ekpo (2015) Issues in sub-national borrowing in Nigeria//The World Bank (2018) Nigeria Public Sector Governance Reform and Development Project//Amah, E. I. (2018) Devolution of Power to Local Government: Appraising Local Government Autonomy under Nigerian Federation. Beijing Law Review, 9, 275–293//UCLGA and Cities Alliance (2018) Assessing the Institutional Environment of Local Governments in Africa//Adejare, A. (2017) The Impact of Personal Income Tax on Government Expenditure in Oyo State//CLGF (2017) The Local Government System in Kenya. Country Profile 2017/18//Adams, P. (2016) State (s) of crisis: Sub-national government in Nigeria. Africa Research Institute//Social Development Integrated Centre//Oluwafemi I. Ajayi (2013) The Nigerian Bonds Market. Central Bank of Nigeria
Rwanda (FY 2016)	Low income	11,980,937	0.37	Fiscal data: IMF Government Finance Statistics Database//Rwanda Ministry of Finance and Economic Planning, 2016–2017 Budget Execution By Sectors//Rwanda Ministry of Finance and Economic Planning, 2016–2017 Earmarked Transfers Guidelines to Decentralized Entities//Rwanda Ministry of Finance and Economic Planning, The Annual Economic Report Fiscal Year 2016/2017//Rwanda Ministry of Finance and Economic Planning, Budget Framework Paper 2018/2019–2020/2021//Rwanda Ministry of Finance and Economic Planning, Updated Macro-Framework Public Dataset by End 2017 (as of 5th April 2018). Other sources of information: Center for Financial Markets (Melken Institute) (2015). Capital Markets in Rwanda: Assessment and Aspirations//CLGF. The Local Government System in Rwanda. Country Profile 2017–18//Embassy of the Kingdom of the Netherlands (2016) Evaluation of the LODA Program to Finance Infrastructure at De-central Level//IMF (2017) Staff Report for the 2017 Article IV Consultation, Seventh Review Under the Policy Support Instrument, and Second Review Under the Standby Credit Facility—Debt Sustainability Analysis Update//International Growth Center (2016) An effective property tax regime for Rwanda//PEFA Assessment Report of Rwanda (2016)//UCLGA and Cities Alliance (2018) Assessing the institutional environment of local governments in Africa

Senegal (FY 2017 for expenditure and revenue, FY 2016 for debt)	Low income	15,419,381	0.37	Fiscal data: Ministry of Economy, Finance and Planning (2018) Document de Programmation budgétaire et économique pluriannuelle (DPBEP) 2019–2021//General balance of consolidated accounts of local authorities. Other sources of information: Ministry of Local Government (2015) Lettre de Politique sectorielle de Développement de la Gouvernance locale, du Développement et de l'Aménagement du Territoire 2015–2020//World Bank (2018) Systematic Country Diagnostic of Senegal//World Bank Group (2017) Economy Profile of Senegal. Doing Business 2018//UCLGA and Cities Alliance (2018) Assessing the institutional environment of local governments in Africa
South Africa (FY 2016)	Upper middle income	57,000,451	5.86	Fiscal data: IMF-Government Finance Statistics//National Treasury (2016). Municipal Finance Data//Statistics South Africa (2017) Financial Statistics of Provincial Government 2015/16. Other sources of information: South African Local Government Association—SALGA (2015). 15 Years of Development and Democratic Local Government//SALGA (2016). About Municipalities//South African Government (2016) Local Government//CLGF (2013) The Local Government System in South Africa//Department of Cooperative Governance and Traditional Affairs (2014) The Integrated Urban Development Framework//UCLGA and Cities Alliance (2018) Assessing the institutional environment of local governments in Africa
Tanzania (FY 2016)	Low income	54,663,906	686.37	Fiscal data: World Bank (2018) The World Bank In Tanzania//Ministry of Finance and Planning (2017) The Budget Execution Report for the Fourth Quarter and Fiscal Year 2016/17 (July 2016–June 2017)//National Audit Office (2017) Annual General Report of the Controller and Auditor General on the Financial Statements for the Year Ended 30th June, 2017. Other sources of information: CLGF (2018) The Local Government System in Tanzania—Country Profile 2017–2018//UCLGA and Cities Alliance (2018) Assessing the institutional environment of local governments in Africa//PEFA (2016) Sub-national (Local Government) PEFA Assessment in Tanzania—Final Consolidated Report//National Bureau of Statistics (2012) Population and Housing Census//Kessy, A. (2011) Local Government Reforms in Tanzania: Bridging the Gap between Theory and Practice, in Democratic Transition in East Africa//Shadrack, B (2010) Local Government Authorities in Tanzania
Tunisia (FY 2016)	Lower middle income	11,433,443	0.68	Fiscal data: Dafflon B. and Gilbert G. (2018), L'économie politique et institutionnelle de la decentralisation en Tunisie, état des lieux, bilan et enjeux—AFD//Tunisian Ministry of Finance//OECD (2018) Subnational Government in OECD countries: key data (brochure and database). Other sources of information: Dafflon B. and Gilbert G. (2018) L'économie politique et institutionnelle de la decentralisation en Tunisie, état des lieux, bilan et enjeux—AFD//S. Yerkes and M. Muasher (2018) Decentralisation in Tunisia: Empowering Towns, Engaging People, Carnegie Endowment for International Peace//World Bank (2018) Tunisie—Evaluation PEFA 2015–2016//OECD (2018) OECD Economic Surveys: Tunisia 2018//Institut Tunisien de la Compétitivité et des Etudes Quantitatives (ITCEQ) and International Labour Organization (ILO) (2017) Tunisian government's 5-year plan//POMED project on Middle East Democracy (2018) Tunisia's Municipal Elections—The View from Tunis//Caisse des Prêts et de Soutien aux Collectivités Locales (2016) Etats financiers arrêtés au 31/12/2016 et extrait du rapport du commissaire aux comptes

(continued)

Table 9.1 (continued)

Country name	Income group (2018)	Population (2016)	Exchange rate in US$ PPP (2016)	Key sources of data used
Uganda (FY 2016)	Low income	41,162,465	0.32	Fiscal data: IMF Government Finance Statistics//Ministry of Finance, Planning and Economic Development, Annual Budget Performance Report, FY 2016/17. Other sources of information: UCLGA and Cities Alliance (2018) Assessing the institutional environment of local governments in Africa//CLGF. The Local Government System in Uganda. Country Profile 2017–18//Ministry of Local Government Factsheet (2017)//Auditor General (2016) Financing of Local Governments in Uganda Through Central Government Grants and Local Government Revenues//Local Public Sector Initiative, Country Profile: Uganda 2010–11//Local Government Finance Commission Review of Local Government Financing (2012) Management and Accountability for Decentralized Service Delivery
Zimbabwe (FY 2014)	Low income	14,236,745	0.51	Fiscal data: World Bank (2017) Local Government Service Delivery—Zimbabwe Public Expenditure Review//Zimbabwe National Statistics (2015) 2009–2015 National Accounts Report.Other sources of information: UCLGA and Cities Alliance (2018) Assessing the institutional environment of local governments in Africa//Chigwata, T.C. and de Visser, J. (2018) Local government in the 2013 constitution of Zimbabwe: Defining the boundaries of local autonomy//CLGF (2013) The constitution of Zimbabwe 2013 as a basis for local government transformation: A Reflection Analysis

9.5 The Fiscal Multilevel Governance Context in Africa

The rest of this chapter unpacks the findings from this work. We have arranged these findings into three key insights. First, we argue that there is limited comparative data on African countries decentralization. Ambitious efforts to consolidate data on subnational finance have usefully demonstrated the incredible diversity and heterogeneity of Africa's experience of fiscal decentralization. Second, these data exercises have foregrounded several challenges related to data and comparison, including limited fiscal transparency and different fiscal accounting and reporting styles. The implications are that such data sets need to be considered in context and triangulated with qualitative information. Finally, there is considerable conflation between local government and city government. This has serious implications for fiscal decentralization, which requires tailored approaches for rural, urban, and metropolitan areas.

9.5.1 Comparing Across Diversity

Ambitious efforts to consolidate data on subnational finance have usefully demonstrated the incredible diversity in African contexts. The data shows that African subnational governments today are highly heterogeneous (Paulais 2012; Clarke Annez et al. 2008). According to the UCLG data sets, there are over 15,000 recognized subnational governments across the 54 countries in Africa. Most African countries have emerged from decentralization reform processes with two levels of subnational government: the regional level and the municipal (local) level, which forms the lowest level of government (UCLG and Cities Alliance 2015). For example, Zimbabwe has two tiers, with 92 local councils at the local (municipal) level and 8 provinces and 2 metropolitan cities at regional level. There are also several cases that have three subnational levels. The level between the regional and the local is often referred to as the intermediate level. For example, Togo has a three-tier system of 5 regions, 39 prefectures at the intermediate level, and 116 municipalities at municipal level. Finally, some African countries only have one level of subnational government. In these cases, it is often hard to tell if this single level should be referred to as a local level or a regional level. For example, Malawi has only 35 councils, and Kenya has only 47 counties (OECD and UCLG 2019b). These differences in subnational architecture have significant impacts on the public finance systems and operations of countries, shaping how money flows.

The extent to which subnational governments control expenditure is an important indication of the level of fiscal decentralization. Table 9.2 outlines the subnational spending per capita, as a proportion of total government spend is divided between current and capital. Unfortunately, the UCLG data set did not differentiate between regional and local government spending for most countries. It is only possible to ascertain this for federal countries, such as South Africa, and countries with only one tier of subnational government (as it can thus be assumed that it is all local spend).

When comparing subnational expenditure levels per inhabitant (see Table 9.2, column 2), the highest level of subnational expenditure per capita is in South Africa. South Africa, however, is a severe outlier, as the next highest country is Botswana, with less than a quarter of the per capital spending of South Africa. The lowest per capita expenditure is in Burundi and Malawi. These extremes show the diversity across the continent. Notably, while there is no clear trend in terms of regions, the upper-middle-income countries have higher levels of subnational spending per capita. In the cases of Botswana and Mauritius, two of the higher spending countries, this can be attributed to a combination of the countries' economic wealth, their small population number (resulting in higher expenditure per capita), and the decentralization of responsibilities in the sectors of general public services, economic affairs and transportation, environmental protection, and culture and recreation. In South Africa, the high level of expenditure per

Table 9.2 Subnational expenditure data for selected countries in Africa (OECD and UCLG 2019c)

Country	Subnational government expenditure PPP per capita (in US$)	Subnational government expenditure as % of total government expenditure	Current expenditure as % of subnational government expenditure	Capital expenditure as % of subnational government expenditure
Angola	81	5.10%	92.60%	7.40%
Benin	29	6.00%	55.00%	44.80%
Botswana	579	10.20%	NO DATA	NO DATA
Burundi	6	3.90%	41.00%	5.60%
Cabo Verde	213	9.80%	74.00%	26.00%
Eswatini	60	2.20%	NO DATA	NO DATA
Ethiopia	152	48.00%	62.40%	37.60%
Kenya	133	15.00%	68.00%	32.10%
Malawi	7	3.70%	15.60%	NO DATA
Mauritius	320	6.10%	85.00%	15.00%
Morocco	273	11.80%	64.20%	35.80%
Mozambique	65	16.80%	NO DATA	1.60%
Namibia	286	6.40%	NO DATA	NO DATA
Nigeria	254	47.90%	76.30%	23.70%
Rwanda	107	18.80%	75.00%	25.00%
Senegal	37	3.90%	74.20%	25.80%
South Africa	2703	49.30%	90.70%	9.30%
Tanzania	40	17.90%	81.50%	18.50%
Tunisia	246	7.90%	40.40%	59.60%
Uganda	57	17.20%	91.60%	8.40%
Zimbabwe	49	9.00%	NO DATA	NO DATA

inhabitant is a direct result of the country's well-established and comprehensive legal fiscal framework at subnational levels and its overall wealth (although unequally distributed).

In terms of total spend, expenditures are highly centralized in Angola, Burundi, Senegal, Malawi, and Eswatini, where subnational expenditure share in the general government is equal to or below 5% (see Table 9.2, column 3). There are different explanations for this in the various countries. Angola and Eswatini experience highly centralized control of all government responsibilities and resources due to the current political regimes (although in Angola there appears to be a move towards allowing subnational governments greater control of their finances). In Malawi and Senegal, a lack of capacity to manage resources at the local level is largely the cause for their low levels of expenditure (OECD and UCLG 2019b). In the case of Burundi, the low levels of total subnational spending are mostly due to a lack of resource allocation from the central government, resulting in subnational governments not being able to fully implement and fulfill their devolved functions. According to a World Bank Public Expenditure Review, an adequate and transparent intergovernmental transfer system is yet to be defined and put in place by the Burundian government (World Bank 2014). In contrast to Nigeria, Ethiopia, and South Africa, subnational expenditure accounts for nearly half of all public expenditures (see Table 9.2, column 3). For Nigeria, the federal structure concentrates fiscal resources with the states. In South Africa, the provinces control big-ticket items, such as health, education, and housing subsidies.[3] The larger

[3]For both South Africa and Nigeria, there was available data on subnational spending between the regional and local governments.

South African metropolitan areas also control large infrastructure grants for urban service delivery. In Ethiopia, the high proportion of subnational expenditure is a direct result of large social sector expenditure undertaken at the subnational level. For example, in 2016, about 60% of the national health on-budget expenditure and 47% of the national education expenditure were incurred by subnational governments (UNICEF 2017). Ethiopia stands out as the most decentralized country among the low-income countries, with 48% of public expenditure occurring at the subnational level. This is mainly due to Ethiopia's public sector-led development strategy, which includes high levels of pro-poor expenditure at subnational level in the sectors of health, education, nutrition and food security, water, energy, electricity, and road construction.

Beyond the question of proportional control, it is important to understand the nature of subnational expenditure. In particular, it is important to understand how much control subnational governments have over capital expenditures and current expenditures. In Africa, subnational governments are important public employers, pushing up current expenditures (see Table 9.2, column 4). In some countries, current expenditure is more than 90% of subnational expenditure. In contrast, capital expenditure across the continent ranges between 7 and 58% of total subnational expenditure, with Benin and Tunisia leading (see Table 9.2, column 5). In the Tunisian case, local governments have few management responsibilities in areas requiring substantial current spending and therefore can play more of an investment role (OECD and UCLG 2019b). These cases, however, are exceptions, and capital investment on the continent generally accounts for a small share of total subnational expenditures (e.g., see Mali, Mauritania, Malawi, Uganda, and Zambia).

Moving from expenditure to revenue, there are similar trends which can be seen (this is because governments can only spend what they have in revenue, except in cases of extreme over or under spending). Revenues include the revenues from own collection and grants. Similar to expenditure patterns, higher-income countries tend to have higher per capita incomes (e.g., South Africa, Botswana, and Mauritius). Most African countries fall in the range of US$ 10 to US$ 2500 per capita revenues (see Table 9.3, column 2). Similarly, the federal states of Nigeria, South Africa, and Ethiopia show the highest proportion of subnational revenue. In all three countries, regional (states, provinces) entities raise a larger proportion of subnational revenues, and local (municipalities) entities tend to rely significantly more on transfers from the federal government compared to regions. Table 9.3 shows the break down in subnational revenues between transfers and own-source revenues. Transfers dominate subnational revenue (see Table 9.3, column 5). Extreme cases include Uganda, Tanzania, and Rwanda, where transfers make up around 90% of subnational revenue. In Zimbabwe, by contrast, transfers are very low, less than 4%. This is largely due to severe political tensions between local and national government in Zimbabwe which result in limited transfers. For instance, Zimbabwe's 2013 Constitution (Article 301), introduced the fiscal transfer to provinces and local authorities of 5% of national revenue raised, but this has not yet been implemented. This is not uncommon, as in many African countries central governments resist transferring money to subnational governments (OECD and UCLG 2019b).

Overall, this data on revenue and expenditures shows that there is incredible diversity among African countries.

9.5.2 Need for Contextualized Understanding

Apart from the vast diversity in fiscal experiences across the continent, there are some serious gaps in the data which make it difficult to both understand what is going on, and to compare across contexts. Everything from effective forward planning to risk calculation relies on being able to understand what is happening, or could happen, in different geographies. Notably, we have significantly lower levels of insights into the fiscal data space in Africa compared to many other

Table 9.3 Subnational revenue data for selected African countries (OECD and UCLG 2019c)

Country	Subnational government revenue PPP per capita (in US$)	Subnational government revenue as % of total government revenue	Tax revenue as % of total subnational government revenue	Transfers (grants and subsidies) as % of total subnational revenue	Tariffs and subsidies as % of total subnational revenue	Property income as % of total subnational	Other source of revenue as % of total subnational
Benin	27	7.90%	32.10%	49.80%	13.20%	1.80%	3.10%
Botswana	621	11.20%	43.50%	56.50%	NO DATA	NO DATA	NO DATA
Burundi	7	7.00%	NO DATA	NO DATA	NO DATA	NO DATA	NO DATA
Cabo Verde	282	13.80%	16.00%	43.80%	0.00%	18.70%	17.40%
Eswatini	67	2.90%	17.50%	22.80%	0.60%	55.50%	0.00%
Ethiopia	173	57.80%	37.20%	62.80%	0.00%	0.00%	0.00%
Kenya	137	20.60%	3.30%	88.70%	0.00%	0.00%	7.90%
Malawi	10	3.70%	13.10%	71.20%	15.20%	0.00%	0.60%
Mauritius	326	6.90%	4.40%	84.30%	10.70%	0.60%	0.00%
Morocco	292	14.70%	25.60%	63.90%	3.60%	3.00%	4.00%
Namibia	287	7.60%	8.20%	20.70%	0.60%	0.00%	70.50%
Nigeria	202	38.30%	30.70%	47.40%	NO DATA	NO DATA	0.00%
Rwanda	108	17.50%	9.50%	89.20%	1.10%	0.30%	0.00%
Senegal	45	5.60%	35.50%	25.90%	0.00%	0.00%	0.00%
South Africa	2496	50.20%	7.00%	70.30%	6.00%	0.50%	18.00%
Tanzania	45	18.90%	4.80%	89.70%	2.10%	1.50%	1.90%
Tunisia	242	6.40%	22.20%	69.20%	4.00%	1.80%	2.70%
Uganda	60	20.90%	1.40%	96.30%	2.20%	0.10%	0.00%
Zimbabwe	39	9.10%	86.60%	3.70%	0.00%	0.00%	9.70%

Note: Angola and Mozambique have been excluded from this table as it was not possible to get any subnational revenue data for these countries (but see Chap. 11 for some subnational data on Mozambique) parts of the world. While this knowledge gap is as much political as it is technical, a well-rehearsed and important part of this story relates to the limited data which is collected in and on Africa. As the possibilities of the digital age emerge, there are many questions about what sort of data can and should be collected in Africa and who should be responsible for analyzing, sharing, and protecting such data. Limited subnational data collection is not particularly surprising considering the many challenges regarding access and availability at the city level (Borel-Saladin 2017). However, the gap in fiscal documentation is more difficult to comprehend as public finance (unlike other aspects, such as the informal economy, corruption, or other more subjective data) is generally part of the financial accounting frameworks of countries. Unfortunately, there are very few contemporary examples of publicly available and disaggregated fiscal data on Africa.

The recently published data set on subnational government finance, developed by UCLG with the support of the OECD, aims to fill some of the gaps in subnational fiscal data collection globally, with a considerable sample of African countries. The observatory/database aims to track data regularly over time, providing insights into trends and patterns regarding subnational finance and fiscal decentralization. As we show above, useful insights can be gained from this exercise. However, while the framework provided a robust method for data collection, the actual data collected on Africa remained limited. Some insights are possible, but there are also gaps which require additional research and contextualization.

Two important data issues are worth mentioning. First, there are many contexts where it is simply impossible to access reliable subnational data. For many countries, data is not published online, or if it is, the information is scattered. This issue is attributed to limitations in terms of capacity, accountability, and transparency (IBP 2017). African countries, except for South Africa, rank low on international budget transparency indexes for their limited budget accessibility and transparency to the public. Fiscal data on expenditure, revenue collection, and debt at subnational levels are largely on a cash basis and remain unknown and inaccessible for 33 countries on the continent, situated mostly in West, East, and Central Africa. Second, a particularly important and underexplored issue related to fiscal data is the incomparability and incommensurability of the data, owing to different systems of public finance accounting.

The differences among African countries stem from diverse fiscal legal regulations and accounting systems used in different countries. Most national governments in Africa have inherited and adopted—entirely or at least partially—the judicial and institutional framework of their respective colonial powers. For example, Anglophone African governments have adopted a mixed legal system of English common and customary law. The Francophone and Arab governments have adopted a mixed legal system of French civil law and Islamic law. The handful of Lusophone countries have inherited the legal system of Portuguese civil law and fiscal policies. When comparing British, French, and Portuguese colonial legacies, there is a greater uptake of accrual financial accounting methods and adoption of international accounting standards and reporting practices among Anglophone-speaking countries with legal systems based on the English common law. Overall, these challenges make comparing fiscal data difficult as the categorization of information is not uniform across the different contexts.

9.5.3 Local vs. City Government

In the context of African urban debates, there is considerable conflation between "local government" and "city government" as there are many different ways in which subnational governments are administratively defined. This has serious implications for fiscal decentralization, which requires tailored approaches for rural, urban, and metropolitan areas. For instance, local governments in Africa are increasingly characterized as either "urban" or "rural." As shown in Table 9.4, examples of countries which categorize local government as urban or rural include Botswana, Burkina Faso, Burundi, Eswatini, Guinea,

Table 9.4 Data on urban differentiation in 25 African countries (Source: compiled by authors)

Country name	Urban differentiation
Angola	None
Benin	Special status given to Cotonou, Porto Novo, and Parakou
Botswana	Differentiation of town councils and city councils
Burkina Faso	Differentiation of urban municipalities
Burundi	Differentiation of urban municipalities
Cabo Verde	None
Eswatini	Differentiation of urban municipalities
Ethiopia	Differentiation between districts and city administration. Special status of "Chartered City" given to Dire Dawa and Addis Ababa
Ghana	Differentiation of metropolitan assemblies
Guinea	Differentiation of urban communes. Special status given to region of Conakry
Kenya	Nairobi and Mombasa granted status as city-counties
Malawi	Differentiation of urban districts
Mauritius	Differentiation of urban councils
Morocco	None
Mozambique	Four different types of urban municipalities, ranging from type "A" to "B" and "C" for provincial capitals and other cities, and type "D" for small towns. Capital city of Maputo singularly classified as type "A"
Namibia	Differentiation of city councils and town councils
Niger	Differentiation of cities and urban municipalities
Nigeria	None
Rwanda	None
Senegal	Differentiation of cities and urban municipalities
South Africa	Differentiation of metropolitan municipalities
Tanzania	None
Tunisia	None
Uganda	Special status given to Kampala City
Zimbabwe	Differentiation of urban municipalities

Malawi, Mauritius, Namibia, Niger, Senegal, Senegal, and Zimbabwe. Further differentiation may be made for larger cities, which can be granted "city" or "metropolitan" status (examples of local governments like this include Mombasa, Johannesburg, Dakar, Kampala, and many others). The definition of "urban" used across African countries is also highly differentiated (Farvacque-Vitkovic et al. 2008). However, the growing attention to rural urban differentiation—in particular, when it comes to the structure of public funding—is a useful recognition of the complex processes of urbanization evident on the continent (Ribot 2002; Pieterse 2008).

While South Africa is the only country with official metropolitan classifications of local government, there are several cases where some level of metropolitan level coordination is being attempted in capital cities, for example, in Senegal (Dakar, established in 1996) and Ivory Coast (Abidjan, established in 2011), both of which have created bodies to coordinate the local government units which make up big cities. In other cases, where there are very large cities, these metropolitan areas more closely align with regional government boundaries, for example, in the case of Nairobi City County, Maputo City province, Addis Ababa [Chartered] City, and Lagos State. There are also cases wherein national bodies have been created, in parallel to the elected local government, to manage key cities. A good example of this is the Nairobi Metropolitan Services (NMS). These different arrangements reflect the efforts that African countries are making to grapple with their largest urban areas.

In the context of rapid, and highly uneven, urbanization in Africa, it is impossible to consider decentralization without accounting for urban dynamics. Rapidly growing urban and metropolitan areas on the continent have the potential to become the engines of inclusive and resilient growth. However, this requires a better understanding of these cities, in terms of how they operate and how they fit into the spatial and fiscal hierarchies of countries and regions. Overall, and particularly in comparison to Western counterparts, data on African cities is a challenge. This is particularly true for fiscal data which may be collected at the local level, but are rarely made public or aggregated for analysis. However, where data has been collected, there are interesting insights about inter-city fiscal differentiation which can be seen. Reflecting on data from 27 cities in Africa, we see several inter-

Table 9.5 City level expenditure and revenue data for selected African cities (UCLGA 2017)

City	Country	Population of city	City revenues per capita (in US$)	Total city expenditure per capita	Current city expenditure per capita	City capital expenditure per capita
Abodo	Ivory Coast	1,450,000	1.67	1.48	1.47	0.10
Antananarivo	Madagascar	1,848,998	2.89	2.51	2.23	0.28
Bambari	Central African Rep.	41,486	2.20	1.57	1.19	0.38
Banjul	The Gambia	31,301	36.06	38.02	38.02	0
Brazzaville	Congo, Republic	1,600,000	19.31	21.94	18.32	3.61
Bujumbura	Burundi	590,297	7.41	9.09	5.57	3.52
Conakry	Guinea	2,164,282	1.60	0.23	0.10	0.13
Dakar	Senegal	1,146,053	43.66	50.92	35.18	15.74
Dar Es Salam	Tanzania	3,700,000	0	0	0	0
Ezulwini	Eswatini	5800	134.09	260.93	134.09	126.84
Gitega	Burundi	184,026	1.59	1.99	1.21	0.77
Gomoa West	Ghana	135,189	6.34	6.09	6.09	0
Kampala	Uganda	1,750,000	22.31	20.92	10.87	10.05
Kigali	Rwanda	835,000	9.97	4.83	1.80	3.04
Koutiala	Mali	151,212	12.1	12.86	12.10	0.75
Lilongwe	Malawi	826,614	4.85	4.85	4.40	0.45
Lomé	Togo	921,563	15.01	13.07	10.45	2.62
N'Djamena	Chad	1,000,000	7.04	8.69	6.71	1.98
Nairobi	Kenya	2,900,000	37.74	42.38	27.46	14.92
Niamey	Niger	1,302,910	15.16	32.98	21.72	11.26
Nouakchott	Mauritania	958,399	3.15	1.19	1.19	0
Ouagadougou	Burkina Faso	2,031,540	11.73	11.66	9.65	2.01
Port Louis	Mauritius	144,894	168.87	178.46	168.48	9.99
Porto Novo	Benin	331,419	3.02	7.05	0.90	6.15
Praia	Cabo Verde	515,320	27.02	31.28	21.19	10.08
Rabat	Morocco	1,884,917	283.64	439.29	228.9	210.39
Sousse	Tunisia	231,484	58.86	55.97	44.45	11.51

esting factors related to city level finances (see Table 9.5).

As with the country data, revenues and expenditures track each other closely. Across African cities, per capita spending is wildly variate, ranging from less than US$ 1 in Dar es Salaam and Conakry to US$ 439 in Rabat (notably, no South African cities were included in this data set). This is a substantial level of variation which would have been further exacerbated by the inclusion of South Africa. Most African cities which were reviewed for this work spent below US$ 50 per inhabitant per year. The nature of this expenditure is also important. In most of the cities, most of the spending is on current expenditure, in other words staff costs and operations (UCLGA 2017). There is limited capital expenditure, ranging from US$ 15 to less than US$ 1 per inhabitant per year. There are exceptions to this, for example, Rabat has a total capital expenditure of US$ 210 per inhabitant per year, and Ezulwini has a total capital expenditure of US$ 126 per inhabitant per year. The extent to which cities are able to spend on capital and development has fundamental implications for the role they play in shaping the urban fabric. Together, Tables 9.4 and 9.5 point not only to the rising importance of urban areas and city finance but also the important role that metropolitan regions are now playing in the fiscal structures of African countries.

9.6 Conclusion

Public finance will be central to any chances of successfully implementing the SDGs. The question is not only about *how much* funding is needed but also *who controls* available funding. Given the limitations in fiscal data, it is very difficult to get a clear picture of the extent to which city governments in Africa control money. And by extension, it is difficult to assess the extent to which city governments are fiscally capacitated to support SDG implementation.

Despite this challenge, the chapter makes three important arguments. First, we show that there is incredible diversity in subnational government control over finance. Making broad-brush statements about the extent to which subnational governments are or are not fiscally capacitated thus requires more detailed interrogation. Second, the data is uneven and incomplete, therefore requiring a more holistic understanding of the complexity of the local government fiscal space. In addition to an overall lack of fiscal data, there is insufficient data on the divide in fiscal control between intermediate and local government. There are also serious challenges related to the commensurability of data. This is owing both to gaps in data collection and very different methods of fiscal and financial accounting.

The vastly differentiated multilevel governance arrangements and urban classifications make homing in on urban fiscal data a challenge. Significantly more work is therefore needed to get a clear understanding of the complex spatial and fiscal dynamics that are emerging through Africa's urbanization process.

References

Becker CM, Hamer AM, Morrison AR (1994) Beyond urban bias in Africa: urbanization in an era of structural adjustment. Heinemann, Portsmouth, NH and James Currey, London

Bekker S, Therborn G (2012) Capital cities in Africa: power and powerlessness. HSRC Press, Cape Town

Bird RM, Vaillancourt F (2008) Fiscal decentralization in developing countries. Cambridge University Press, Cambridge

Boadway R, Shah A (2007) Intergovernmental fiscal transfers: principles and practices. World Bank Publications, Washington DC

Borel-Saladin J (2017) Data dilemmas: availability, access and applicability for analysis in sub-Saharan African cities. Urban Forum 28:333–343

Buchanan JM (1989) Richard Musgrave, public finance, and public choice. Public Choice 61(3):289–291

Clarke Annez P, Gwenaelle H, Peterson GE (2008) Lessons for the urban century: decentralized infrastructure finance in the World Bank. World Bank, Washington DC

Engberg-Pedersen L (2016) The Addis Ababa Action Agenda: breaking new ground, incremental changes, or neoliberal business as usual? In: Fejerskov AM, Funder, M, Engberg-Pedersen L, Jiang, Y, Ravnborg, HM, Webster, N (eds) Financing sustainable development—actors, interests, politics. Danish Institute for International Studies, Copenhagen, pp 19-31

Faguet JP (2014) Decentralization and governance. World Development 53(C):2–13

Farvacque-Vitkovic C, Raghunath M, Eghoff C, Boakye C (2008) Development of the cities of Ghana: challenges, priorities, and tools. Africa region working paper series no. 110. World Bank. http://documents.worldbank.org/curated/en/318351468256521007/Development-of-the-cities-of-Ghana-challenges-priorities-and-tools. Accessed 12 October 2020

Farvacque-Vitkovic C, Godin L (1998) The future of African cities: challenges and priorities for urban development. World Bank Publications, Washington DC

International Budget Partnership (IBP) (2017) Open budget index ranking 2017. https://www.internationalbudget.org/wp-content/uploads/open-budget-survey-2017-OBIrankings.pdf. Accessed 12 October 2020

Lincoln Institute and World Bank (2016) Municipal finance and local fiscal systems. Lincoln Institute, Cambridge, Massachusetts and World Bank Publications, Washington, DC

Moeti K, Khalo T, Mafunisa J (eds) (2007) Public finance fundamentals. Juta, Cape Town

Move Humanity (2018) Closing the SDG budget gap. https://movehumanity.org/wp-content/uploads/2018/10/FINAL-2018-10-18_Closing-the-SDG-Budget-Gap.pdf. Accessed 15 May 2020

Oates WE (2008) On the theory and practice of fiscal decentralization. In: Auerbach AJ and Shaviro DN (eds) Institutional foundations of public finance: economic and legal perspectives. Harvard University Press, Cambridge, pp 165–76

Organisation for Economic Cooperation and Development (OECD) (2020) The territorial impact of COVID-19: managing the crisis across levels of government. http://www.oecd.org/coronavirus/policy-responses/the-territorial-impact-of-covid-19-managing-the-crisis-across-levels-of-government-d3e314e1. Accessed 12 October 2020

Organisation for Economic Cooperation and Development (OECD) and United Cities and Local Governments (UCLG) (2019a) SNG-WOFI Database. World Observatory on Subnational Government Finance and Investment. https://stats.oecd.org/viewhtml.aspx?datasetcode=SNGF_WO&vh=0000&vf=00&l&il=blank&lang=en&vcq=1111. Accessed 12 October 2020

Organisation for Economic Cooperation and Development (OECD) and United Cities and Local Governments (UCLG) (2019b) 2019 report of the World Observatory on Subnational Government Finance and Investment—country profiles. http://www.sng-wofi.org/reports/SNGWOFI_2019_report_country_profiles.pdf. Accessed 12 October 2020

Organisation for Economic Cooperation and Development (OECD) and United Cities and Local Governments (UCLG) (2019c) 2019 report of the World Observatory on Subnational Government Finance and Investment—key findings. http://www.sng-wofi.org/reports/2019_SNG-WOFI_Report_Key_Findings.pdf. Accessed 12 October 2020

Paulais T (2012) Financing Africa's cities: the imperative of local investment. World Bank Publications, Washington DC

Pieterse E (2008) City futures: confronting the crisis of urban development. Zed Books, London

Ribot JC (2002) African decentralization: local actors, powers and accountability. United Nations Research Institute for Social Development, Geneva

Smoke P (2001) Fiscal decentralization in developing countries: a review of current concepts and practice. United Nations Research Institute for Social Development, Geneva

Smoke P (2003) Decentralisation in Africa: goals, dimensions, myths and challenges. Public Administration and Development 23(1):7–16

Tanzi V (2016) Public finance in developing countries: an introduction. In: Erdoğdu M and Christiansen B (eds) Handbook of research on public finance in Europe and the MENA region. IGI Global, Hershey, PA, pp 1–10

United Cities and Local Governments Africa (UCLGA) (2017) Africa review local finance 2016, 2nd edn. United Cities and Local Governments of Africa, Rabat

United Cities and Local Governments of Africa (UCLGA) and Cities Alliance (2015) Assessing the institutional environment of local government in Africa. United Cities and Local Governments of Africa, Rabat and Cities Alliance, Brussels

United Nations (UN) (2015) Addis Ababa Action Agenda. https://sustainabledevelopment.un.org/content/documents/2051AAAA_Outcome.pdf. Accessed 12 October 2020

United Nations (UN) (2017) New Urban Agenda. https://uploads.habitat3.org/hb3/NUA-English.pdf. Accessed 1 February 2021

United Nations (UN) (2018) Financing for development: progress and prospects 2018. Report of the Interagency Task Force on Financing for Development. United Nations, New York

United Nations (UN) (2020) Financing for development: progress and prospects 2020. Report of the Interagency Task Force on Financing for Development. United Nations, New York

United Nations Conference on Trade and Development (UNCTAD) (2014) World investment report. Investing in the SDGs: an action plan. United Nations Conference on Trade and Development, New York and Geneva

UN-Habitat (2009) Guide to municipal finance. Human settlements financing tools and best practices series. United Nations-Habitat, Nairobi

United Nations International Children's Emergency Fund (UNICEF) (2017) Ethiopia 2017 national budget brief. A Synopsis analysis of 2006-2016 national trends and the 2017/2018 federal budget proclamation. https://www.unicef.org/esa/sites/unicef.org.esa/files/2018-09/UNICEF-Ethiopia-2017-National-Budget-Brief.pdf. Accessed 12 October 2020

World Bank (2014) Burundi public expenditure report. Republic of Burundi fiscal decentralization and local governance: managing trade-off to promote sustainable reforms. http://documents.worldbank.org/curated/en/970171468014351438/pdf/929140WP0P14530English0web00PUBLIC0.pdf. Accessed 12 October 2020

Wunsch JS (2001) Decentralization, local governance and 'recentralization' in Africa. Public Administration and Development: The International Journal of Management Research and Practice 21(4):277–288

10 Activating Urban Planners for Fostering Urban Integrity: An Inroad into Curbing City Level Corruption

Laura Nkula-Wenz, Gilbert Siame, and Dieter Zinnbauer

Abstract

Corruption is one of the major hurdles to achieving the Sustainable Development Goals (SDGs) globally. Money is lost due to illicit financial flows and the lack of progress in updating policies and improving the technical capabilities of key departments at the national level. In addition to that, many African countries face the challenge of reforming colonial-era public administrations, making them more accessible and accountable to the people they serve. In turn, successfully localizing the SDGs requires a competent and ethical public service, including at the local government level, which is the primary state-citizen interface. This chapter sets out to connect the aims of SDG 11 and SDG 16 through exploring the potential of the urban planning profession as a key custodian of transparency, accountability, and, ultimately, urban integrity. Drawing on research among urban and regional planners in Zambia and South Africa, done as part of the Cities of Integrity project and funded by the Department for International Development (DFID), we argue that planners are among the key stakeholders to engage when addressing corruption and maladministration at the local level. We further maintain that it is worthwhile to move from a narrow focus on legalistic compliance with anti-corruption measures towards a more proactive promotion of professional integrity and collective accountability mechanisms. This is especially true in the African context. Finally, using a case study involving planning professionals in Lusaka, we explore the opportunities and challenges faced when addressing corruption and promoting integrity among the local community of practice.

Keywords

SDG 11 · SDG 16 · Local government · Corruption · Integrity · Urban planning · South Africa · Zambia

L. Nkula-Wenz (✉)
African Centre for Cities, University of Cape Town, Cape Town, South Africa
e-mail: laura.nkula@uct.ac.za

G. Siame
Centre for Urban Research and Planning, Department of Geography and Environmental Studies, University of Zambia, Lusaka, Zambia
e-mail: Siamegilbert@yahoo.co.uk

D. Zinnbauer
Copenhagen Business School, Copenhagen, Denmark
e-mail: dz.msc@cbs.dk

10.1 Introduction

Across the globe, critical work towards achieving the Sustainable Development Goals (SDGs) is being hampered by corruption (Murshed and

S. Croese, S. Parnell (eds.), *Localizing the SDGs in African Cities*, Sustainable Development Goals Series, https://doi.org/10.1007/978-3-030-95979-1_10

Mredula 2018). At the same time, SDG 16.5 commits countries to "substantially reduce corruption and bribery in all its forms." These forms are many and span different scales, from grand corruption like state capture and international money laundering syndicates to the so-called "petty" corruption involving the solicitation of small bribes and favors by often underpaid public officials. This makes corruption a "wicked problem" because it requires context-specific, multiscale, and whole-of-society interventions.

Corruption and illicit financial flows are also hard to measure, which is why there is a lively debate around how to effectively devise and monitor corruption-related indicators within SDG 16 (Khan 2018; Razafindrakoto and Roubaud 2010). While some scholars see Africa as leading the way when it comes to adopting ways to measure SDG 16, a commitment to monitoring corruption-related indicators remains a blind spot (Cling et al. 2018).

Ultimately, SDGs cannot be implemented at a local level without paying attention to the pervasive issue of corruption. This chapter argues that SDG 11, in particular—which aims to make cities and human settlements inclusive, safe, resilient, and sustainable—can benefit substantially from new perspectives on city level corruption, and the role of professional integrity in curbing it.

Generally, cities are high-stakes environments for development interventions. By the same token, cities offer unique opportunities to make critical headway in localizing the SDGs. Cities can be key drivers of local sustainable development if coherent urban development policies are in place that identify and exploit systematic synergies between the local economy, sociocultural and environmental parameters, and long-term infrastructural planning (Parnell 2016). At the same time, cities are also high-risk zones for corruption, for a number of interconnected reasons (Zinnbauer 2019):

- Population growth in cities increases competition for resources and infrastructure provision, while curtailing the enforcement of rules and regulations.
- In cities with a significant informal economy, poor and marginalized populations are more vulnerable to extortion and becoming victims of corruption (see also Zinnbauer 2020).
- The highest urban growth is projected for countries that already face high corruption risk, where "(e)ight of the ten countries forecast to experience the highest urbanization rates between 2014 and 2050 are perceived to be saddled with high levels of public sector corruption, and rank in the bottom half of countries globally when it comes to expert perceptions of the severity of public sector corruption" (Zinnbauer 2019, p. 3).
- The continuous influx of people looking for opportunities in cities, particularly from poor and marginalized groups, has historically lent itself to bolstering political patronage networks. This links political support to the supply of essential services and skews critical infrastructure development through political favoritism.

We argue that to effectively localize the SDGs, it is critical to tackle corruption at the urban scale. However, to do so, and to devise effective strategies to respond to it, we need a better understanding of the problem and how it presents itself in different contexts. Unfortunately, the city remains vastly under-researched in anti-corruption scholarship, which is traditionally situated in disciplines like public administration, political science, and behavioral economics. Urban studies have furthermore not engaged sufficiently with questions of urban corruption, even though it is a significant force shaping contemporary urban development across the globe (Chiodelli and Moroni 2015; Kim 2020). Over the past decade, the limited success of traditional legalistic anti-corruption measures has prompted a substantial reckoning in the anti-corruption advocacy space. This initiated an ongoing paradigm shift which Zinnbauer (2019, p. 6) summarizes as follows:

> A narrow, legalistic focus on direct anti-corruption measures and a largely punitive approach to step up monitoring, legal sanctions, compliance, and related reporting requirements are not sufficient and at times even counterproductive. Instead, what holds more promise is a more encompassing approach that embraces the broader ambition of strengthening integrity, rather than reducing corruption as its main guiding principle.

Recognizing this research gap and heeding the need to shift the paradigm to focusing on doing the right thing rather than just avoiding the wrong thing, the Cities of Integrity project was initiated as part of UKAid's Global Integrity Anti-Corruption Evidence Program. It focused on investigating the nexus between corruption, urban development, and professional integrity among urban and regional planners through in-depth research in South Africa and Zambia. The data collection ran from March 2019 to December 2020. It encompassed over 50 semi-structured interviews and focus groups with planning professionals in both public and private practice, as well as a Qualitative Action Experiment in the form of an integrity training for 40 planners in Lusaka, Zambia (see also Siame et al. 2020). Between February and September 2020, we also administered a detailed quantitative online survey among planning professionals in both countries in cooperation with the respective national professional association, which assisted in distributing the survey among their members. The survey solicited a sample of 211 complete responses (African Centre for Cities and Centre for Urban Research and Planning 2020).[1] The main research objectives were to examine the current integrity landscape in local urban development practice. It sought to expand on existing mechanisms and practices to develop, promote, and enforce integrity as a professional norm and to identify public roles and responsibilities that practicing planners in South Africa and Zambia see for themselves. It further considered the pressures they face at the intersection of the public and private sector and in light of high urbanization pressures in both countries.

The following arguments draw from this rich set of empirical data, and we will proceed as follows. First, we provide a brief overview of the dynamics of urban corruption in Africa before discussing the current state of the planning profession on the continent. In the second part of the chapter, we then hone in on the case of Zambia, discussing the specific pressures and repercussions planners face when trying to carry out their job in the face of widespread urban corruption pressures. We conclude by reflecting on some of our interventions that have sought to actively promote professional integrity in urban planning and what we regard as productive pathways to further engage with this pertinent issue.

10.2 Brief Overview of Urbanization and Urban Corruption in Africa

Corruption, and by extension urban corruption, is in no way a uniquely African problem. Indeed, it is a global phenomenon that is profoundly shaping cities across the globe. A case in point are the "million dollar ghost buildings" (Boles 2017) in high-end neighborhoods of London or New York, where all-cash purchases of luxury real estate have yielded ample opportunities for money laundering, while leaving behind underused houses and lifeless neighborhoods (Atkinson 2019; Fernandez et al. 2016; Glucksberg 2016). Across the world, urban real estate is one of the most coveted asset classes for money laundering, as a number of red flags indicate: in London more than 36,000 plots (5.7 km^2) are owned by shell companies (de Simone 2015); 30% of worldwide confiscated criminal assets are real estate (FATF 2013); in a recent World Bank study, a Kampala business representative divulged that "land prices have gone up very fast—it is overvalued because there's a scramble for land with the influx of migrants and corrupt officials who have a lot of cash and need to put it somewhere" (World Bank 2018).

[1]The full dataset, questionnaire and additional survey materials have been deposited with DataFirst at the University of Cape Town. They can be accessed and reused under a CC-BY-SA license by following this link: https://www.datafirst.uct.ac.za/dataportal/index.php/catalog/871.

Corruption in Africa usually only makes global news in connection to large-scale state capture and national tender fraud, seen most notably with the investigation in South Africa by the Judicial Commission of Inquiry into Allegations of State Capture (2018–2021)—the so-called Zondo Commission, named after its chairperson, Deputy Chief Justice Raymond Zondo—in connection with the tenure of former president Jacob Zuma. However, corruption on the continent is felt most acutely at the local scale. More than half (55%) of all African citizens think that corruption increased in the previous 12 months, and only 23% think it declined. More than one in four African citizens who accessed public services, such as health care and education, paid a bribe in the previous year (Pring et al. 2019). This is equivalent to approximately 130 million people. Finally, more than 60% of people polled in 34 African countries in 2017 thinking it is very likely that rich people can get away with paying bribes to falsely register land that does not belong to them, a number that rises to almost 67% in South Africa and 72% in Zambia (Afrobarometer 2018). This is particularly indicative when we consider the rapidly rising value of urban land and the large windfalls that can be generated by re-zoning rural land for urban expansion.

Tenure insecurity—measured as the perceived likelihood for eviction within the forthcoming 5 years—can be another red flag for corruption, indicating high levels of impunity when it comes to land grabbing. These rates of perceived insecurity are highest in sub-Saharan Africa and North Africa/Middle East, where 26 and 28% of respondents respectively feel in danger of being evicted. Tenure insecurity tends to be more pronounced in urban areas as compared to rural areas, with sub-Saharan Africa posing the starkest difference: 27% of respondents in cities feel insecure, compared with 22% in rural areas. A particularly strong indicator for the presence of corruption with impunity in relation to tenure insecurity is that even the possession of formal documentation does not offer protection or significantly increase the perceived level of security: in sub-Saharan Africa, 70% with formal documentation feel secure compared with 65% without formal documentation, the smallest difference of all regions (Prindex 2020).

Finally, there is widespread suspicion in the general public when it comes to the question of who benefits from urban development and policy initiatives: 77% of survey respondents in African cities and 61% in Asian cities believe public office holders benefit most from urban reforms due to corruption (UN-Habitat 2010). As two detailed case studies from South Africa illustrate (Olver 2017, 2019), this impression is either corroborated by direct party political meddling and graft, or by close relationships between local officials, political circles, and property developers, fanning suspicions of undue influence and favoritism.

Thus, the premise of the Cities of Integrity project is twofold: first, that urban development is not only a particular node for corruption but can also be a fulcrum for fostering integrity and ethical conduct and, second, that urban planners present a key professional group that could spearhead innovative anti-corruption and integrity practices. This is because planners sit at the nexus of complex public and private interests and must consistently navigate this treacherous terrain to safeguard the public good while also meeting developmental goals.

The aim of our research has therefore been to understand how to best sensitize and activate urban planners for tackling corruption. Before we share a few insights from our research in Zambia, we provide a brief overview of the African planning profession and why we consider planners as potentially potent custodians of urban integrity.

10.3 The Urban Planning Profession in Africa

Broadly speaking, planning deals with the location of rights and services within a legal spatial framework that is ideally committed to safeguarding the public good, as well as the long-term interests of the city and its inhabitants. This public interest focus is an essential pillar of formal urban planning education across the globe, and in contrast to other spatial disciplines such as architecture, most planning graduates spend at least some of their professional life in public service (Sager 2009; Johnson 2010).

10.3.1 Urban Planning Education Programs

Standalone urban planning programs only developed in the early twentieth century in response to accelerating rates of urbanization, incorporating strands of civil engineering, architecture, landscape architecture, public health, and law. However, by 2017, over 600 planning education programs were offered in half of the world's countries (Frank and Silver 2017, p. 338). While Frank and Silver note that there remain gaps in some regions, the pronouncement of a now majority urban world and, in consequence, the recognition of urbanization as a crucial focal point for sustainable development—codified in SDG 11—is likely to only increase the demand for planning education. Internationally, the long-term need for capacity building in the urban planning sector has also been foregrounded in both SDG 11 and the New Urban Agenda within the broader commitment of signatory countries to building stronger and more inclusive urban governance institutions (Parnell 2016).

With regard to planning education, the recognition of widespread urban informality and socio-spatial inequality as primary markers of urbanization in most low- and middle-income countries has started to bring practical changes to the curriculum. Straddling both technical engineering and social science influences, planning theory, planning pedagogy, and planning practice are becoming more closely intertwined through joint discussions relating to how to implement participatory and inclusive planning practices and realize more just and equitable cities. Following the imperative that "planning educators and their students need 'to get their shoes dirty'" (Watson 2013, p. 9), many degrees are now offering regular practice courses and applied research studios from undergraduate level onwards. This allows students to hone both their technical and social problem-solving skills through real-life engagement with contemporary planning challenges, more often even in cooperation with local authorities, non-governmental organizations (NGOs), or private planning firms (Balassiano 2011).

10.3.2 Urban Professional Associations

Alongside planning education programs, national professional associations have also proliferated. Among the largest and most influential are the British Royal Town Planning Institute (RTPI), established in 1914 with currently about 22,000 members (RTPI 2019), and the American Planning Association, established in 1978 with currently 46,398 members, of which 44% are also certified through its statutory arm, the American Institute of Certified Planners (American Planning Association 2019). Often, these professional associations also function as legal bodies at a national level, requiring planners to register and subscribe to professional codes of conduct before they can practice. In line with these codes, many associations have also defined self-regulatory mechanisms geared towards addressing professional misconduct, including corruption.

As town planning historically constituted a key practice of British colonial expansion, the RTPI has also been a model for many national planning bodies in its former colonies and across the Commonwealth. However, to be able to better respond to the planning challenges across the rapidly urbanizing continent, African planners adopted the so-called Durban Declaration in 2002, leading to the formation of the African Planning Association (APA) (UN-Habitat and APA 2013). Intended to strengthen the public voice of the profession and act as a platform to link African planners, by 2012 a total of 26 planning institutions had signed up to be part of the APA, though the organization has been largely dormant over the past 5 years, due to lack of

	Population (million) 2011	No of accredited planners	No of planners per 100,000	Year of Estimate
APA Countries				
Burkina Faso*	16,970,000	14	0.08	2011
Ghana	24,970,000	150	0.60	2011
Nigeria*	162,500,000	2,333	1.44	2011
Mali*	15,840,000	50	0.32	2011
Kenya*	41,610,000	194	0.47	2011
Uganda	34,510,000	90	0.26	2011
South Africa*	50,800,000	1,690	3.33	2011
Malawi	15,300,000	30	0.20	2011
Mauritius	1,286,000	27	2.10	2011
Tanzania*	46,200,000	158	0.34	2011
Zambia	13,400,000	60	0.45	2011
Zimbabwe	12,700,000	262	2.06	2011
Other countries				
United Kingdom	61,126,832	23,000	37.63	
United States	304,059,724	38,830	12.77	2010
Australia	18,972,350	4,452	23.47	2009/10
Pakistan	173,593,383	755	0.43	2010
India	1,210,193,422	2,800	0.23	2011

*Countries that regulate the registration of planning at a national level

Fig. 10.1 Ratio of planners to country population (UN-Habitat and APA 2013)

funding.[2] Moreover, many African countries do not yet have professional planning associations or dedicated higher education courses. Even in many of the countries that do, the number of professional planners is too low and thus insufficient to respond to the growing needs of their rapidly urbanizing populations (see Fig. 10.1).

For example, it is estimated that Burkina Faso has only about 14 accredited planners for a population of nearly 17 million people, while in Nigeria a mere 2333 planners serve 162.5 million people (UN-Habitat and APA 2013, p. 23). This sits in contrast to the United Kingdom which has 23,000 planners for 61.1 million people. Moreover, there is a lack of data regarding the share of female planners across the continent, but according to Olufemi (Olufemi 2008, p. 412), "their numbers remain negligible and their output unrecognized." This speaks volumes about the detrimental lack of and continued disregard for female perspectives and gender-based struggles in African cities.

Partly because of this scarcity of planning capacity, planners in African cities often alternate between working on publicly tendered projects and private development commissions, which increases the possibility for conflicts of interest to arise. The issue of so-called private jobs is particularly illustrative of the "conflicting rationalities" (Watson 2003). African planners are confronted within their daily practice. Our survey data[3] from Zambia confirms this (see Fig. 10.2): compared with South African respondents—of whom almost 80% have never found themselves working on two sides of a matter—almost half of all surveyed planners in Zambia indicate having

[2] This stands in contrast to the thriving Association of African Planning Schools (AAPS), established in 1999, which links planning programs across the continent and has a very active membership base, hosts a biennial conference as well as regular forums on curriculum development and academic exchange (Watson and Agbola 2013).

[3] These figures have been generated through an online survey targeted at practicing planners in both Zambia and South Africa.

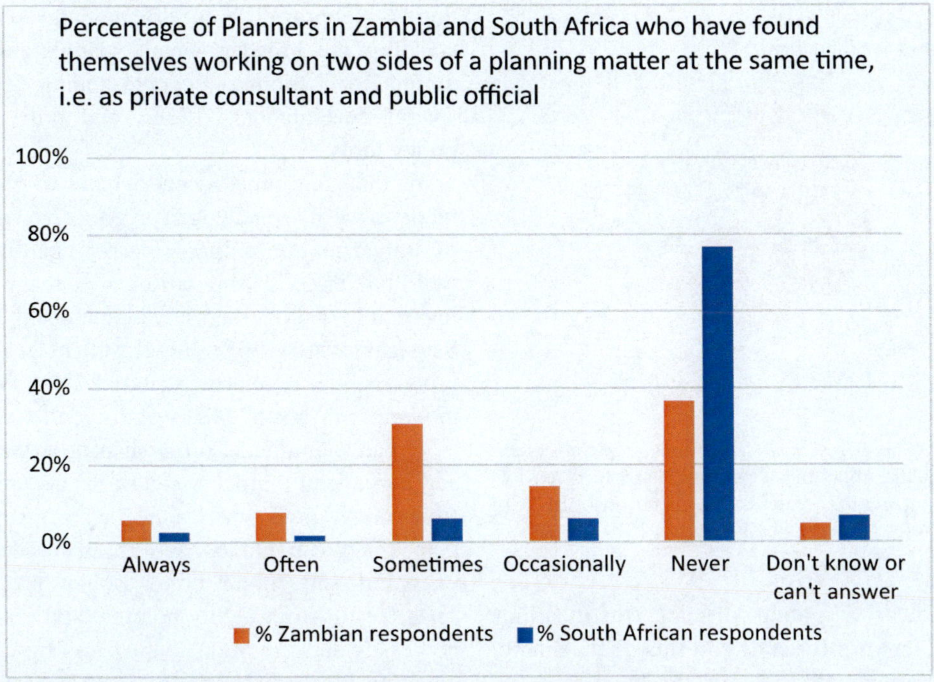

Fig. 10.2 Survey responses to the question whether planners in South Africa and Zambia ever found themselves working on two sides of a planning matter at the same time (In % of $n = 211$, Cities of Integrity online survey 2020)

worked on the same job as both a private consultant and a public official, at least on occasion. Private jobs are typically done on tasks involving the preparation of site plans, building designs, reviews of designs, and lodging of document. Often the same planners who assisted with the private submission are then involved in assessing it and making recommendations for final decisions in their capacity as civil servants. Two areas of conflict emerge. The first area is regarding the question of whether the private job is done during the planner's personal off-duty time, or when the planner is on-duty and expected to do work in their official capacity. The second area is the conflict of interest that arises when one planner works on two sides of the same job.

Furthermore, our data also shows that early career planning professionals with less than 10 years of experience are more likely to engage in private jobs than older and more experienced planners. From our interview data, we can deduct several reasons for this. Firstly, young professionals are more likely to be directly exposed to developers, the public, and other clients, making them more easily noticed and targeted by potential clients. Secondly, more than 62% of Zambian planners are young, with less than 10 years' experience. This cohort also confided that they often receive orders from senior officials, such as directors of heads of local planning authorities, to achieve predetermined planning outcomes. This makes them more susceptible to private jobs and the associated integrity dilemmas.

A fierce debate is currently taking place among Zambian planners on how to address issues surrounding private jobs and potential conflicts of interest. The Zambia Institute of Planners (ZIP) currently seems unclear on how to guide planners on this emotive professional issue. ZIP acknowledges that private jobs clearly create a conflict of interest and it has sought to include guidelines in the draft Practice Manual for Planners in Zambia. Zambian planners are divided on the legality and morality of private jobs as Fig. 10.3 shows. About one third of the surveyed planners indicate that it is justifiable to work on the same job as both a private consul-

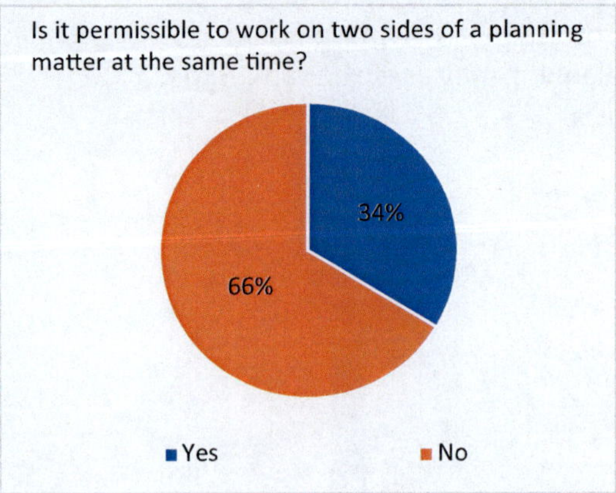

Fig. 10.3 Survey responses from Zambian planners indicating whether they find it permissible to work on two sides of a planning matter at the same time (In % of $n = 98$, Cities of Integrity online survey 2019)

tant and as a public official. Justifications brought forward for acting in this dual capacity include among other things the belief that, as trained professionals, they are able to maintain objectivity on the matter, even when working on both sides simultaneously; the critical shortage of qualified planners in Zambia, especially in rural municipalities where private planners are particularly scarce; and low and irregular salary payments, resulting in those planners in lower salary brackets feeling forced to accept private jobs to make ends meet. Some have even argued that maintaining a steady income through private jobs-on-the-side helps them resist the temptation of bribes and maintain their professional integrity.

These issues and dilemmas resonate across the African continent, with both planning education and professional associations representing spheres where collective ethics and professional value systems are forged and vividly debated. It is widely accepted that "[p]lanning is fundamentally an ethical activity as it raises questions about what should be done, for whom and by whom, and with what benefits or losses" (Watson 2003, p. 404). Like medical doctors whose fiduciary duty is helping their patient understand different treatment options and trade-offs between them, planners—through their specialized expertise coupled with their legal public engagement duties—support different urban actors in understanding the broader social, spatial, and economic implications of specific zoning decision, spatial development visions, and other urban policy tools.

At the same time, recent debates in planning theory have also challenged preconceived notions of universal ethics, particularly regarding the multitude of conflicting rationalities that practicing planners find themselves subjected to (de Satgé and Watson 2018), not least in places where urban realities regularly exceed ideals of Western modernity (Watson 2002; Robins et al. 2008).

Urban planning is often seen as a technical and bureaucratic process, decisions about where rights are granted, where and when infrastructure is installed or upgraded, as well as broader city-visioning and future urban policy processes. These result from highly political and contested processes that are often shaped less by rational planning ideals and more by local contexts and historic path dependencies.

10.4 Tackling Urban Corruption: Why It Matters

In practice, spatial and urban planning fundamentally support important political and jurisdictional decision-making on the location of key public infrastructure, such as schools, hospitals, roads, public transport facilities, and so forth. Drawing on their specialized expertise, planners are deeply involved in implementing land-use regulations, thus shaping the allocation of development rights, and are entrusted with monitoring the supply of developable land through spatial development frameworks and other legislated plans. In turn, Chiodelli and Moroni (2015, p. 437) argue that "[c]orruption in the planning field is largely tied to the opportunities that land-use planning generates by allocating development rights and land uses (following a discretionary and differentiated logic)."

In other words, planning entails a constant balancing of diverse interests. In day-to-day practice, planners are forced to make trade-offs among conflicting objectives, such as environmental protection, social provisions, and economic development. Their decisions thus directly impact the lives and livelihoods of different social groups, particularly in contexts marked by high levels of informality and inequality (e.g., in terms of gender, income, race, and ethnicity). Operating at the nexus of different state systems, private sector interests, citizen groups, and the public good, planners often feel the pressure from developers, land owners, political party representatives, other departments, and other urban actors to compromise the principles of their planning practice in favor of specific interests.

To make matters more complex, in many postcolonial contexts antiquated planning legislation and cumbersome regulatory frameworks clearly do not serve the public good (Watson 2009, 2011; Berrisford 2011). This makes it harder to ascertain a basis for just and fair decision-making and uncover potential corruption. While planning decisions that benefit or disadvantage particular groups or identities thus might not always be clear-cut cases of corruption, they nonetheless tend to further entrench existing patterns of socio-spatial inequality within the city.

10.5 Building Cities of Integrity in Zambia

Zambia records both high rates of urbanization and high numbers of alleged cases of corruption in urban development. The country ranked a lowly 150th in the world for effectiveness in registering property, and enforcing property rights through the courts is an often cumbersome and uncertain process (US Department of State 2017). There are numerous past and ongoing urban development projects that have been investigated by Transparency International Zambia, other civil society organizations, and public authorities for alleged corruption, fraud, and financial mismanagement (Chulu 2019). Moreover, the media is replete with reports that implicate the planning profession in compromised urban development and land administration processes (Phiri 2019). Thus, planners in Zambia are increasingly exposed, and sometimes also alleged as party, to corruption in local urban development processes.

Figure 10.4 illustrates how different factors affect the performance of Zambian planners on the job. A key impediment is the low level of funding for planning work. Typically, fieldwork is negatively affected, and clients often have to offer additional resources such as transport, fuel, allowances to facilitate site inspections, and final determination on planning submissions. When clients take charge of funding official planning work, the independence of the profession and objectivity of the decision-making process tends to be compromised. Planners have also indicated that they do not always have sufficient and guaranteed support from planning department directors and senior institutional managers. The lack of support is visible when directors or heads of planning authorities do not respect or adopt specific planning recommendations by planners and, in certain circumstances, might even request the planner arrives at a predetermined planning outcome. A further factor is that there is a widespread view among the public that planners are corrupt and demand bribes to do their job. Those planners who resist and reject bribes from the public are seen to be indirectly asking for bigger bribes. This results in mistrust and lack of public support, hence making it harder for planners to navigate the thin line between harsh public perception and fast-moving urban developments. Finally, with most planners being relatively young and inexperienced, there is a lack of familiarity with the latest planning laws and newly promulgated planning guidelines. Both our survey and interview data indicate that these factors, taken together, negatively impact the professional performance of planners in Zambia.

Zambia is a signatory of the three most important global anti-corruption conventions, the UN Convention Against Corruption, the African Union (AU) Convention on Preventing and Combating Corruption, and the Organization for Economic Cooperation and Development (OECD) Convention Against Foreign Bribery, which has identified corruption as one of the major impediments to local growth and development (OECD 1997). Choosing Zambia as the

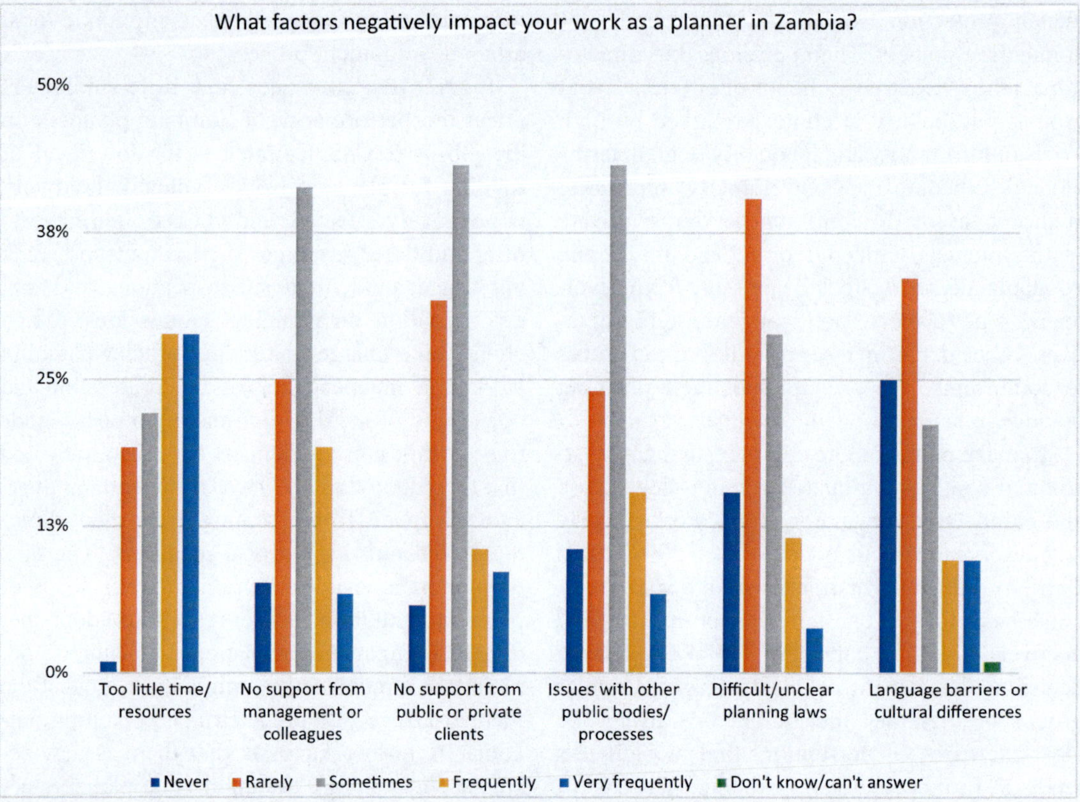

Fig. 10.4 Survey responses from Zambian planners regarding the factors that most negatively impact their performance at work (In % of $n = 98$, Cities of Integrity online survey 2020)

location for our research intervention was thus not primarily spurred by the mere prevalence of corruption in urban development but also—and perhaps more importantly—by the variety of local efforts to create an enabling environment for planners to make a positive contribution to national development. Taken together, these conditions create an exciting window of opportunity for administering integrity training, leveraged for research in the form of a Qualitative Action Experiment (QAE).

The practice of urban and regional planning in Zambia is regulated by the Urban and Regional Planners Act of 2011. The Act provides for the establishment of the Zambia Institute of Planners (ZIP), whose mandate is to register planners and planning firms, to regulate their professional conduct, and to provide for matters connected with or incidental to town planning. Thus, every planner in Zambia is required by law to be registered and regulated but is also entitled to be professionally supported by ZIP.

Until recently, ZIP failed to perform in accordance with its national goals, objectives, and vision. Poorly organized and lacking funding, the profession was performing very poorly (Taylor and Thole 2015), and practicing as a planner in Zambia was not sanctioned by any formal accreditation. In 2011, only 60 accredited planners were registered as members.

After the enactment of the Urban and Regional Planners Act of 2011, Zambia took further reforms, including the Urban and Regional Planning Act in 2015, and the introduction of a capacity building-focused Master's program in Spatial Planning at the University of Zambia and its support research center, the Centre for Urban Research and Planning (CURP). As a result, urban planning has recorded remarkable growth and assumed an important role and voice on matters of urban development in Zambia. As of 2019, the Institute records show over 800 accredited practicing planners (personal communication, Treasurer of ZIP, 2019).

Following these reforms, ZIP has achieved a powerful position from which to influence the collective development and sociopolitical positioning of the profession. Through its mandate to formally register planners, it regulates access to the profession, thus holding the opportunity to control quality, sanction irregular planning practice, and promote integrity in the profession. From 2017 onwards, ZIP has promulgated a Code of Ethics as a reference document for ethical and professional conduct of planners in Zambia. Further, ZIP is currently working to formulate the Planners' Practice Guide to operationalize the Code of Ethics. Used together with the provisions in the Urban and Regional Planners Act of 2011 and the Urban and Regional Planning Act of 2015, ZIP is a key stakeholder in developing and enforcing professional accountability mechanisms and proactively promoting integrity as a shared collective of planners in Zambia.

Unlike in the past, only qualified planners with full accreditation can now practice as planners in both the public and private sector in Zambia. ZIP has also recorded an increase in the number of private planning firms, with numbers currently standing at one international and seven national firms. There are now accredited planning firms offering planning services from consultancy and education, to community empowerment and civic development in the urban land and urban development sectors. Other reform efforts include the launch of an innovative forward-looking Association of African Planning Schools (AAPS) model planning program at the University of Zambia. This seeks to support the implementation of the new national efforts to reform and enhance the role of planning in Zambia and serves as a model for other planning schools in Africa to reform their planning curriculum and pedagogy. Watson and Agbola (2013, p. 7) characterize this program as "embodying content" and one that seeks to foster a pedagogy that can respond to current and future urbanization challenges and promote transparency in urban planning and urban development in Africa.

An overwhelming majority of 86% of Zambia planners surveyed indicate that planning institutes like ZIP have a direct role in promoting integrity in planning and urban development. Figure 10.5 shows that Zambian planners expect

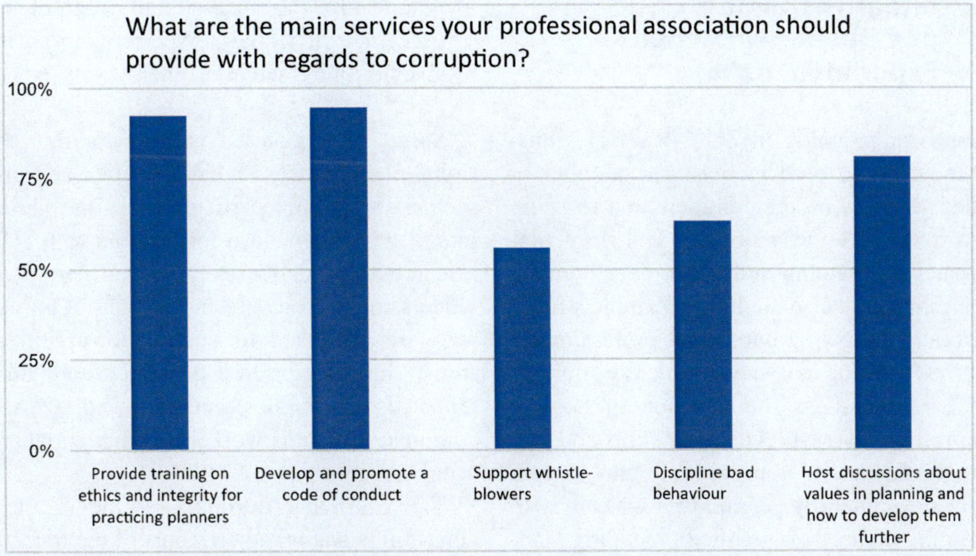

Fig. 10.5 Survey responses from Zambian planners regarding the services that should be provided by their professional association regarding curbing corruption and promoting integrity (In % of n = 98, Cities of Integrity online survey 2020)

the ZIP to train planners in how to avoid corruption and uphold integrity. This is crucially important, as the majority of interviewees and survey participants indicated that their university training—mostly at bachelors level—did not adequately address corruption and integrity, making them feel ill-prepared for facing the harsh reality of corruption in the planning and urban development sector. The planners equally hold the view that ZIP needs to support moral professional behavior and punish wrongdoing. To be able to do this, the planners believe a practice manual must be formulated and promoted so that all planners are well-guided and that ZIP can effectively sanction corrupt and unethical conduct.

In sum, the outlook for planning in Zambia presents an opportunity to embed integrity both in the growth of the planning profession and in the profession's role in influencing the nature of urban development in Zambia. Presenting itself as a case with numerous opportunities for interventions to transform and enhance the role of planners in national development, Zambia struck the research team as a unique case that warranted the implementation of an action-driven research agenda.

10.6 Integrity Training as a Qualitative Action Experiment (QAE)

The status of planning in Zambia clearly illustrates the need to both increase the number of qualified planners on the continent and to equip them with the soft power skills to deal with complex issues in planning and urban development. The current status of planning in Zambia—characterized by a growing number of professionals, sometimes working in severely ethically compromised contexts—begs the question of how to equip professionals with adequate skills and values that champion transparency and ensure greater accountability. Similarly, measures to increase the number of qualified practicing planners must be matched by efforts to strengthen their professional integrity.

The Qualitative Action Experiment (QAE) in Lusaka, which took place in October 2019, sought to leverage professional pride and collective acumen, channeling it towards countering potentially corrupt practices and presenting an opportunity to empower Zambian planners to detect, report, and avoid corrupt practices (for more details on the research methodology, see Siame et al. 2020).

Taking the form of an intensive 2.5-day training workshop, the aim of the intervention was to offer a safe space for local planners from across the country to discuss their day-to-day encounters with corruption, share personal responses, think through possible collective support strategies together with the professional body, and build a community of practice around the positive notion of integrity. Translating our research principles into practice, this meant that the training:

- Placed a strong focus on practical mechanisms of lived professional integrity, rather than a narrow focus on anti-corruption compliance
- Adopted a systemic approach that takes seriously the positive influence of collective identities, social norms, and shared organizational cultures
- Allowed for the discussion of different and, at times, controversial perspectives in an inclusive non-judgmental manner

Some 38 registered and practicing professional planners from both the public and private sectors in Zambia participated in the Cities of Integrity QAE. In close cooperation with ZIP, the research team deliberately invited certain individuals considered to be in positions of influence, who are subjected to intense corruption pressures, and also ensured equal representation in terms of age, gender, seniority, and size of the administrative unit worked in (municipal, rural, or urban centers).

The following introduces some of the key aims and pedagogical elements of the training:

Raising awareness about the scale, scope, and consequences of corruption and debunking common corruption myths. Developed to kickstart the discussion, this component honed in on the detriments of corruption and set out to debunk common myths about the causes and effects of corruption. For example, it is an often-repeated truism that corruption is only about a few bad apples that spoil the barrel. This has been challenged in multiple behavioral studies that show that we overestimate our ethical agency and that certain contexts might override our moral reasoning (Ashforth et al. 2008).

Sharing context-specific information about corruption risks in urban development and planning. Beyond general corruption dynamics, there are specific processes that are important to pay attention to when considering corruption in urban development, e.g., around re-zoning and building application and the financial windfalls that are often attached to them. Participants took the opportunity to compare their own experiences in the Zambian context, what common issues they face in their daily practice, and what specific consequences they or their colleagues have faced when calling out professional misconduct.

This conversation was further deepened through creating awareness about the practical challenges of ethical agency. This introspective session illustrated the subtle psychosocial mechanisms that can even lead well-intentioned people down a slippery slope of corruption. In interactive sessions, participants were asked to "make the call" on common moral challenges in the planning sector, with scenarios covering issues like the "revolving door" and cases of undue political interference with planning decisions.

Furthermore, in line with the action-oriented ethos of the research project, the QAE also aimed to equip participants with practical tools and support mechanisms for asserting integrity. Here, we drew on the expertise of Transparency International's Zambia chapter, whose president Rueben Lifuka joined the workshop to discuss local anti-corruption instruments and reflect on the potential of leveraging ZIP as a crucial integrity infrastructure in the profession. Lifuka's input also created the space for a heated discussion around issues of private planning jobs fulfilled by public servants—a widespread phenomenon in the Zambian context—which raised questions around how planners deal or fail to deal with conflict of interest.

Taken together, the aim of our QAE workshop was to catalyze action for behavior change and foster peer involvement, as well as commitment to champion institutional reforms in local planning departments. Thus, on the final day of the workshop, all participants were invited to commit to one or several activities (see Table 10.1). These exemplary commitments were derived from some of the micro-tactics and broader strategies for integrity promotion brought forward by

Table 10.1 Integrity commitments of Qualitative Action Experiment (QAE) participants (The number in brackets refers to the number of participants who committed to the specific activity. Participants were able to commit to more than one activity. Source: Siame et al. 2020)

- Meet with colleagues/mentees/subordinates to share the results of the workshop (24)
- Draft/develop service charter (9)
- Make service charter and procedures more publicly accessible/visible (9)
- Draft/develop checklists (7)
- Improve personal efforts and practices to be a person of integrity (7)
- Strengthen integrity of colleagues/workplace (6)
- Put sign up with integrity-focused message in office or visible public place of work/vehicle (4)
- Take steps to educate clients regarding processes/procedures (4)
- Be a resource on or champion for integrity and anti-corruption (4)
- Call out the corruption of others (2)
- Praise/reward demonstrations of integrity by others (2)
- Introduce regular work meetings to reflect on integrity (2)

participants over the course of the workshop and were further informed by a growing empirical literature on the efficacy of behavioral nudges (Thaler and Sunstein 2009). To fully retain the voluntary aspect and avoid undue public pressure, participants made commitments to themselves in writing, sharing them with the course convenors in anonymous form. In our ongoing engagement with the QAE participants, the list serves as a reminder to participants of their individual promises and commitments, making them aware of the collective promises of the group and inviting them to reflect on potential challenges and progress made over the course of the research project.

While follow-up research with QAE participants suggests that integrity challenges have by no means eased after the training—with two-thirds stating that they were asked to approve a predetermined planning outcome—there is a prevailing commitment to champion integrity in local work environments, e.g., by drafting a departmental service charter and lobbying for the creation of integrity committees (Siame et al. 2020, p. 15).

10.7 Conclusion

In this chapter, we have argued that urban development is a particularly prolific turf for corruption, but by the same token, urban planning can be an equally potent node for combatting it and fostering professional integrity. This makes the sector particularly interesting for integrating SDGs 11 and 16 at the city level. Early impacts from our multifaceted, action-oriented research are promising, showing that promoting professional integrity among urban planners is an important lever for tackling urban corruption. At the same time, it is a piece of a larger puzzle when it comes to the quest of building cities of integrity, particularly in the African context.

At first glance, the big picture does not bode well. Rapid urbanization coupled with late industrialization presents a challenge to create economic opportunities for the growing, young urban populations on the continent (Pieterse 2008). Furthermore, climate change puts enormous strain on habitats and livelihoods, exacerbating existing corruption and creating new corruption risks (Transparency International 2011; Fredriksson and Neumayer 2016). Finally, the global COVID-19 crisis that hit the world full steam in 2020 pushed governments and populations to the brink of economic collapse, generating news about cases of procurement fraud in several African counties (Merten 2020; Schipani et al. 2020). In sum, meeting both SDGs 11 and 16 has become ever more challenging.

At the same time, there remain reasons for holding out hope. Cities have emerged as trail blazers for new standards of openness and transparency, for instance in local procurement and public contracting, with 56 new local municipalities particularly from Africa and Asia joining the global partnership to promote open government (Open Government Partnership 2019). While a universal leapfrog narrative for the continent has been rightly questioned (Alzouma 2005; Srinivasan et al. 2019), there are undeniably growing pockets of digital and technological innovation, not least in the civic tech and social innovation bracket. It is particularly the cities with a critical mass of young and well-educated entrepreneurs that are at the cutting edge of new data-sharing models and platform economies, as well as novel approaches for local democratic participation in Africa.

Finally, while urban life continues to present itself as messy, pluralistic, and experimental, this experimentation and "hustle" also continues to yield new political openings for local action and social justice (Simone and Pieterse 2018). In order to build cities of integrity and craft a sustainable urban future, it is therefore crucial that the process takes a context-sensitive approach that works closely with local actors, takes seriously their lived realities, and is open to experimentation.

References

African Centre for Cities, Center for Urban Research and Planning (CURP) (2020) Cities of Integrity online survey 2020 [dataset], version 1. African Centre for Cities, CURP [producers]; Cape Town, Lusaka; DataFirst [distributor], Cape Town. https://doi.org/10.25828/sy2r-y323

Afrobarometer (2018) Data from 2016/2018 survey (round 7). http://www.afrobarometer.org/online-data-analysis/analyse-online. Accessed on 20 October 2020

Alzouma G (2005) Myths of digital technology in Africa. Global Media and Communication 1(3):339–356

American Planning Association (2019) Annual Report 2019, APA Dashboard. https://www.planning.org/dashboard/. Accessed 21 October 2021

Ashforth BE, Gioia DA, Robinson SL, Treviño LK (2008) Re-viewing organizational corruption. Academy of Management Review 33(3):670–684

Atkinson R (2019) Necrotecture: lifeless dwellings and London's super-rich. International Journal of Urban and Regional Research 43(1):2–13

Balassiano K (2011) Tackling "wicked problems" in planning studio courses. Journal of Planning Education and Research 31(4):449–460

Berrisford S (2011) Why it is difficult to change urban planning laws in African countries. Urban Forum 22(3):209–228

Boles JR (2017) Million dollar ghost buildings: dirty money flowing through luxury real estate markets. Real Estate Law Journal 476. https://papers.ssrn.com/sol3/papers.cfm?abstract_id=3518771. Accessed 20 October 2021

Chiodelli F, Moroni S (2015) Corruption in land-use issues: a crucial challenge for planning theory and practice. Town Planning Review 86(4):437–455

Chulu C (2019) Transparency International Zambia's statement on leaders' acquisition of land on Forest 27. https://tizambia.org.zm/transparency-international-zambias-statement-on-leaders-acquisition-of-land-on-forest-27. Accessed 21 October 2020

Cling JP, Razafindrakoto M, Roubaud F (2018) SDG 16 on governance and its measurement: Africa in the lead. Working paper DT/2018-2. https://horizon.documentation.ird.fr/exl-doc/pleins_textes/divers18-03/010072517.pdf. Accessed 21 October 2020

Fernandez R, Hofman A, Aalbers MB (2016) London and New York as a safe deposit box for the transnational wealth elite. Environment and Planning A: Economy and Space 48(12):2443–2461

Financial Action Task Force (FATF) (2013) Money laundering and terrorist financing vulnerabilities of legal professionals. https://www.fatf-gafi.org/media/fatf/documents/reports/ML%20and%20TF%20vulnerabilities%20legal%20professionals.pdf. Accessed 21 October 2020

Frank AL, Silver C (2017) Urban planning education: beginnings, global movement and future prospects. Springer Publishing, New York

Fredriksson PG, Neumayer E (2016) Corruption and climate change policies: do the bad old days matter? Environmental and Resource Economics 63(2):451–469

Glucksberg L (2016) A view from the top: unpacking capital flows and foreign investment in prime London. City 20(2):238–255

Johnson BJ (2010) City planners and public service motivation. Planning Practice & Research 25(5):563–586

Khan MH (2018) Corruption and illicit financial flows. https://new.unctad.org/system/files/non-official-document/stat2018_em_iff0620_session5_khanandreoni.pdf. Accessed 21 October 2020

Kim H (2020) Corruption as infrastructure: rendering the new Saigon global. International Journal of Urban and Regional Research 44 (6): 1057-1071

Merten M (2020) SIU probes R5.08-bn in questionable Covid-19 tenders, while lists of PPE contracts emerge. Daily Maverick, 20 August. https://www.dailymaverick.co.za/article/2020-08-20-siu-probes-r5-08-bn-in-questionable-covid-19-tenders-while-lists-of-ppe-contracts-emerge. Accessed 21 October 2020

Murshed M, Mredula F (2018) Impacts of corruption on sustainable development: a simultaneous equations model estimation approach. Journal of Accounting, Finance and Economics 8(1):109–133

Olufemi O (2008) Experiential and emotional encounters of women planners in Africa. Development in Practice 18(3):412–419

Olver C (2017) How to steal a city: the battle for Nelson Mandela Bay: an inside account. Jonathan Ball Publishers, Johannesburg and Cape Town

Olver C (2019) A house divided: the feud that took Cape Town to the brink. Jonathan Ball Publishers, Johannesburg and Cape Town

Open Government Partnership (2019) Global report volume 1: democracy beyond the ballot box. https://www.opengovpartnership.org/wp-content/uploads/2019/09/Global-Report_Volume-1.pdf. Accessed 21 October 2020

Organisation for Economic Cooperation and Development (OECD) (1997) Convention on combating bribery of foreign public officials in international business transactions. Signed 17 December 1997, entry into force 15 February 1999. http://www.oecd.org/corruption/oecdantibriberyconvention.htm. Accessed 21 October 2020

Parnell S (2016) Defining a global urban development agenda. World Development 78:529–540

Phiri C (2019) The "mysterious" 48 apartments. Zambia Reports, 11 August. https://zambiareports.com/2019/08/11/mysterious-48-apartments. Accessed 21 October 2020

Pieterse E (2008) City futures: confronting the crisis of urban development. Zed Books, London and New York; UCT Press, Cape Town

Prindex (2020) A global assessment of perceived tenure security from 140 countries: comparative report. https://www.prindex.net/documents/537/PRINDEX-Comparative-2020.pdf. Accessed 21 October 2020

Pring C, Vrushi J, Afrobarometer (2019) Global corruption barometer Africa 2019: citizens' views and experiences of corruption. https://www.transparency.org/files/content/pages/2019_GCB_Africa.pdf. Accessed 21 October 2020

Razafindrakoto M, Roubaud F (2010) Are international databases on corruption reliable? A comparison of expert opinion surveys and household surveys in sub-Saharan Africa. World Development 38(8):1057-1069

Robins S, Cornwall A, von Lieres B (2008) Rethinking "citizenship" in the postcolony. Third World Quarterly 29(6):1069–1086

Royal Town Planning Institute (2019) The UK planning profession in 2019: statistics on the size and make-up of the planning profession in the UK. https://www.rtpi.org.uk/media/5889/theplanningprofessionin2019.pdf. Accessed 21 October 2020

Sager T (2009) Planners' role: torn between dialogical ideals and neo-liberal realities. European Planning Studies 17(1):65–84

de Satgé R, Watson V (2018) Urban planning in the Global South: conflicting rationalities in contested urban space. Springer Publishing, New York

Schipani A, Cotterill J, Munshi N (2020) Africa's Covid-19 corruption: "theft doesn't even stop during a pandemic." Financial Times, 1 September. https://www.ft.com/content/617187c2-ab0b-4cf9-bdca-0aa246548745. Accessed 21 October 2020

Siame G, Nkula-Wenz L, Zinnbauer D (2020) Methodological considerations for researching corruption and building urban integrity: a qualitative action experiment in Zambia. Cities of Integrity working papers, no. 2. African Centre for Cities. https://www.africancentreforcities.net/wp-content/uploads/2020/07/GI_ACE_WORKING_PAPER_2_3.pdf. Accessed 21 October 2020

Simone AM, Pieterse E (2018) New urban worlds: inhabiting dissonant times. Polity Press, Cambridge

de Simone M (2015) Corruption on your doorstep: how corrupt capital is used to buy property in the UK. Transparency International UK, London

Srinivasan S, Diepeveen S, Karekwaivanane G (2019) Rethinking publics in Africa in a digital age. Journal of Eastern African Studies 13(1):2–17

Taylor TK, Thole CB (2015) Re-thinking town and country planning practice in Zambia. Developing Country Studies 5(10):34–43

Thaler RH, Sunstein CR (2009) Nudge: improving decisions about health, wealth, and happiness. Yale University Press, New Haven

Transparency International (2011) Global corruption report: climate change. Sweeney G, Dobson R, Despota K, Zinnbauer D (eds). https://images.transparencycdn.org/images/2011_GCRclimatechange_EN.pdf. Accessed 21 October 2020

UN-Habitat (2010) State of the world's cities report (2010/11): bridging the urban divide. https://sustainabledevelopment.un.org/content/documents/11143016_alt.pdf. Accessed 20 October 2020

UN-Habitat, African Planning Association (2013) The state of planning in Africa. http://unhabitat.org/the-state-of-planning-in-africa. Accessed 21 October 2020

United States Department of State (2017) Investment climate statement: Zambia. https://www.state.gov/reports/2017-investment-climate-statements/zambia. Accessed 11 October 2020

Watson V, Agbola B (2013) Who will plan Africa's cities? Africa Research Institute. https://www.africaresearchinstitute.org/newsite/wp-content/uploads/2013/09/ARI-Counterpoint-Who-will-plan-Africas-cities1.pdf. Accessed 21 October 2020

Watson V (2013) The ethics of planners and their professional bodies: response to Flyvbjerg. Cities 32:167–168

Watson V (2011) Changing planning law in Africa: an introduction to the issue. Urban Forum 22(3):203–208

Watson V (2009) "The planned city sweeps the poor away…": urban planning and 21st century urbanisation. Progress in Planning 72(3):151–193

Watson V (2003) Conflicting rationalities: implications for planning theory and ethics. Planning Theory & Practice 4(4):395–407

Watson V (2002) The usefulness of normative planning theories in the context of sub-Saharan Africa. Planning Theory 1(1):27–52

World Bank (2018) From regulators to enablers: the role of city governments in economic development of greater Kampala. https://openknowledge.worldbank.org/bitstream/handle/10986/28459/119806-REVISED-PUBLIC-ADD-AUTHORS-P158599-GreaterKampalaFinalOutput.pdf?sequence=5&isAllowed=y. Accessed 21 October 2020

Zinnbauer D (2019) Leveraging the role of the urban planning profession for one of the central policy challenges of our times. Cities of Integrity working papers, no. 1. African Centre for Cities. https://www.africancentreforcities.net/wp-content/uploads/2019/07/GI_ACE_WORKING_PAPER_1_DZ_.pdf. Accessed 20 October 2020

Zinnbauer D (2020) Urbanisation, informality, and corruption: designing policies for integrity in the city. U4 issue 2020:6. https://www.u4.no/publications/urbanisation-informality-and-corruption. Accessed 6 October 2020

Part IV

Collaborative Experiences from the Frontier of Practice

11 Mozambique's Voluntary Local Review: SDG Localization, Decentralization, and the Role of Local Governments and Associations

Pedro Laíce and Sylvia Croese

Abstract

The National Association of Mozambican Municipalities (ANAMM) has been undertaking a series of activities since 2016 that aim to raise awareness of Agenda 2030 and the Sustainable Development Goals (SDGs) among its members. These activities have contributed to increased ownership of the SDGs at the municipal level, the start of the incorporation of the SDGs in local planning processes, and the completion of the country's first Voluntary Local Review (VLR) in 2020, with contributions from 16 out of the country's 53 municipalities. This chapter draws on the experience of ANAMM to highlight the opportunities and challenges for SDG localization in urban Africa. The case of Mozambique shows that while the SDGs are high on the central government's agenda, municipalities lack the resources and capacity to fully contribute to meeting the SDGs as a result of uneven levels of decentralization. Nevertheless, the VLR revealed high levels of local awareness and diverse examples of initiatives and good practices that can be built on in order to strengthen and support SDG implementation at the local level.

Keywords

National Association of Mozambican Municipalities (ANAMM) · SDG localization · Voluntary local review (VLR) · Decentralization · Mozambique

11.1 Introduction

The National Association of Mozambican Municipalities (ANAMM) represents the united voice of municipalities of Mozambique. Its core mandate is to represent Mozambican municipalities, advocate for their rights and interests among the government and development partners, and inform and train its members on matters related to municipal development. Since 2016, the association has developed a series of activities with the aim of promoting Agenda 2030 and the Sustainable Development Goals (SDGs) among its members. These activities have included awareness raising, dissemination, and capacity building initiatives on the SDGs among elected

P. Laíce (✉)
National Association of Mozambican Municipalities (ANAMM),
Maputo, Mozambique
e-mail: pedro.laice@yahoo.com

S. Croese
University of the Witwatersrand, Johannesburg, South Africa

African Centre for Cities, University of Cape Town, Cape Town, South Africa
e-mail: sylvia.croese@wits.ac.za

S. Croese, S. Parnell (eds.), *Localizing the SDGs in African Cities*, Sustainable Development Goals Series, https://doi.org/10.1007/978-3-030-95979-1_11

officials and technical staff in different municipal development fora. Taken together, these actions have contributed to increased ownership of the SDGs at the municipal level, the start of the incorporation of the SDGs in local planning processes, and the completion of the country's first Voluntary Local Review (VLR) in 2020 with contributions from 16 out of the country's 53 municipalities.

This chapter draws on the experience of ANAMM to highlight the opportunities and challenges of SDG localization at local government level in a context of uneven levels of decentralization. It starts by reviewing the emergence of VLRs and their purpose, contribution, and importance in the urban African context. It then builds on the Mozambican case to highlight the importance of local governments for the implementation of the SDGs. While the SDGs are high on the central government's agenda, municipalities lack the resources and capacity to fully contribute to meeting the SDGs. Nevertheless, the VLR revealed high levels of local awareness and diverse examples of initiatives and good practices that can be built on in order to strengthen and support SDG implementation at the local level.

11.2 Voluntary Local Reviews and the African Context

As part of the review process of Agenda 2030 for Sustainable Development, member states are encouraged to undertake regular and inclusive reviews of their progress towards the SDGs and present these to the annual United Nations High Level Political Forum (UN HLPF). Reporting is processed under the auspices of the Economic and Social Council (ECOSOC).

In the first 5 years following 2015, a total of 168 countries reported on their efforts (Kindornay 2018). Despite the COVID-19 pandemic and its impacts, global interest in preparing a Voluntary National Review (VNR) remained high in 2020, with 47 countries preparing and virtually presenting their VNRs, including 16 from Africa (UNDESA 2020a).

While the involvement of local and regional governments in VNR processes was limited in the first VNR cycles, this has gradually increased. According to the global local government association United Cities and Local Governments (UCLG), one of the main voices and advocates for SDG localization, local and regional governments' involvement in VNR processes increased to 55% in 2020, up from 42% in the 2016–2019 period. Additionally, UCLG has noted an increase in the elaboration of Voluntary Local and Regional Reviews, understood as publications of local and regional governments that "assess and present advances on the fulfilment of the 2030 Agenda from a local standpoint and through a locally-developed narrative" (UCLG and UN-Habitat 2021, p. 11). Some countries have started to include the results of these subnational reports in national reviews (UCLG 2020a), while the UN has started to record available VLRs on the UN Department for Economic and Social Affairs website (UNDESA 2021).

The city of New York was the first to submit a VLR on the SDGs to the UN HLPF in 2018, and the year after, 22 cities signed a Voluntary Local Review Declaration, committing themselves to submit a VLR of local progress in achieving the SDGs to the UN HLPF (NYC Mayor's Office for International Affairs 2019). Since then, the number of local governments that has completed a VLR has exponentially increased, from over 50 as of July 2020 to at least 110 VLRs either already published and publicly available or to be published in 2021 and 2022. This also includes a growing number of what have come to be referred as Voluntary Subnational Reviews (VSRs), which represent country-wide bottom-up reports that assess the state of localization processes at subnational levels (UCLG and UN-Habitat 2020, 2021) (see Fig. 11.1).

Due to the novelty, and in many cases spontaneous emergence of VLRs and VSRs, many of them vary in terms of their content, scope, purpose, methodology, template, and format. For instance, VLRs vary in terms of the number of SDGs that are covered, the kind of data used and reported on, the number and extent of stakeholder input, and the actors involved in putting the VLR

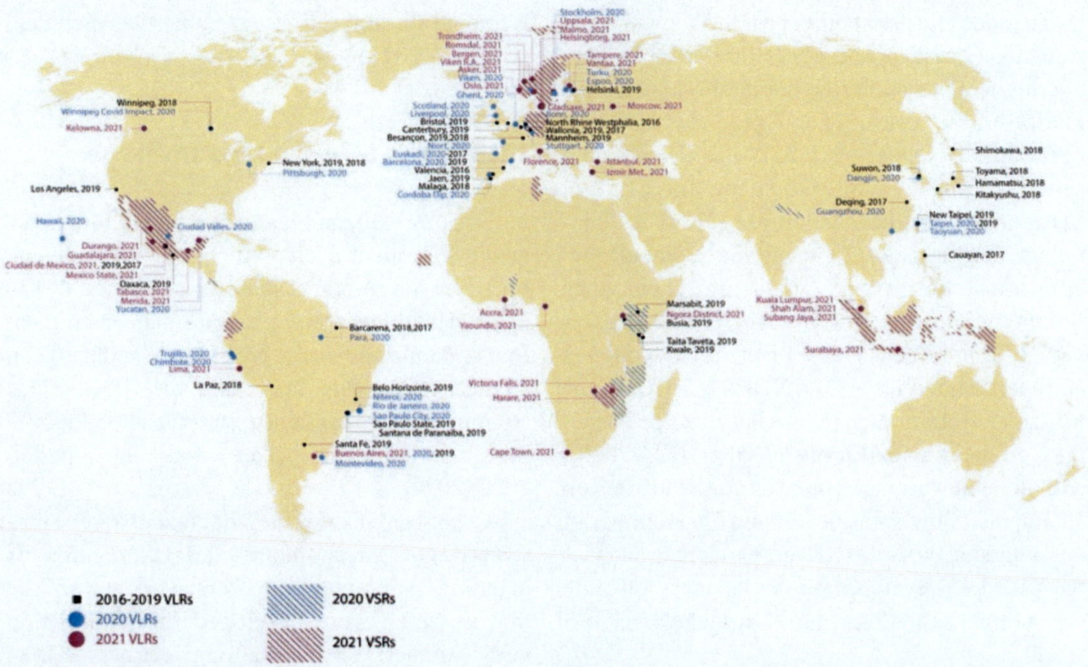

Fig. 11.1 Map of currently completed Voluntary Local Reviews (VLRs) and Voluntary Subnational Reviews (VSRs). (Source: elaborated by UCLG based on the VLR and VSR repository, See https://gold.uclg.org/report/localizing-sdgs-boost-monitoring-reporting#field-sub-report-tab-2)

together, with approaches ranging from more internal to external collaborations and partnerships (Pipa and Bouchet 2020; Ortiz-Moya et al. 2020). Many of the first-generation VLRs have been completed in developed countries and relatively very few in Africa. Support for the development of VLRs in African cities is therefore particularly important, considering that VLRs represent a useful instrument and opportunity to connect local strategies to the SDGs, strengthen evidence-based policymaking, identify gaps and priorities, as well as mobilize new policies, partnerships, and resources (Pipa and Bouchet 2020). As such, VLRs offer a practical way to participate in a growing global movement and accelerate local awareness and action for sustainable urban development.

The development of VLRs is especially relevant in the African context as efforts to include local and regional governments in national SDG reviews remain limited across the continent, in spite of the rapid growth of cities and their central contribution to national development. Although most African countries have integrated the SDGs in their national development plans and reported on their SDG progress to the UN HLPF, very few have actively included local governments or their associations in coordination or reporting processes. More specifically, only about half of the countries that reported to the UN between 2016 and 2019 involved local or regional government associations in their VNR processes, while only about a third of these countries reported having involved local or regional governments in national coordination mechanisms on the SDGs (UCLG 2019).

The limited involvement of local governments reflects the uneven levels of decentralization across the continent and the limited extent to which institutional political, administrative, and fiscal environments are favorable for the action of cities and regions (UCLGA and Cities Alliance 2018). Because of such unfavorable environments, there are large gaps in terms of the data required to monitor and implement development policies at the local level, resulting in inadequate urban policies and interventions, reduced accountability, and reduced social inclusiveness.

To guide and assist other cities, policy makers, researchers, and practitioners interested in conducting a VLR, numerous guidelines, tools, and platforms for sharing good practices are now becoming available. UNDESA has developed a series of Global Guiding Elements for VLRs of SDG implementation which are meant to "provide a starting point for local and regional governments considering producing their own Voluntary Local Review and aim to give an overview of useful focus areas for the review process and the eventual report" (UNDESA 2020b). The European Commission in turn has developed specific guidelines for European cities (EU 2020), and there are various online sharing platforms on VLRs, including some developed by a number of cities themselves, as well as networks, such as the VLR Lab developed by the Japanese Institute for Global Environmental Strategies (IGES 2021).

To contribute to such sharing and learning, UCLG has developed a range of training tools, guidelines, and modules and started a VLR Community of Practice in collaboration with UN-Habitat in order to facilitate and support the development of VLRs as well as key practical knowledge, lessons, and learnings (UCLG and UN-Habitat 2020). Important in this regard is UCLG's support to a number of bottom-up, country-wide pilot experiences on subnational reporting in collaboration with local government associations. In Africa, these include experiences in Kenya, Benin, and Mozambique, the latter which we report on in this chapter (UCLG 2020b).[1] Importantly, the United Nations Economic Commission for Africa (UNECA) has also started discussions around the preparation of a template for peer-reviewed VLRs in Africa that is aligned with both Agenda 2030 and Agenda 2063 reporting, following requests made from African member states to UNECA at the Africa Regional Forum on Sustainable Development (UNECA 2021).

11.3 The SDGs in Mozambique

In 2015, the Mozambican government initiated a new government cycle, which was accompanied by a new development plan for the period 2015–2019. This plan represents the main instrument for economic and social governance and has been aligned to Agenda 2030 and the SDGs. This alignment was maintained with the adoption of a new development plan for the period 2020–2024.

A number of other key documents are representative of Mozambique's efforts to align its planning and statistical frameworks with the SDGs, which were developed in collaboration with national and international partners. These include a number of interrelated studies, namely:

- A study mapping government capacity for SDG implementation, which assesses institutional, planning, and statistical capacity (MEF 2016a; FAO 2018)
- A study on the viability of SDG indicators, which presents existing indicators relevant for SDG monitoring (MEF 2016b)
- A study on the "mozambicanization" of the SDGs through the 5-year development plan (MEF 2017)

These studies were followed by the development of an Initial Report on the implementation of the SDGs in Mozambique in 2018, which outlines the status of Agenda 2030 implementation, with a particular focus on SDG 2, 3, 4, 6, 13, and 14 (MEF 2018). The report also includes a proposal for a National Indicator Framework on the SDGs in Mozambique, which was updated in 2020 (INE 2020).

To coordinate the country's efforts around the SDGs, in 2017 the government also created a National Reference Group for the implementation, monitoring, evaluation, and financing of the SDGs, as well as Agenda 2063. The Reference Group is presided over by the Ministry of

[1] Note that in line with more recent definitions by UCLG, the Mozambique exercise should perhaps more appropriately be referred to as a VSR, however in this chapter we follow the term VLR as this is how it was initially conceived and is referred to locally.

Economy and Finances and further includes representatives from the Ministry of Foreign Affairs and Cooperation, the National Institute of Statistics, the National Assembly, civil society, ANAMM, the private sector, and bilateral and multilateral development partners and UN agencies.

Members of the National Reference Group worked together towards the development of the country's first VNR, which was presented to the (virtual) UN HLPF in 2020 (Republic of Mozambique 2020). In this report, the government outlines its progress and the challenges involved in SDG implementation and the importance of multi-stakeholder collaboration, including local governments, which have been instructed to integrate the SDGs into their local strategies and operations.

However, while the National Reference Group on SDGs is meant to be replicated at the provincial and district government level, notably there are no provisions in place for its replication at the municipal government level. This illustrates the limited space afforded to municipal governments when it comes to driving local development or engaging autonomously with multilateral processes, in spite of important progress made in decentralization over the past decades.

11.4 Local Governments and SDGs

Decentralization reforms in Mozambique followed the end of the civil war in the early 1990s and culminated in the creation of municipal authorities in 1997. Under Law 8/97 of 31 May, municipal authorities were administratively classified into five different categories, ranging from categories "A" to "D", in addition to the category of small "towns" (*vilas*). This categorization takes different criteria into account, including political, economic, social, cultural factors, as well as population density, number and type of industries, degree of development of trade activities, education, and sanitation. Based on these criteria, the capital city of Maputo is accorded a special status and is singularly classified as type "A". The adjoining municipality of Matola and provincial capitals of Nampula and Beira are classified as type "B", while type "C" includes all other provincial capitals, as well as other cities. Type "D" are small districts that play an important role in the process of local development (ANAMM and World Bank 2009).

This classification has guided the country's gradual approach to decentralization, limiting the local governments that could participate in the country's first local elections in 1997 to 33 urban cities and towns, with another 10 added in the elections of 2008, followed by 10 more in 2018. In addition, Mozambique has 128 districts and 11 provincial governments, which represent two additional tiers of subnational government. As part of the most recently adopted package of decentralization laws in 2018 and 2019, the government has introduced a new model of "decentralized provincial governance" (Forquilha 2020). This includes the creation of a new figure of the Provincial Secretary of State, who is appointed by the president of the republic and functions alongside the elected provincial governor. Under a new legislation approved in 2020 by Resolutions 21/2020 and 22/2020 of 26 March, the Mozambican government has updated the classification of cities, districts, and towns as well as introduced the classification of category "A" towns, some of which coincide with existing municipal authorities.

In this complex legal and administrative context, powers to municipal governments have only gradually been devolved, first under the local government laws adopted in 1997 which were revised and expanded between 2006 and 2008 and more recently in 2018. Currently, the following competencies fall under the exclusive responsibility of local governments: land management, urban roads, solid waste collection and treatment, markets and fairs, cemeteries, public parks, gardens and culture, sport and local law enforcement (see Fig. 11.2). Other functions are shared with central or provincial governments, including education, health, public illumination and distribution of electricity, water and sanitation, public transport and roads, social housing, and environmental conservation and protection, although

1. Urban and Rural Infrastructures
- Green spaces, including gardens and nurseries
- Roads, including sidewalks
- Housing and economy
- Public cemeteries
- Public municipal offices
- Markets and fairs
- Firefighters

2. Basic Sanitation
- Municipal water supply systems
- Sewerage systems
- Waste and cleaning collection and treatment systems

3. Energy
- Electricity distribution
- Public, urban and rural illumination

4. Transport and Communication
- Urban road network
- Municipal public transport systems

5. Education
- Pre-school education centers
- Public primary schools
- School transport
- Equipment for elementary adult education
- Other activities complementary to educational action

6. Culture, Leisure and Sports
- Cultural centers, libraries and museums
- Cultural, landscape and urban heritage
- Camping parks
- Sport installations and equipment

7. Health
- Primary health care units

8. Social Assistance
- Assistance to vulnerable parts of the population
- Social housing

9. Environmental Management
- Environmental protection or rehabilitation
- Afforestation, tree planting and conservation
- Establishment of municipal nature reserves

Source: elaborated by authors based on Law No 2/97 of 18 February and Law No 1/2008 of 16 January as amended by Law No 6/2018 of 3 August.

Fig. 11.2 Mozambican municipal government competencies

some of these are set to be transferred to the municipal level.

Most of these responsibilities and competencies are directly linked to the actions that are needed to meet the SDGs, particularly SDG 11, highlighting the importance and relevance of municipal governments as key actors when it comes to making progress on Agenda 2030. However, a number of factors constrain the extent to which municipal governments are able to fulfill their mandate.

This is especially important considering the rapidly growing levels of urbanization in the country, which are mainly driven by high birth rates. In 2019, 36% of the country was urbanized compared to little over 10% in 1980 (Hansine and Arnaldo 2019). The UN has estimated growth to increase to 60% by 2030, but this growth is highly uneven with growth levels ranging from less than 1% to over 10% in a selected number of municipalities (ANAMM and World Bank 2009, p. 297). As a result of this growth, the majority of urban dwellers live in unplanned "informal" areas marked by limited access to basic services and high levels of urban poverty. Moreover, many urban areas are situated in ecologically vulnerable locations, making them prone to natural disasters and other impacts of climate change. These include repeated floods, most catastrophically in 2000, as well as cyclones such as Kenneth and Idai in 2019, which led to the destruction of thousands of human lives, infrastructure, and local economies with long-lasting effects (Oxfam 2020). Local governments have also had to deal with the effects of on-going political tensions, resulting in repeated skirmishes between national

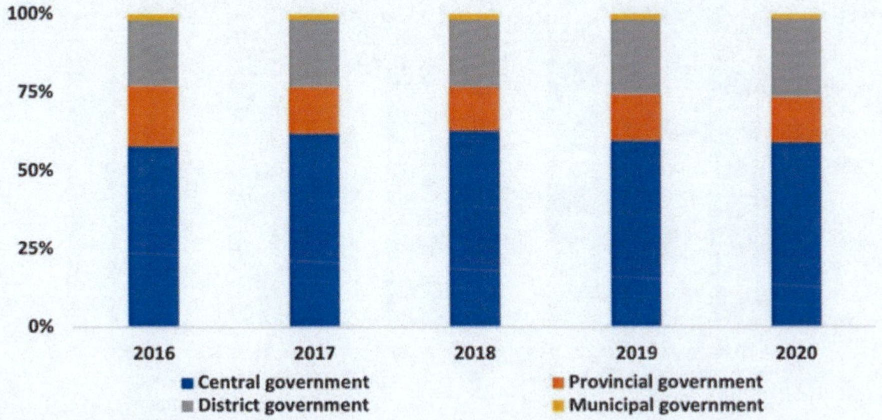

Fig. 11.3 Mozambican state budget allocation (operational) per government tiers (ANAMM 2020)

government and opposition forces, particularly after the general government elections of 2014, as well as rising acts of violent terrorism in the north of the country since 2019.

Overall, the capacity for service delivery, combined with the need to prepare for and respond to the impact of natural calamities and political conflict, is limited. This results from historically weak organizational structures and poorly maintained infrastructural systems. Moreover, municipal tax bases are limited,[2] which also derives from historically weak but centralized tax systems (Canhanga 2009; ANAMM and World Bank 2009). Importantly, there is a large gap between the increase in spending responsibilities that has come with the devolution of competencies to the municipal level and the revenues available for spending, also referred to as "unfunded mandates" (Raich n.d.). To a large extent, this is the result of the minimal levels of central government transfers to municipal government. These mainly consist of funds transferred under the Municipal Compensation Fund (*Fundo de Compensação Autárquica* or FCA), the Local Initiative Investment Fund (*Fundo de Investimento de Iniciativas Locais* or FIIL), or transfers destined for a specific objective, such as the Road Fund or the Fund for the Reduction of Urban Poverty. In spite of the increase in the number of elected municipalities over the years, the level of intergovernmental transfers to municipalities has not increased correspondingly and is also much lower compared to transfers to provincial or district governments. For instance, between 2007 and 2010 on average only about 1% of the general state budget was allocated to municipal governments, compared to 4% to district governments and 26% to provincial governments (Raich n.d.). While budget allocations to provincial governments have reduced to the benefit of district governments in recent years, funding for municipal governments remains largely the same, as can be seen in Fig. 11.3. A similar trend applies when looking at the central state budget for public investments (Fig. 11.4).

Although the transfers allocated to the municipal level are low, a study conducted among six Mozambican municipalities shows that they represent a very important part of municipal budgets, with on average almost half of total local revenues consisting of intergovernmental transfers (48%), against 36% of own revenues (divided between 26% of tax

[2] In 2006, the total budget for all municipalities (excluding aid) was of MT 1.1 billion, equivalent to about US$46 million. In total per capita terms, this means that on average, municipal spending for each resident was little less than MT 300, equivalent to about US$4.7 at 2021 exchange rates, with significant variation from MT 90 to MT 900, or US$1.4 to US$14 at 2021 exchange rates (ANAMM and World Bank 2009, p. 99).

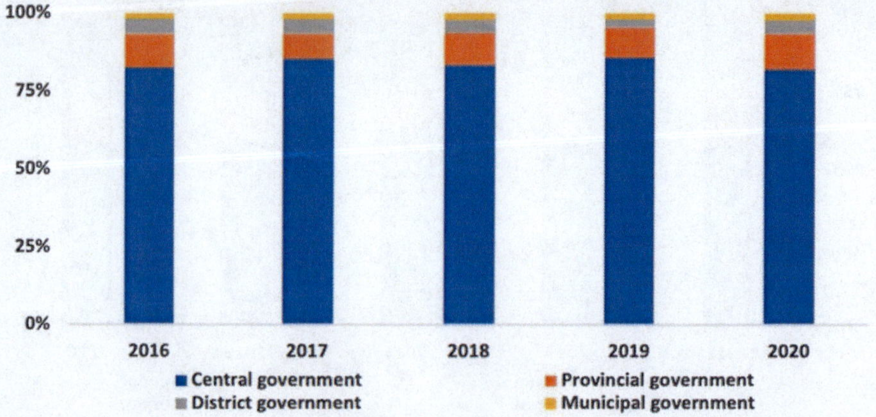

Fig. 11.4 Mozambican state budget allocation (investment) per government tiers (ANAMM 2020)

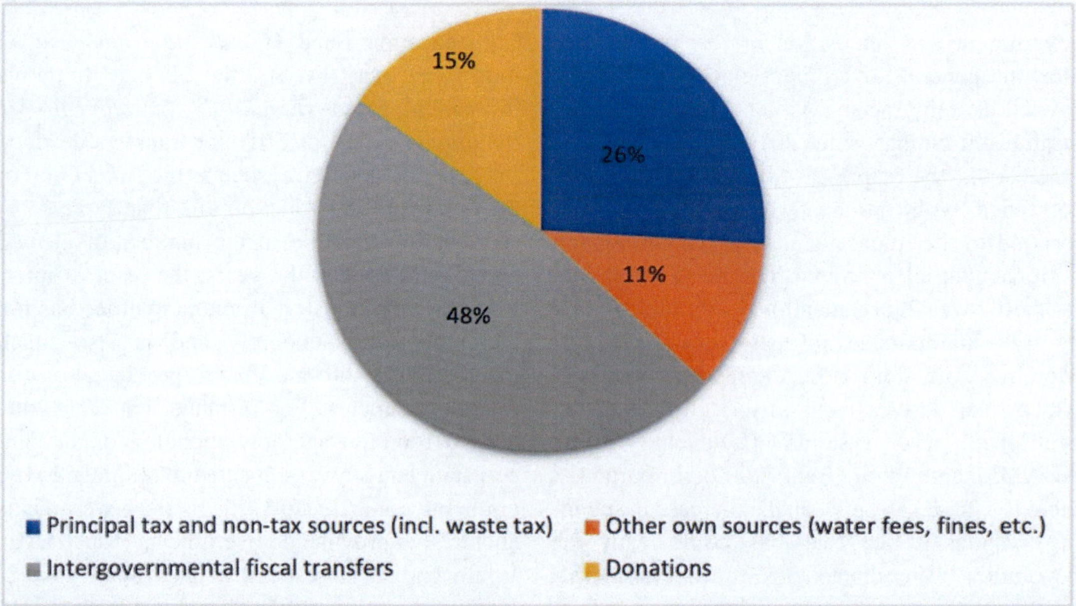

Fig. 11.5 Municipal aggregated revenue in six selected Mozambican municipalities in 2009. (Adapted from Weimer 2012)

and non-tax and 11% other own revenue sources such as fines) and 15% of aid (see Fig. 11.5) (Weimer 2012, pp. 364–365; see also Ilal and Weimer 2018). As in many other African cities (see Chap. 9), most of these revenues are spent on wage bills and other operational costs, even in larger cities with more expansive tax bases such as the capital city of Maputo or cities such as Nampula or Quelimane (ANAMM 2017), with little left to spend on capital investments required for improved service delivery.

In spite of these common trends, there is a lot of diversity in the performance of municipal governments across the country, with the VLR recording many innovative actions undertaken by municipal authorities in spite of local barriers and constraints. These illustrate the importance of recognizing local government diversity and the ways in which this diversity originates from local histories, geographies, leadership, and administrative competency and capacity (also Weimer 2012, p. 366).

11.5 Voluntary Local Review Experience in Mozambique

In November 2019, ANAMM conducted its first capacity and awareness training about the SDGs with municipal officials from across the country. The training was supported by UCLG, UNDP, the Municipal Council of Maputo, and the German cooperation (GIZ/BGF). The objectives of the training were to raise awareness and to train local authorities on the implementation and localization of the SDGs, territorial planning, monitoring and evaluation of the global agendas, and providing a space for the sharing and exchange of knowledge and experiences on SDG implementation at the city level (UCLGA 2019).

The development of VLRs was also discussed in the training, sparking the interest of ANAMM to contribute to this growing global movement. The Mozambican VLR received financial support from UCLG, as well as technical support from GIZ/BGF under a tripartite Memorandum of Understanding (ANAMM 2020).

One of the aims of the VLR was to assess the extent to which municipalities were already contributing to the SDGs in Mozambique. This required an analysis of levels of poverty and access to basic services in each municipality. Census data is essential for such an analysis as the census is based on household surveys, with the most recent census in Mozambique having been conducted in 2017. However, these data are not disaggregated to the municipal level. As such, the analysis of the VLR was focused on the identification of programs, projects, actions, and their outcomes at the municipal level. Through the assessment of local perceptions, the review also sought to identify the most critical challenges for service delivery at the municipal level (ANAMM 2020).

Key instruments to collect this information included interviews and four types of questionnaires with open-ended and multiple-choice questions. These included the following:

- **General questionnaire:** The objective of this questionnaire was to assess the level of awareness of the SDGs among local authorities and officials, as well as representatives of civil society, and the private sector. It also sought to measure the extent to which the SDGs are already incorporated in municipal actions and policies and which SDGs are considered to be most relevant. This questionnaire was also meant to contribute to the development of a dissemination and training strategy on the SDGs by ANAMM.
- **Municipal questionnaire:** The principal purpose of this questionnaire was to collect quantitative data on municipal management and SDG implementation. It included questions on (1) general municipal data; (2) municipal officials according to gender; (3) information on municipal plans, budgets, and financial resources; (4) implemented programs and projects with data on beneficiaries; (5) spaces for participation and accountability; and (6) the municipality's institutional partners.
- **Questionnaire for members of Municipal Assembly:** This questionnaire was aimed at assessing SDG-related regulations and accountability mechanisms as issued or exercised by the Municipal Assembly.
- **Questionnaire on social perceptions and participation:** This questionnaire aimed to assess the perception and experience of citizens, the extent to which municipalities promote participatory decision-making and accountability, and the extent of the involvement of civil society in municipal management by assessing the extent to which citizens are familiar with municipal programs and projects that are being implemented and the extent to which these contribute to a better quality of life among citizens.

Initially, the VLR sought to collect local data in and with 17 out of the country's 53 municipalities in the period February to June 2020.

However, the spread of the COVID-19 pandemic meant that it was only possible for ANAMM staff members to physically visit little over 40% of the targeted municipalities. In other municipalities, information questionnaires and interviews had to be conducted remotely through the use of digital communication platforms, while secondary sources were used to validate the collected information. One municipality (Maputo) ended up not being included in the report as the data provided could not be included in time (ANAMM 2020).

The municipalities that participated in the VLR represented both large and small cities, covering administrative categories "A" to "D", as well as municipal authorities from the south to center and northern part of the country (see Fig. 11.6 and Table 11.1). They also included the municipalities that had taken part in the SDG training in Maputo in November 2019.

11.5.1 SDG Awareness and Relevance

The general questionnaire on SDG awareness and relevance showed that 66% of participants in the study had heard of the SDGs, but had only limited knowledge of the goals, while 26% knew of the SDGs and were able to explain what they were. Only 8% had no knowledge at all of the SDGs (see Fig. 11.7). The main source of information on the SDGs was television and to a lesser extent social media, showing the potential of these and complementary communication channels and activities, such as public events, workshops and trainings, to disseminate information about the SDGs among city officials (see Fig. 11.8).

Some 76% of participants in the study consider the SDGs to be very relevant for municipal management, which implies that even in the absence of widespread detailed knowledge about the SDGs, global development agreements are considered to have local relevance (see Fig. 11.9). Moreover, according to 46%, the SDGs are aligned to the priorities of the municipality (Fig. 11.10). Notably, SDG 6 on water and sanitation was considered to be the most important and relevant goal at the municipal level, which was corroborated in various interviews, followed by SDG 3 on health, SDG 4 on education, SDG 1 on poverty, SDG 8 on employment, and SDG 5 on gender (ANAMM 2020).

11.5.2 SDG Implementation

The data collected on municipal management indicate that when it comes to the implementation of municipal programs and projects, four key SDGs are addressed. These include SDG 6 on water and sanitation, SDG 3 on health, SDG 11 on sustainable cities and communities, and SDG 1 on poverty. Drilling down into this information illuminates the local challenges and governance complexities of implementing sustainable development in a low-income African context.

Interventions related to SDG 11 are part of the regular services that fall under the responsibility of municipalities and include the issuing of property and land-use rights, waste collection, road maintenance, public transport, and environmental protection. However, many other municipal projects are implemented in response to local needs and demands and do not always fall within municipal mandates. Such emergency interventions also often require resources that were not planned for, such as food assistance for vulnerable communities affected by natural disasters. Many municipalities also report using their own resources for the construction of health centers to meet the demand for local health care, even if the transfer of this function has not been completely devolved to the municipal level. Municipalities also organize public events to promote preventive health measures, for instance, related to family planning and the prevention of unwanted pregnancies. Similarly, while the responsibility for education has also not been completely devolved to the municipal level, many municipalities report to be building classrooms to meet local needs as some schools have in excess of 100 learners per class. Municipalities also report to provide students with desks and other school supplies.

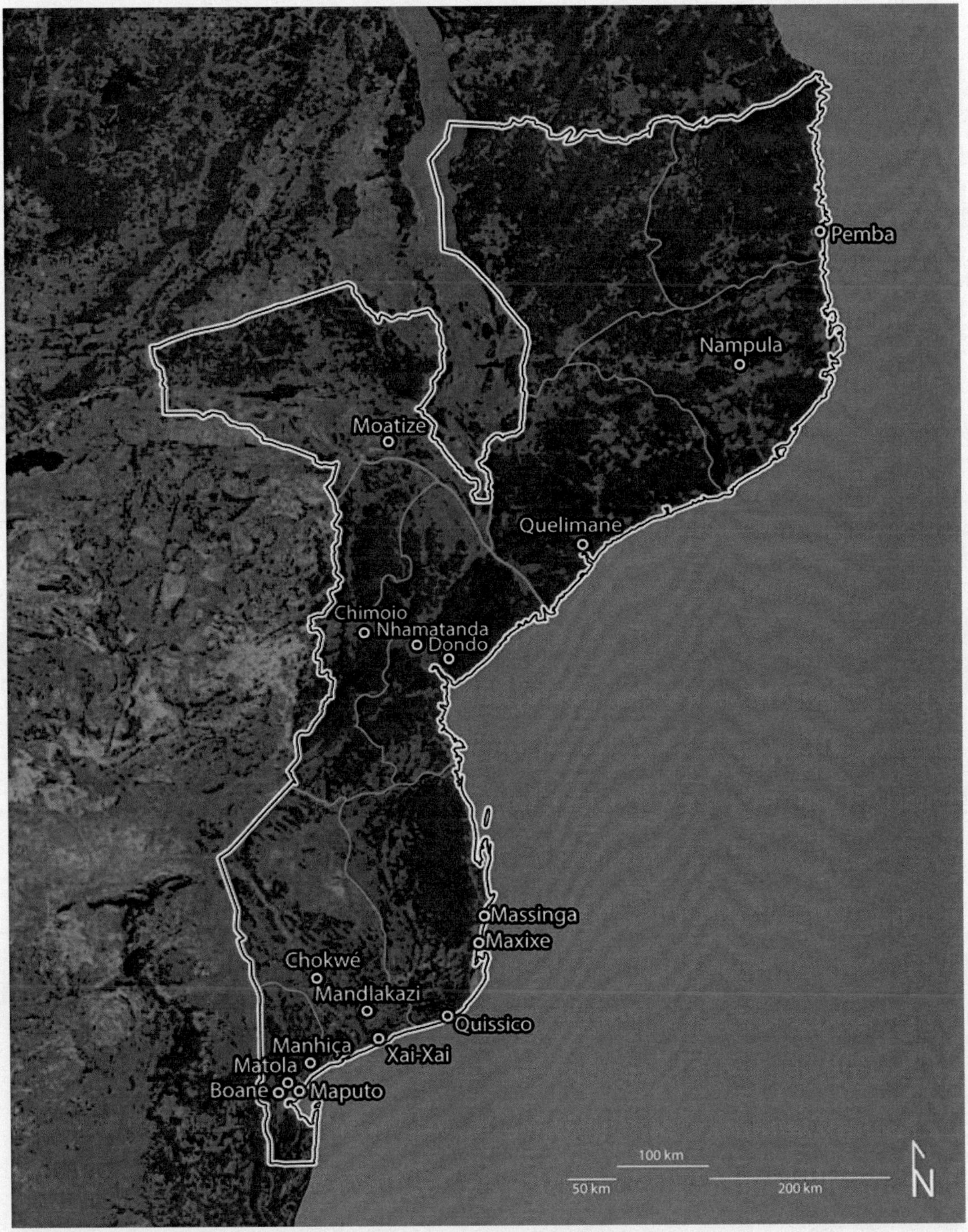

Fig. 11.6 Map of Mozambique with participation VLR cities. (Source: Authors)

Table 11.1 Profiles of Mozambican Voluntary Local Review (VLR) municipalities (ANAMM 2020)

	Nampula	**Chimoio**	**Dondo**	**Nhamatanda**
Population	653,961	363,336	103,493	62,362
Number of neighborhoods	18	33	10	12
Year of creation	1997	1997	1997	2013
Category	City "B"	City "C"	City "D"	Town "B"
	Mandlakazi	**Quelimane**	**Manhiça**	**Pemba**
Population	59,714	246,915	77,592	200,529
Number of neighborhoods	15	52	21	13
Year of creation	1997	1997	1997	1997
Category	Town "B"	City "C"	Town "A"	City "B"
	Moatize	**Massinga**	**Quissico**	**Maxixe**
Population	64,398	49,635	25,000	129,993
Number of neighborhoods	8	14	11	17
Year of creation	1997	2008	2013	1997
Category	City "C"	Town "A"	Town "C"	City "C"
	Chokwe	**Xai**	**Boane**	**Matola**
Population	70,100	140,000	106,000	1,032,197
Number of neighborhoods	8	15	33	42
Year of creation	1997	1997	2013	1997
Category	City "C"	City "C"	Town "A"	City "B"

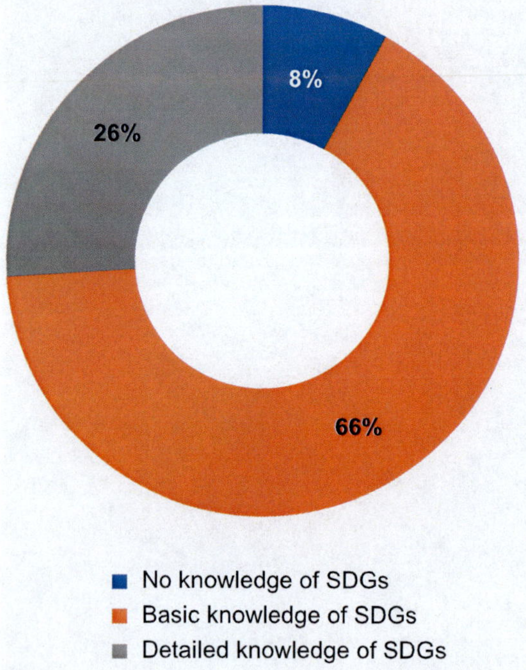

Fig. 11.7 Levels of SDG awareness (ANAMM 2020)

Poverty reduction is a major concern for municipalities and is addressed by initiatives promoting economic development and activities supporting agriculture, fishing, and livestock aimed at creating both employment opportunities as well as improving food security.

Nevertheless, for the majority of municipalities water provision represents the main priority, as public water networks managed by central state-owned utilities often only partially reach the most centrally located neighborhoods and communities, with peripheral neighborhoods requiring additional systems and sources such as fountains, wells, or informal water service providers (Matsinhe et al. 2008). This has important implications for the state's ability to recover costs and further invest in water provision (Farolfi and Gallego-Ayala 2014). There are also three municipalities that are not covered at all by public water systems and where water provision is of the exclusive responsibility of the local municipality. In the municipality of Nhamatanda, this was addressed by drilling 61 wells and constructing three small water systems, two in the period after Cyclone Idai. However, 9 out of the town's 12 neighborhoods still do not have regular access to water, and there are only 200 residential water connections for a population of more than 62,000 inhabitants (ANAMM 2020).

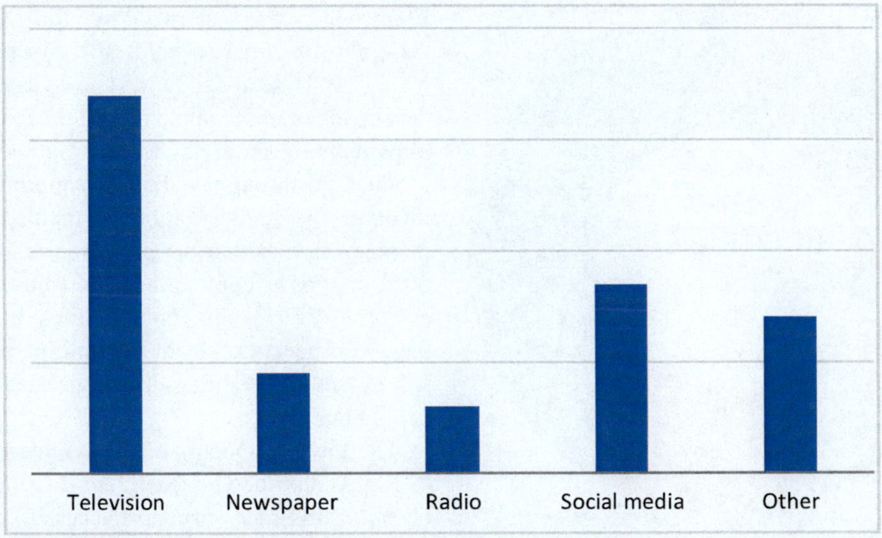

Fig. 11.8 Main source of information on SDGs (ANAMM 2020)

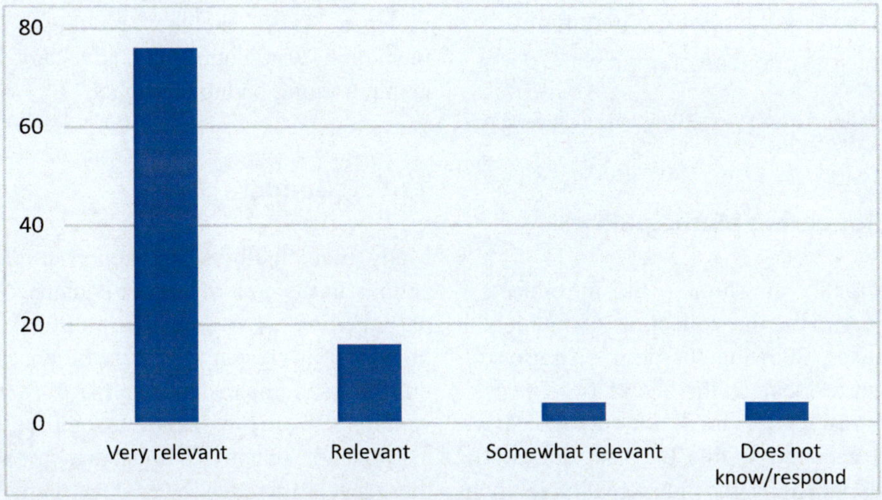

Fig. 11.9 SDG relevance for municipal management (ANAMM 2020)

11.5.3 Good Practices

The municipal programs, projects, and initiatives outlined above indicate that many municipalities in Mozambique are actively involved in improving the lives of local residents, in spite of the many financial and governance limitations. The complexity of the local context is the recurrence of natural disasters, representing a major barrier to long-term planning and development (Koivisto and Nohrstedt 2017). But even amidst adversity, there is innovation. Among the 16 municipalities covered in the VLR, a number of them stand out for being particularly innovative. This section outlines a number of good practices developed by these municipalities with the aim of offering opportunities for learning and action by other municipalities. This is followed by cross-cutting examples around gender and partnerships.

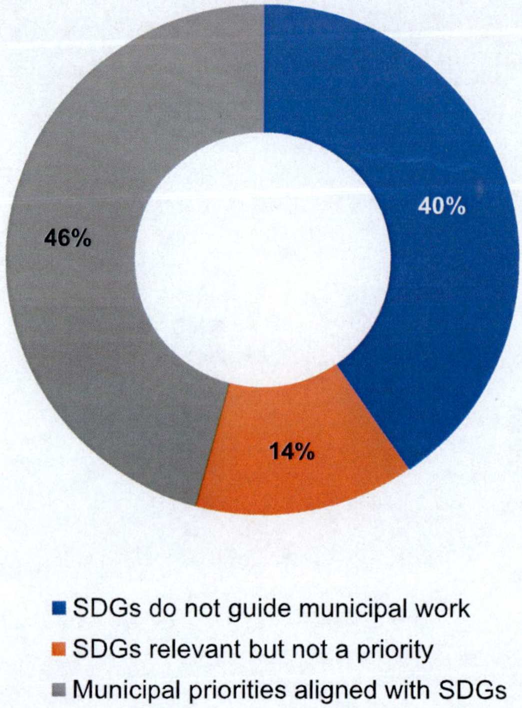

Fig. 11.10 Alignment between SDGs and municipal priorities (ANAMM 2020)

11.5.4 Pioneering Municipalities

The municipality of Chimoio has introduced a new IT system for the collection of municipal fees and taxes, allowing for the registration of 60,000 local residents in the first phase. The system allows residents to track not only their payments and accounts online but also the public works undertaken by the municipality in their neighborhoods and therefore how their taxes contribute to local development. Information is further available on the timeframes of each project, as well as contact details of the contractors, allowing citizens to monitor municipal works.

The city of Dondo was one of the main cities affected by Cyclone Idai in 2019. In order to improve disaster preparedness in the aftermath of the cyclone, the city created a City Resilience Framework for Action with support of UN-Habitat. This involved working together with local communities to create a map of risk and vulnerability of the city of Dondo. Importantly, after its finalization in March 2020, the Action Framework was adopted by the Municipal Assembly in order to guide the city going forward (UN-Habitat and DiMSUR 2020).

The cities of Nampula and Maxixe have been implementing programs to promote agricultural production through a holistic approach. This involves the regularization of tenure for low-income families and the provision of seeds and other inputs to communities as well as farmer cooperatives. Fishing communities in Maxixe have also benefitted from training in sustainable fishing and the planting of mangroves to protect fishing populations.

The town of Quelimane has created a House of Rights and the Citizen (*Casa do Direito e do Cidadão*) in order to support access to free legal assistance to citizens of Quelimane, especially those in vulnerable situations. It also has developed preventive strategies to address (domestic) violence. These include activities of conflict mediation, legal counseling, advocacy, and leadership training on human rights.

11.5.5 Gender

Many municipalities also report implementing actions in the area of gender equality. While the participation of women in political life has increased in recent years, with an increasing number of female deputies (37.6%), civil servants (39%), provincial assembly members (35%), and Council of Ministers members (41%), these levels are much lower at the municipal level (Republic of Mozambique 2020).

After the local elections of October 2018, only six women (11%) out of a total of 53 were elected President of a Municipal Council, an important position in the Municipal Executive, responsible for the management and implementation of municipal programs and projects. The percentage of female Executive Council members is higher and reached 24%. However, female council members tend to occupy positions in the area of social affairs such as health or culture and less in areas such as administration or finances.

A similar trend emerges when assessing the number of female Presidents of the Municipal

Assembly of which there are only 8 (15%) out of a total of 53, while 21 (40%) occupy the post of Vice President of the Municipal Assembly. Members of the Municipal Assembly also only include 34% women against 66% men.

The reduced participation of women in local decision-making spaces can be explained by a mix of socio-cultural factors, as well as a lack of incentives and political support. While there are no laws or quota to guarantee the participation of women in political life, a number of municipalities such as Chimoio, Nampula, and Mandlakazi have taken it upon themselves to develop initiatives aimed at women empowerment, as well to adopt measures to increase the number of female public servants, both in technical and decision-making positions.

11.5.6 Partnerships

Because municipal financial resources are often limited, many municipalities have promoted the consolidation of partnerships with external actors, such as development agencies but also private entrepreneurs, churches, and other actors that provide technical and financial support.

Such partnerships represent an important contribution to the implementation of mid- to long-term development plans and objectives. Important partners include international organizations and foundations such as the NGO Engineering for Human Development (ONGAWA), the Lay Volunteers International Association (LVIA) and related Association of Volunteers in International Service (AVSI), national non-governmental organization (NGOs) such as the Mozambican Association for the Development of Female Heads of Household (MUCHEFA), and the Mechanism for Civil Society Support (MASC) Foundation, as well as international development agencies such as USAID, GIZ-German Cooperation, the European Union, and UN agencies such as UN-Habitat. The initiatives that count with most external support are those in the area of water, urban planning, infrastructural development, and actions aimed at the protection of children and vulnerable communities. A number of municipalities have also benefitted from financial support from multilateral organizations such as the World Bank and the German KfW Development Bank.

The private sector also represents a (potentially) important partner in the implementation of municipal development initiatives. However, it is still relatively weak, especially when it comes to the support of the implementation of SDG-related initiatives. However, municipalities such as Chimoio are rapidly advancing in the consolidation of public-private partnerships as an important part of local development initiatives is co-financed by local businesses. This has been the result of persistent efforts by the municipal leadership, including efforts to restructure and improve its services. Such efforts in turn have contributed to building confidence and increased perceptions of municipal management as efficient and transparent, important in attracting private sector investment. Notably, few municipalities reported on the details of possible partnerships with researchers and academic institutions.

11.6 Lessons

As the world enters into the decade of action, counting down to the year 2030, many challenges remain when it comes to the space that local governments have for SDG localization and implementation. A lack of local awareness, resources, and capacity represent major barriers, as illustrated by the case of Mozambique. Nevertheless, the VLR process also reveals important lessons and opportunities.

Firstly, local leadership is crucial for successful SDG implementation. This involves political will to commit to the SDGs and incorporate these goals as an integral part of local development agendas. The VLR process showed that in municipalities where presidents were more aware of the importance of the SDGs, there was a better integration and visibility of the SDGs in either on-going or projected local plans and efforts.

On the other hand, while the SDGs remain a new concept, most municipalities are already implementing the SDGs in their day-to-day actions and initiatives. This means that there is important scope for awareness raising among local elected officials, technical staff, and citizens around the SDGs in order to build ownership and strengthen the integration of the SDGs in local development plans.

Decentralized cooperation represents an important mechanism for strengthening the human, technical, and material capacity of municipalities, but also for further alignment and integration of the SDGs with local development plans. This in turn requires a greater recognition by national government of the important role of local governments for SDG implementation, more attention to the capacity and resources this requires at the subnational level, as well as the importance of social responsibility. Addressing these issues would require stepping up awareness and communication campaigns around the SDGs and local government and capacity building among a wide range of city officials.

Municipalities need to be supported when it comes to the elaboration of SDG localization strategies, for instance, by building on the current collective VLR to develop individual reviews that use the same methodological instruments. In this regard, monitoring and evaluation efforts will need to be strengthened, for instance, through the creation of locally generated indicator frameworks, in order to support follow-up and implementation (Croese et al. 2021).

In order to strengthen implementation, there is a need to recognize the importance of supporting specific projects that can work as accelerators, but also projects that address the impact of extreme natural calamities. This requires technical and financial support of both the national government, cooperation partners, the private sector, and academia.

Existing as well as additional or alternative financial mechanisms and means of implementation also need to be reviewed, both at the level of local municipalities, national government as well as external development partners. This includes exploring ways to improve local tax collection, reconsidering existing intergovernmental transfers to the municipal level as well as expanding private sector support.

In sum, while there are still many challenges and barriers to SDG localization and implementation at the local level, proactive and effective leadership and coordinated and committed partnerships represent an important contribution to local progress.

11.7 Conclusion

Voluntary Local Reviews on the SDGs are emerging as an important instrument for local governments to affirm the importance of their actions for SDG progress. In spite of their usefulness as an instrument and opportunity to connect local strategies to the SDGs, strengthen evidence-based policymaking, identify gaps and priorities, as well as mobilize new policies, partnerships, and resources at the local government level, few African cities so far have engaged in VLR processes.

In part this reflects slow, complex, and uneven processes of decentralization across the continent, as illustrated by the case of Mozambique. Decentralization in Mozambique has been guided by the principle of gradualism, meaning that power has only been gradually geographically and functionally been devolved to the local level. Responsibilities have increased but so have structural as well as extraordinary challenges and limitations, arising from limited financial resources combined with the effects of climate change that force municipalities to adapt outside of established or projected plans and policies.

However, the VLR shows that municipalities in Mozambique implement the SDGs in various and often locally innovative and creative ways. Moreover, a number of municipalities have developed cross-cutting initiatives to improve gender equity and build partnerships with a range of partners and stakeholders.

There is much scope to strengthen and support these on-going efforts and share lessons and learnings. Local leadership and commitment are crucial in this regard, but so is the importance of tracking and evaluating ongoing work through the improvement of local statistical, monitoring, and data management capacity. Local government associations can play an important role in supporting and harmonizing these processes and making sure these are fed back to national government.

References

Associação Nacional dos Municípios de Moçambique (ANAMM) (2020) Relatório Local Voluntário: avanços e desafios de 16 municípios de Moçambique. ANAMM, Maputo. https://gold.uclg.org/sites/default/files/mozambique_2020_portuguese.pdf. Accessed 30 December 2020

Associação Nacional dos Municípios de Moçambique (ANAMM) (2017) Empenho e potencial tributários dos municípios de Nampula e Quelimane. ANAMM, Maputo

Associação Nacional dos Municípios de Moçambique (ANAMM), World Bank (2009) Municipal development in Mozambique: lessons from the first decade. https://doc.rero.ch/record/258327/files/03-Municipal_Development_Study_Long_Version_English.pdf. Accessed 12 July 2021

Canhanga NJV (2009) Descentralização fiscal, transferências inter-governamentais e dinâmicas da pobreza nas autarquias locais. Comunicação submetida no âmbito da Conferência sobre "Dinâmicas da pobreza e padrões de acumulação económica em Moçambique" organizada pelo Instituto de Estudos Sociais e Económicos (IESE). IESE, Maputo

Croese S, Dominique M, Raimundo IM (2021) Co-producing urban knowledge in Angola and Maputo: towards meeting SDG 11. npj Urban Sustainability. https://doi.org/10.1038/s42949-020-00006-6

European Union (EU) (2020) European handbook for SDG voluntary local reviews. Publications Office of the European Union, Luxembourg

Farolfi S, Gallego-Ayala J (2014) Domestic water access and pricing in urban areas of Mozambique: between equity and cost recovery for the provision of a vital resource. International Journal of Water Resources Development 30(4):728-744

Food and Agriculture Organization (FAO) (2018) Mapping the capacity of Mozambique to report on Sustainable Development Goals. Food and Agriculture Organization of the United Nations

Forquilha S (2020) Decentralization reforms in Mozambique: the role of institutions in the definition of results. WIDER Working Paper 2020/132 (October 2020). United Nations World Institute for Development Economics Research (UNU-WIDER). https://www.wider.unu.edu/publication/decentralization-reforms-mozambique. Accessed 12 July 2021

Hansine R, Arnaldo C (2019) Natureza demográfica e consequências do crescimento urbano em Moçambique. Presentation given at IESE conference "Desafios para Moçambique: dez anos pensando no país", Maputo, 19 September 2019. https://www.iese.ac.mz/wp-content/uploads/2019/09/Conf2019_RHansineCArnaldo.pdf. Accessed 12 July 2021

Ilal A, Weimer B (2018) Urbanização, serviços públicos e recursos fiscais—Desafios principais para as autarquias moçambicanas. Uma análise com enfoque às transferências fiscais intergovernamentaihas. In: Forquila S (ed) Desafios para Moçambique 2018. IESE, Maputo, pp 65-107. https://www.iese.ac.mz/wp-content/uploads/2019/05/artigo_ai-bw.pdf. Accessed 14 July 2021

Institute for Global Environmental Strategies (IGES) (2021) Online Voluntary Local Review (VLR) Lab. https://www.iges.or.jp/en/projects/vlr. Accessed 14 July 2021

Instituto Nacional de Estatística (INE) (2020) Progress towards Sustainable Development Goals: National Voluntary Review 2020 (statistical annex). Instituto Nacional de Estatística, Maputo

Kindornay S (2018) Progressing national SDGs implementation: an independent assessment of the Voluntary National Review reports submitted to the United Nations High-level Political Forum on Sustainable Development in 2017. Canadian Council for International Co-operation, Ottawa. https://www.partners-for-review.de/wp-content/uploads/2020/03/Progressing-national-SDG-implementation2018_3rd-Edition.pdf. Accessed 29 November 2020

Koivisto JE, Nohrstedt D (2017) A policymaking perspective on disaster risk reduction in Mozambique. Environmental Hazards 16(3):210-227

Matsinhe NP, Juízo D, Macheve B, dos Santos C (2008) Regulation of formal and informal water service providers in peri-urban areas of Maputo, Mozambique. Physics and Chemistry of the Earth, parts A/B/C 33(8–13):841-849

Ministério de Economia e Finanças (MEF) (2016a) Transformando o nosso mundo: a Agenda 2030 para o desenvolvimento sustentável. Ministério de Economia e Finanças and Direcção Nacional de Planificação e Orçamento; Maputo, República de Moçambique

Ministério de Economia e Finanças (MEF) (2016b) Mapeamento dos Instrumentos e sistemas de Monitoria sobre os Objectivos de Desenvolvimento Sustentável. Ministério de Economia e Finanças, Direcção Nacional de Planificação e Orçamento and Direcção Nacional de Monitoria e Avaliação; Maputo, República de Moçambique

Ministério de Economia e Finanças (MEF) (2017) Moçambicanização dos Objectivos de Desenvolvimento Sustentável através do Plano Quinquenal do Governo para alcançar o crescimento inclusive e sustentabilidade ambiental. Ministério de Economia e Finanças and Direcção de Estudos Económicos e Financeiros; Maputo, República de Moçambique

Ministério de Economia e Finanças (MEF) (2018) Relatório Inicial da Implementação dos Objectivos de Desenvolvimento Sustentável em Moçambique. Ministério de Economia e Finanças, Direcção Nacional de Planificação e Orçamento and Direcção Nacional de Monitoria e Avaliação; Maputo, República de Moçambique

NYC Mayor's Office for International Affairs (2019) NYC and world cities unite to join Voluntary Local Review declaration. https://www1.nyc.gov/site/international/programs/voluntary-local-review-declaration.page. Accessed 30 December 2020

Ortiz-Moya F, Koike H, Ota J, Kataoka Y, Fujino J (2020) State of the Voluntary Local Reviews 2020. Local action for global impact in achieving the SDGs. Institute for Global Environmental Strategies, Kanagawa

Oxfam (2020) After the storm: barriers to recovery one year on from cyclone Idai. https://reliefweb.int/sites/reliefweb.int/files/resources/After%20the%20storm%20-%20barriers%20to%20recovery%20one%20year%20on%20from%20Cyclone%20Idai%20-%20Media%20brief%20-%20Embargoed%2010%20March%202020.pdf. Accessed 13 July 2021

Pipa T, Bouchet M (2020) Next generation urban planning: enabling sustainable development at the local level through Voluntary Local Reviews (VLRs). The Brookings Institution, Washington, DC

Raich U (n.d.) Finanças autárquicas em Moçambique. World Bank presentation. https://www.citiesalliance.org/sites/default/files/Presentation_WB_Uri%20Raich.pdf. Accessed 30 December 2020

Republic of Mozambique (2020) Voluntary National Review of Agenda 2030 for Sustainable Development. https://sustainabledevelopment.un.org/content/documents/26314VNR_2020_Mozambique_Report.pdf. Accessed 12 July 2021

United Cities and Local Governments (UCLG) (2020a) Towards the localization of the SDGs: how to accelerate transformative actions in the aftermath of the COVID-19 outbreak. https://www.uclg.org/sites/default/files/report_localization_hlpf_2020.pdf. Accessed 13 July 2021

United Cities and Local Governments (UCLG) (2020b) UCLG annual retreat report. https://www.uclg.org/sites/default/files/m15_uclg_retreat2020_informe.pdf. Accessed 13 July 2021

United Cities and Local Governments (UCLG) (2019) Fifth global report on decentralization and local democracy. The localization of the global agendas: how local action is transforming territories and communities. https://www.gold.uclg.org/sites/default/files/ENG-GOLD-V-2020.pdf. Accessed 13 July 2021

United Cities and Local Governments Africa (UCLGA) (2019) Sessão de formação e conscientização sobre: adaptar planos locais e assegurar o monitoramento e a apresentação de relatórios para localizar Objetivos de Desenvolvimento Sustentável (ODS) [conceptual note]. ANAMM-Maputo, Moçambique, 18-19 November 2019. https://uclgafrica-alga.org/wp-content/uploads/2019/09/Nota-conceitual_AteliêODS_17-18nov2019-22.08.19.pdf. Accessed 30 December 2020

United Cities and Local Governments Africa (UCLGA), Cities Alliance (2018) Assessing the institutional environment of local governments in Africa, 3rd edn. UCLG Africa, Rabat; Cities Alliance, Brussels

United Cities and Local Governments (UCLG), UN-Habitat (2020) Guidelines for Voluntary Local Reviews, volume 1: a comparative analysis of existing VLRs. UCLG, Barcelona

United Cities and Local Governments (UCLG), UN-Habitat (2021) Guidelines for Voluntary Local Reviews, volume 2: towards a new generation of VLRs. Exploring the local-national link. UCLG, Barcelona

United Nations Department of Economic and Social Affairs (UNDESA) (2021) Voluntary Local Reviews: reports from local authorities. https://sdgs.un.org/topics/voluntary-local-reviews. Accessed 14 July 2021

United Nations Department of Economic and Social Affairs (UNDESA) (2020a) 2020 Voluntary National Review synthesis report. UNDESA, New York. https://sustainabledevelopment.un.org/content/documents/27027VNR_Synthesis_Report_2020.pdf. Accessed 13 July 2021

United Nations Department of Economic and Social Affairs (UNDESA) (2020b) Global guiding elements for Voluntary Local Reviews (VLRs) of SDG implementation. https://sdgs.un.org/sites/default/files/2020-10/GlobalGuidingElementsforVLRs_FINAL.pdf. Accessed 14 July 2021

United Nations Economic Commission for Africa (UNECA) (2021) Africa regional guideline & template for Voluntary Local Reviews. Prepared for UNECA by Lusungu Kayani, 27 February 2021. https://www.uneca.org/sites/default/files/TCND/ARFSD2021/Presentations/MGoS/Africa%20Regional%20Guideline%20%26%20Template%20for%20Voluntary%20Local%20Reviews%20-%20Lusungu%20Kayani.pdf. Accessed 6 July 2021

UN-Habitat, Disaster Risk Management, Sustainability and Urban Resilience (DiMSUR) (2020) CityRAP tool: city resilience action planning tool. United Nations-Habitat, Nairobi; Sub-Regional Technical Centre for Disaster Risk Management, Sustainability and Urban Resilience, Maputo. https://unhabitat.org/sites/default/files/2020/05/cityrap_tool_booklet_2020.pdf. Accessed 13 July 2021

Weimer B (2012) A base tributária das autarquias Moçambicanas: características, potencial e economia política. In: Weimer B (ed) Moçambique: descentralizar o centralismo: economia política, resursos e resultados. IESE, Maputo, pp 359-500

12 Localizing the SDGs in Complex Metropolitan Structures: Lessons and Insights from eThekwini Municipality, South Africa

Puvendra Akkiah

Abstract

Over the past years, eThekwini municipality (also known as Durban) has emerged as an exemplar city for biodiversity and climate change action and also for the localization of the Sustainable Development Goals (SDG) in an African city context. Key to the SDG localization approach has been the use of its medium-term Integrated Development Plan to align the work of the municipality with the SDG goals and targets. This chapter looks at some of the factors that have shaped eThekwini's approach to SDG localization. Chief among them is the city's longstanding track record of engagement and participation in global sustainability networks and agendas through an active leadership that has fostered the emergence of local sustainability champions within the city. While eThekwini shares many challenges with other metropolitan African cities around multilevel governance, active local leadership has made the city carve out its own path toward the practical implementation of global policy in a way that is grounded in local processes and through priority actions that are suited to its specific context.

Keywords

SDG localization · eThekwini · Integrated Development Plan · City leadership · Metropolitan governance

12.1 Introduction

eThekwini Municipality is the government authority with oversight over South Africa's third-largest urban hub—Durban—that is also an important development hub on the east coast of Southern Africa. Like many other African cities, it has experienced rapid urban growth over the past years. Covering a vast, mostly rural territory of some 2297 km^2, its estimated 3.3 million population is mostly settled in and around the city's urban core. High levels of formal unemployment put pressure on its ability to sustainably manage, finance, and deliver services (eThekwini Municipality 2021).

The municipality has aimed to tackle the challenges of growth, poverty, and environmental integrity through innovative policy reform and through an integrated approach to strategic city planning. Although predating the SDGs, integrated development planning is a local government process adopted across South Africa that aligns with the shift towards a more holistic understanding of sustainable development of the post-2015 development agenda (Binns and Nel

P. Akkiah (✉)
Senior Manager Programmes: Integrated Development Planning, eThekwini Metropolitan Municipality, Durban, South Africa
e-mail: puvendra.akkiah@durban.gov.za

S. Croese, S. Parnell (eds.), *Localizing the SDGs in African Cities*, Sustainable Development Goals Series, https://doi.org/10.1007/978-3-030-95979-1_12

2002; Madzivhandila and Asha 2012). This chapter directly explores the value of the Integrated Development Plan (IDP) as a tool for SDG localization, using the experience of Durban, and indirectly highlights the critical role of the support and activities of individual officials in building a global agenda from the bottom up.

As might be expected of an African city that is habitually used as a case study of urban change in Africa, eThekwini Municipality has been the subject of considerable practice-based and academic reflection. There are both critical and complimentary assessments of the council's performance. Some of the most interesting and reflective work on the localization of global agendas comes from Debra Roberts, who aside from being the co-chair of the United Nations International Panel of Climate Change (IPCC) is also a local government official from Durban and has published extensively on the challenges of working at the local level (Roberts 2008, 2010; Roberts and O'Donoghue 2013; Cockburn et al. 2016; Boon et al. 2016). Building on her invaluable insights as a city insider, this chapter is also written by a Durban city official. It reflects from within local government on the factors, lessons, and challenges that have shaped the city's own unique approach to SDG localization in eThekwini.

The story and the position of its storyteller are important because Durban has embraced localizing global policy since the endorsement of the 2030 Agenda in 2015. This started with the use of its IDP as the basis for capital expenditure in alignment with SDG goals and targets. Early efforts around IDP/SDG harmonization fed into ongoing efforts around monitoring and reporting in preparation for its first Voluntary Local Review (VLR) on the SDGs. Key to this IDP-based approach to pushing for SDG enhancement is that local priorities draw from and reinforce international agendas. This global-local articulation has been a feature of Durban governmentality and is based on a longstanding engagement with global sustainability agendas, going back to the Rio Earth Summit of 1992. Since then, there has been extensive participation by Durban officials in numerous global city networks. International engagement has been led by an active city leadership, which in turn has fostered the emergence of local sustainability champions.

12.2 eThekwini's Planning Approach in Context

eThekwini's key entry point for global policy alignment at the city level is its 5-year Integrated Development Plan (IDP) process, through which the SDGs are now filtered to reinforce sustainability planning and implementation (eThekwini Municipality 2021). For those less familiar with South African context, it is helpful to know that the IDP is a municipal-led strategic planning tool that for almost two decades has been used across South Africa to ensure cross-department alignment and to allow for medium-term prioritization (Harrison 2001; Todes 2004). Its origins lie with the end of apartheid and the ambition for radical change.

In 1994, South Africa became a newly democratic country after decades of white minority rule. Apartheid left behind an entrenched legacy of social and spatial inequality, which the new post-apartheid government sought to address through the introduction of a new legal, policy, and governance framework. Key to this framework was the restructuring of local government, translating into a period between 1994 and 2000, during which cities were governed through a transitional local government. The adoption of the Municipal Systems Act of 2000 and the Municipal Structures Act of 2000 established South Africa's current local government structure with three categories of municipalities: metropolitan, district, and local (Cameron 2001).

The basis for this structure was the new Constitution of 1996, which assigned specific service delivery responsibilities to national, provincial, and local government under a nationally sponsored Reconstruction and Development Program (RDP). Under this program, local governments were to have an explicit developmental mandate, with key mandates and responsibilities in the area of basic service delivery (Parnell et al. 2002). A key instrument in this regard is the Integrated Development Plan (IDP). The IDP was designed "as an instrument to assist local authorities in transformation, and in fulfilling the objectives of the nationally sponsored Reconstruction and Development Programme (RDP)." It combines a strategic plan, which has a vision and a set of strategies and projects, and a meta-plan, which links a series of separate sectoral-based plans and planning processes (Harrison 2001, pp. 176–177).

During the transitional period for local government, most local authorities in South Africa prepared some kind of IDP with guidance of the national government. After the first local elections of 2000, a new round of IDPs was prepared for the new local municipalities. However, many local authorities in South Africa were unable to complete a meaningful integrated and coordinating planning process. To date, many continue to struggle with the most basic functions of management and service delivery, in a context in which economic performance and efficiency and short-term delivery have increasingly come to prevail over social, participatory, and long-term planning prerogatives (Harrison 2001; see also McDonald and Smith 2004; Palmer et al. 2017).

These challenges are shared by the city of eThekwini, which in recent years have been compounded by global drivers of change such as rapid urbanization, globalization, and climate change. Three distinctive characteristics are important in understanding Durban and its complex challenges. Firstly, approximately 68% of the municipal area is peri-urban or rural in nature and partly governed by traditional authorities. Secondly, Durban is unusual in that it is located in a global biodiversity hotspot (1 of only 36 worldwide), making the protection and management of natural ecosystems a priority. Thirdly, due to apartheid, Durban has a legacy of structural and social inequity which is apparent in all aspects of city life and functions. Durban has a Gini coefficient (measuring the level of inequality) of 0.63, which is among the highest in the world, with an estimated 41% of the population experiencing conditions of poverty (eThekwini Municipality 2021) and a housing backlog of approximately 389,000 units (eThekwini Municipality 2017a, p. 35).

In comparison with many African local governments, the fiscal power of eThekwini may seem great. However, this has to be put against the reality that in this context it is local government that is on the front line of delivering the services that underpin progress towards the SDGs. Like all local governments in South Africa, the city of eThekwini is responsible for the delivery of basic services such as water, sanitation, electricity, and solid waste collection. However, as a metropolitan municipality, created as part of the municipal reforms of 2000, it represents a Category "A" municipality, holding constitutionally protected powers such as the ability and the technical capacity to generate revenue. This gives it more powers than smaller category "B" and "C" municipalities, representing intermediate and small towns, although primary responsibility for social services (education, health, and welfare), the built environment (human settlements), and water and electricity (energy) infrastructure still lies with the provincial and/or national government departments. Over the years, the city of eThekwini has skillfully sought to manage these challenges and responsibilities through long-term evidence-based development planning and active participation in global sustainability organizations and networks. The power of localization in Durban rests in part on the fact that it is the local government that is tasked with action to enhance basic conditions of residents. In this regard, what planners do to harmonize competing imperatives and come up with catalytic action that will drive sustainable development really matters. Experiences in eThekwini, which has been doing this for some time now, are illuminating for any city wanting to push the SDGs from the local scale.

12.3 Planning in eThekwini: From the Local to the Global

During the transitional period of the late 1990s, the city of eThekwini was mainly governed in line with the logic of "Reconstruction and Development," delivering a range of services, including water and transport, and even some primary health. However, the city had little evidence of the impact of this demanding political agenda on its service delivery. The gap between the political aspirations of the post-apartheid transition and local change led to the decision to conduct a Quality of Life survey in Durban in 1998. The survey measured the perception of local residents of municipal service performance and showed that in spite of increased service delivery, the perceived quality of life had decreased. Clearly in breach of the hope of the new democracy, the slip in standards represented a major political concern, sparking a search for more sustainable service delivery mechanisms within the municipality and a focus on process and impact rather than merely on outputs.

In 1999, Durban adopted a methodology of outcome-based planning, which was used to develop the city's Long-Term Development Framework (LTDF) in 2000, which was adopted in 2001, providing an overarching long-term sustainability framework for the midterm national government mandated municipal IDPs (eThekwini Municipality 2001). The kinds of action initiated in South Africa might be thought of as the sort of intervention envisaged by SDG 17 on strengthening the institutional capacity (of local government) to deliver the global agenda (Palmer et al. 2017). In the South African context, Durban was a pioneer of the bold rethink of the role and practices that local government needs to achieve large scale transformation—not just nationally but also internationally.

In 2008, the city partnered with an international organization called the Sustainable Cities Initiative to review the LTDF. This organization promotes the use of community participation and engagement for the development of city strategies and previously had worked in cities like Vancouver and Chicago. Building on knowledge that was coming out of the recently adopted Millennium Development Goals (MDGs) and the World Sustainable Development Summit, held in Johannesburg in 2002, the jointly constructed plan was grounded in sustainable development theory, while also including extensive active public participation processes. The result of this process—called "Imagine Durban"—was a consolidated vision for the future of the city with associated timeframes, roles, and responsibilities, as well as targets that spanned a 50-year sustainable development path. Moreover, the plan recognized that its implementation needed to involve all spheres of government, in consultative and inclusive processes that involved all stakeholders (Roberts 2008).

In recognition that there are multi-scale issues to localization, the city in 2015 started reviewing its alternative strategy through its City Planning Commission (CPC), which existed alongside the National Planning Commission and the Provincial Planning Commission for KwaZulu-Natal, the province that eThekwini is part of. The members of the CPC are experts in their field and provide oversight for the component parts of the city's strategy (eThekwini Municipality 2015). The CPC highlighted three areas that needed attention: urban governance, spatial development, and economic development. Based on these areas, a set of implementation plans were developed to feed into the city's IDP, but also to align with the requirements or indicators of the newly endorsed global development agendas (from the IPCC, Sendai Framework for Disaster Risk Reduction and also the 2030 Agenda).

The overt focus on global development agendas emerged from eThekwini's active participation in a number of international networks and organizations. Apart from the city's membership of a number of national local government associations such as the South African Local Government Association (SALGA) and the South African Cities Association (SACN), eThekwini is an active member of global city networks such as United Cities and Local Governments (UCLG), the Commonwealth Local Government Forum (CLGF), Metropolis, the Global Parliament of Mayors, as well as the African Union's NEPAD Cities Program. Several eThekwini mayors, senior officials, and council members have been active members of the executive leadership of these organizations, hosting numerous high-level meetings and summits in Durban, such as the United Nations Climate Chance Conference (COP17) in 2011, to the UCLG World Congress and Global Parliament of Mayors in 2019.

Working with its commitment to addressing its substantial environmental challenges, the municipality has been particularly active in networks with a focus on climate change, such as the 100 Resilient Cities Initiative, ICLEI Africa, and the C40 Climate Leadership Group. For instance, in the area of climate change, the city contributed to the global climate change negotiations that resulted in the adoption of the Paris Climate Agreement in 2015. A key person in this regard was the city's head of the Environmental Planning and Climate Protection Department between 1994 and 2016. Dr. Debra Roberts was a member of United Nations International Panel of Climate Change (IPCC) working groups, as well as the South African UN Framework Convention on Climate Change (UNFCCC) negotiating team. Roberts, in 2015 elected as IPCC co-chair, is the most high-profile official to combine a local and global role. However, Durban residents were also visible in the area of urban safety through the city's participation in the African Forum for Urban Safety and its partnership with the UN-Habitat Safer Cities Program (UN-Habitat 2016). Key to the city's role in these municipal, regional, and global networks is not just its active membership at the highest political level but also the use of these connections and senior positions in these organizations to advocate and lobby for the inclusion, importance, and capacitation of subnational governments in global sustainable development conversations and implementation. There is an additional value for others of the Durban presence—that of having a strong African voice with a credible base to push for forms of localization that work on the continent.

Individuals engagements in city-to-city and international urban organizations and networks opened the way for broader global advocacy work. This included the city's contributions to the Sustainable Development Solutions Network (SDSN) Thematic Group on Sustainable Cities, a group involved in mobilizing support for the creation of a stand-alone urban SDG in the run-up to the adoption of Agenda 2030 in 2015 (Acuto et al. 2018). The city also started chairing the UCLG Urban Strategic Planning Committee, while participating in the Global Taskforce of Local and Regional Governments, a coordination mechanism that brings together the major international networks of local governments to undertake joint advocacy work relating to global policy processes. As part of this advocacy work, the city participated in numerous important global meetings and events, including the New Urban Agenda working groups, which contributed to the adoption of the New Urban Agenda at Habitat III in 2016 (Moodley 2016).

Participation in global engagements allowed the city not only to advance subnational government interests in South Africa but also to learn from other pioneering cities across the world and the different ways in which they have approached SDG localization. As a result, internal levels of knowledge and awareness of global development agendas were high, contributing to a commitment to align local planning agendas with these global conversations. In practice, this resulted in a bold and independent attitude to forging a bottom-up approach to SDG localization, where the city developed its own approach without much guidance from the South African national government, which has been mainly focused on SDG domestication at the national level (Mthembu and Nhamo 2021).

12.4 Localizing the SDGs in eThekwini

SDG localization in eThekwini has been led by the office of the IDP, which has a strong local political base from the Mayor. The Mayor holds local political leadership of the city's mayoral committee, which coordinates the work of the City Council. However, there is also a strong external political impetus that derives from the role and experience of the city's IDP manager as technical chair of the UCLG Urban Strategic Planning Committee.[1] The power of the IDP as a conduit for SDG localization in eThekwini is not just political, but it is also technical. In any

[1] More information on this committee see the UCLG website: https://www.uclg.org/en/organisation/structure/committees-working-groups/urban-strategic-planning.

South African municipality, the IDP represents the city's guiding document. The IDP office therefore represents a key office or hub in the city. As the IDP requires a holistic approach to strategic development within the municipality, making links across various city units and departments to ensure that the city remains on the correct trajectory is its major purpose. This makes the IDP office a strategic entry point for sustainability planning, over time also becoming a hub for information and knowledge sharing on the processes of implementation and SDG localization with actors inside as well as outside of the city.

Work on SDG localization in Durban actively started upon the completion of the city's municipal IDP for the period of 2017–2022, which is defined according to an eight-point plan: (1) Develop and sustain our spatial, natural, and built environment. (2) Develop a prosperous, diverse economy and employment creation. (3) Create a quality living environment. (4) Foster a socially equitable environment. (5) Support organization design, human capital development, and management, (6) Build a vibrant and creative city—the foundation for sustainability and social cohesion. (7) Provide good governance and responsive local government, and (8) Build a financially accountable and sustainable city (eThekwini Municipality 2017b, p. 226).

Broadly, these eight points reflect the three key components of sustainability: environmental, economic, and social, underpinned by good governance practices, and the correlation with the values of the SDGs is clear. Important to note in this regard is the additional contribution of the New Urban Agenda, which brings a more specifically human rights dimension to the social, economic, and environmental aspects of sustainable development through its focus on the right to the city (UN 2016). Taken together, eThekwini's premise for SDG localization is that there are different starting points for SDG localization and that a city can use whatever model for sustainable development that it chooses and find alignment between the goals and intention of global agendas and local sustainability agendas. The SDG relevance of the normative platform of the eThekwini IDP is only really evident when its incorporation into the work of the Council is assessed.

12.4.1 Mapping Responsibilities

The first step towards SDG localization in eThekwini consisted of an exercise, conducted in 2017, to find out which SDGs fell under the responsibility of the metropolitan municipal government. The starting point for this exercise was a review of all of the 169 SDG indicators that were mapped in accordance with the role and responsibilities of the different tiers of government, as per the South African Constitution, specifically section four and five which outline the powers and functions of each sphere of government.

Through this mapping exercise, 98 of the indicators were found to respond to South African municipal mandates, while 23 and 48 indicators responded to provincial and national responsibilities respectively (see Fig. 12.1). However, of the projects that were implemented in eThekwini, 60 projects did not align with any of SDG indicators, meaning that although the city played a role in facilitating their implementation, these projects did not align with its direct responsibilities as local government. This number of projects was reduced to 48 in the following year, 2018. Conversely, 30 of the 98 indicators associated with local government aligned with a particular SDG target or indicator in 2017, which in 2018 increased to 50 and in 2019 to a further 56 indicators which had associated projects. Overall then, there was a gradual harmonization of SDG and IDP priorities and project selection, highlighting the important influence of the local on the global and the global on the local.

Based on this review of SDG targets and associated projects that the city was implementing as local government, the IDP office listed the number of projects for each relevant SDG target. This allowed it to calculate the capital expenditure for each target as well as average spend, creating an understanding, and therefore awareness, of what capital expenditure was going towards in terms of SDG achievement.

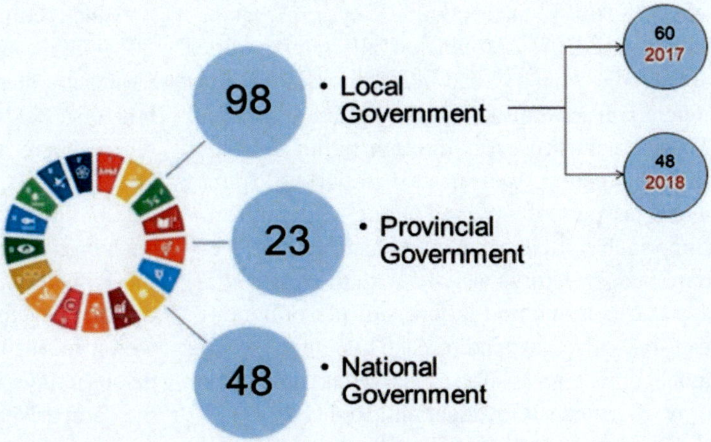

Fig. 12.1 SDG indicators and South African government responsibilities. (Source: author)

For example, the city's highest capital outlay corresponded to projects related to SDG targets 16.7 (to ensure responsive, inclusive, participatory and representative decision-making at all levels), 6.2 (achieve access to adequate and equitable sanitation and hygiene for all and end open defecation, paying special attention to the needs of women and girls and those in vulnerable situations), and 11.2 (provide access to safe, affordable, accessible, and sustainable transport systems for all, improving road safety, notably by expanding public transport, with special attention to the needs of those in vulnerable situations, women, children, persons with disabilities, and older persons).

It also showed that in terms of housing or human settlement-related programs, which is actually a provincial and national mandate, the city was implementing in the region of 700 projects with a total capital budget of ZAR 2.8 billion (about US$ 180 million at 2021 exchange rates) and a highest spend of around ZAR 238 million (about US$ 15 million at 2021 exchange rates). Having this overview therefore gave the city an important snapshot of its expenditure in relation to each of the SDG targets and indicators and an important basis for the alignment of its IDP at a capital budget level. It also highlighted the need for more engagement and active partnership with other levels of government to deliver on the SDGs, as well as more efficient financial flows to the local government level.

12.4.2 Budget Alignment

The IDP consists of a number of goals that provide the basis for a Service Delivery and Budget Implementation Plan (SDBIP) and associated budgets, thereby ensuring that spending matches political planning objectives (Duminy and Parnell 2020). Effectively, through the addition of the SDBIP process, the city looked at how the municipal capital budget implementation and capital budget investment aligned to the SDG goals and targets. Practically, the method used was to align each of the capital projects to a relevant SDG and associated targets. In cases where projects can align to multiple goals and targets, the city used the "best match" principle, aligning the project with a target that matched best with the original intent of that capital delivery program. For example, if a project revolved around the development of new infrastructure to deliver better health services, the city aligned that project to the associated health SDG and not the infrastructure SDG.

Source documents used in this exercise were the city's SDBIP capital projects, as included in its 2017/2018 Municipal Service Delivery Implementation Plan (eThekwini Municipality 2017c). In 2017/2018, the city looked at capital and operational projects, but in the following years of 2018/2019 only looked at capital projects. What contributed to this process of strategic budgetary review was the implementation of the Municipal Standard Chart of Accounts (mSCOA) system. This is an accounting system that has standardized the accounts of all municipalities across the country and aims to change how municipalities transact by standardizing financial management processes through policy formulation, budgeting, and in-year reporting frameworks as part of a wider public sector reform. Importantly, it contributes to ensuring that every item budgeted for is aligned with the city's strategy and is monitored in terms of not only delivery but also expenditure. This then assists management to ensure effectiveness and economy of projects (Govender and Reddy 2019).

The city used the mSCOA to go through and list all IDP goals and outcomes at the program level. The departmental accountants loaded all of the capital projects for a year, which could total between 3000 and 5000. Once that was completed, each project in the list was linked to a respective development agreement and respective goals, including global agreements such as the SDGs, but also other global agendas such as the New Urban Agenda and Sendai Agreement, as well as the KwaZulu-Natal Provincial Growth and Development Strategy (Provincial Planning Commission Kwazulu-Natal 2012) and South Africa's 2030 National Development Plan (Republic of South Africa 2012). This multiscale alignment was done centrally in the IDP office to ensure consistency in the calibration process. However, the ambition is to create enough awareness and knowledge around those global agendas so that line departments can conduct these alignment exercises themselves in the future.

Through the alignment process at a capital budget level, the city started looking at how the IDP plans aligned to the SDGs and respective capital spend. The exercise revealed that the highest capital spend corresponded with projects related to SDG 6 (water and sanitation) and 11 (cities), which in turn aligned with IDP plans 3 (creating a quality living environment) and 6 (a vibrant and creative city). In other words, this allowed the city to see and track which of its plans have an impact in terms of the SDGs, again using the best-match principle using the intent of the initial project.

When comparing the budgets of the years 2017–2018 and 2018–2019, there were some variances. The largest variance concerned capital budgets that align to SDG 9 (infrastructure), where there was a significant decrease from 2017–2018 to 2018–2019 of approximately ZAR 2.1 billion (about UD$ 136 million at 2021 exchange rates). This can be explained through the completion of one of the large parts of the capital development program for 2017/2018, which meant that the city did not have to make those outlays in 2018/2019. On the other hand, the comparison revealed a misallocation in terms of water-related projects that align with SDG 6, leading to an increase in capital budget allocated to this sector from ZAR 516 million (about US$ 33 million at 2021 exchange rates) in 2017–2018 to ZAR 1.4 billion (about US$ 90 million at 2021 exchange rates) the following year. Hence, aligning the capital budget to the SDGs allowed the city to have a better insight into what it was, and should be, spending to meet relevant SDG goals and targets, contributing to supporting the financing of SDGs. The work also represented an important way of building awareness. For instance, through the process of budgeting and business planning, we discovered that 60 projects did not align with the city's direct responsibilities as local government. This represented an opportunity to engage with line departments in order to align projects more directly to the city's responsibilities and respective SDG targets. This work also contributed to the start of creating a benchmark for monitoring by using the city's financial model as the benchmark.

12.4.3 Champions and Capacity Building

The alignment of the city's responsibilities and budget to the SDGs through the IDP gave the city a strategic and high-level entry point into SDG localization. Implementation was in turn supported by institutional reform, which involved the appointment of the city's IDP manager, together with its chief learning officer and the deputy city manager in the Human Settlements, Engineering, and Transportation cluster as champions for SDGs in the municipality. The IDP manager is responsible for strategy development, long-term development planning, IDPs, as well as assisting with the monitoring and evaluation. It therefore makes sense to allocate the functions of SDG localization to this position. The task of the chief learning officer, on the other hand, is to build awareness and advocacy around the SDGs within the organization itself and to start building up capacity at the local level. The deputy city manager for the Human Settlements, Engineering, and Transportation administrative cluster is largely responsible for the implementation of SDG 11 across the municipality by virtue of the departments that report to this office. The IDP manager reports to the chief strategy officer which in turn reports directly to the city manager, while the chief learning officer reports to the deputy city manager in the Human Resources and Capacity Building administrative cluster. Together, these three officers are responsible for supporting and translating the city's political leadership advocacy role from the global to the city level.

The work of the IDP Office on prioritization and alignment of activities with the SDGs was located in a wider change management approach. Importantly, this involved awareness raising among the executive leadership that included the municipal manager, who heads up the municipal administration, as well as the members of the city's Executive or Mayoral Committee, who together with the mayor oversee the work of the municipal manager and department heads. To this effect, the city has also put various reporting mechanisms in place, while linking reports on the alignment of the SDGs to current IDPs and capital projects to its Executive Committee. A series of workshops was also conducted with members of the city's political leadership so that when senior councilors engage in international forums and networks, they do so with full knowledge of the SDGs and the latest information available on the city's activities related to SDG localization. Reference to the SDGs is also made by speechwriters and communication teams in the offices of the mayor and city manager, contributing to raised awareness on the SDGs within, as well as outside, the city.

Awareness and capacity have to go beyond a select group of officials at a senior management level and cascade down to the rest of the organization in order to ensure effective implementation of the SDGs. The main targets of capacity building in eThekwini are members of middle management because that is where the majority of project implementation actually takes place. It is therefore important to build awareness and capacity on the content and intents of the SDGs among operational officials in order to contribute to an understanding of the important role they have to play in SDG implementation through the delivery of efficient and effective basic services. Showing how their efforts impact at the global level also contributes to an improved sense of ownership of their departmental work.

To facilitate capacity building and learning, the city hosts master classes in partnership with the Municipal Institute of Learning (MILE). MILE represents the first-ever local government-driven, practitioner-based Institute of Learning that was created in 2009 to promote learning for and by local government professionals.[2] In partnership with MILE, the city has reviewed the content of existing master classes so that they link to sustainable development and the SDGs. Whether the classes are about urban strategic planning, wastewater sanitation, recycling, or performance management, they all link to the SDGs so as to raise awareness about them among the participants of these master classes and promote the incorporation of the SDGs in their work in their respective municipalities. MILE has not only trained local councilors and officials from eThekwini but it has also functioned as a hub for train-the-trainer courses on the SDGs with officials from other African countries such as Malawi, Namibia, Maldives, Seychelles, and Morocco. Key to these courses has been the development of training modules and toolkits for SDG localization developed together with UCLG and UN-Habitat. eThekwini staff were actively

[2] See MILE website: http://www.mile.org.za/.

involved in developing the first three manuals on the SDGs: the first relates to awareness, the second addresses planning and implementation, while a third manual covers SDG monitoring, evaluation, and reporting (UCLG 2017, 2019a, b). Reflecting the way Durban has influenced international ideas about localization is the fact that the last module was launched at the global UCLG congress hosted by eThekwini in November 2019. Since then, a fourth manual was completed on SDG localization through decentralized cooperation (UCLG 2020).

Localization is not only about implementing in situ—it is also about sharing ideas about how to localize, and eThekwini has also started working together with local universities to build the technical skills and workforce that will be required to implement the SDGs through the development of a sustainable development research agenda, through guest lectures, the review, and external examination of student papers, and to ensure that curricula are relevant to the municipal environment.

12.5 Challenges to SDG Implementation

Building awareness and capacity on the SDGs at the city level is important but should extend to all levels of government in order to facilitate a whole-of-government approach, which is crucial for effective SDG localization. This applies especially in cross-cutting policy areas such as public spaces and urban safety, issues that the city of eThekwini has been actively working on in collaboration with external partners but that require multi-level policy making and implementation, involving provincial as well as national government.

Currently, the lack of policy coherence and collaboration between the different spheres of government in South Africa hinders implementation. This is exacerbated by the lack of political leadership on the SDGs at the national government level. After the adoption of the SDGs in 2015, Agenda 2030 did not immediately have an institutional home. Some saw it as falling within the Department of Planning, Monitoring, and Evaluation (DPME), while others saw it as a mandate for the national statistics office, Stats SA, which is responsible for national reporting. Only a peripheral role was accorded to national government agencies such as the Department of Human Settlements, which is signatory to the New Urban Agenda, or the Department of Cooperative Governance and Traditional Affairs (CoGTA), which is responsible for local governments (Croese et al. 2021). This meant that provincial governments, which hold important mandates over crucial areas such as health, education, and human settlements, were also left out of the conversation. It was only after the development of the country's first Voluntary National Review, presented to the UN High Level Political Forum in 2019 (Republic of South Africa 2019), that different national government departments started engaging more structurally together around the SDGs.

The Durban experience suggests that multiscale cooperation at the local level is not the only impediment to effective SDG localization. Budget constraints, exacerbated by the impact of the 2020–2021 COVID-19 pandemic, have reduced any prospects for additional national government funding for the implementation of the SDGs. This means that local governments will have to use their own internal revenues to fund SDG localization and implementation. While municipal resources may be limited, existing funding and public sector expenditure might be used and coordinated more efficiently. A new integrated planning model launched by the South African national government in 2019, the District Development Model, which counts eThekwini as one of its pilot cities, may contribute to advancing such an all-of-government approach as it gives cities an opportunity to better coordinate across the planning, budgeting, and implementation process across all spheres of government. The review of the Intergovernmental Framework Act would further assist capital investment to align with the co-created vision and strategic direction of local government.

However, there are other specific challenges that eThekwini faces when it comes to the implementation of the SDGs. Chief among them are increasing levels of informality (Patel 2013; Lund and Skinner 2004; Moyo and Gumbo 2021). In recent years, the city has experienced an influx of a large number of residents from surrounding provinces and countries. Many of these are semi-skilled and unskilled residents, while a large number of skilled and newly graduated professionals are leaving the city in search of employment. This means that the disposable income of the city's residents, compared to the country's main economic centers such as Cape Town or Johannesburg, is low. This has an impact on the cost of servicing as the ratio between disposable income and the cost of rates is high, contributing to the need to find alternative ways to generate own revenue without further burdening local rates and taxpayers, for instance, by appealing to higher equitable share transfers and other grant funding from the national government.

The realities of a larger than average informal sector in Durban are compounded by the fact that eThekwini is the only city in the country's nine provinces where some land within its jurisdiction is communally owned and therefore managed by traditional authorities, through the Ingonyama Trust, a trust set up by provincial law in 1994 to oversee land management for the benefit, material welfare, and social well-being of the members of the tribes and communities (Phakathi 2020). The complexities of who owns and runs the poorest parts of the periphery of the city make Durban a difficult, but by no means unique context in urban Africa for implementing the SDGs (Marrengane et al. 2021; Resnick 2021).

Many parts of eThekwini's urban edge have, in the past few decades, merged with communities that have built properties on land that is managed by the Ingonyama Trust Board (ITB), which decides how land will be released and considers applications for use of the land. In fact, between 60 and 70% of the city's area is in rural areas governed by the ITB. This results in challenges relating to the fact that there are large areas of informality within the city's jurisdiction, which are becoming increasingly densified. The influence of traditional authorities on matters of everyday settlement management impacts the city's service delivery mandate, its budgets for service delivery, and its ability to generate revenue from delivering those services (Beall et al. 2015). It also has a fundamental impact on how the SDGs are to be realized in African cities.

The Constitution requires that the city delivers services such as water and electricity to those in its jurisdiction. However, the existing policy framework does not allow the city to charge rates for services given to people living on government land. This means that the city must somehow pay for the cost of rolling out electricity and water services to communal areas but cannot draw revenue from the delivery of these services. This requires more hybrid solutions to governance and planning (Sim et al. 2018). In this particular case, the city has taken the interpretation of ownership of ITB land to the Constitutional Court of South Africa. In June 2021, the court ruled that the ITB is merely a notional owner of customary land, while those living on the land are the "true and beneficial owners" (Cousins 2021). This outcome is important in light of ongoing efforts of the city towards the use of smart metering and a pay-as-you-go system to enable the delivery and payment of basic services. It also highlights the extent of the interplay of politics, technology, and basic service delivery implied by the SDG localization agenda. Not all of these dynamics are easily reflected in the reporting that is associated with local action for 2030.

12.6 Monitoring, Reporting, and Evaluation

Most of the city's efforts to achieve the IDP/SDG objectives so far have centered around planning for alignment and implementation and building the necessary awareness and capacity to support these processes. The next step involves institutionalizingreporting on SDG implementation within the municipality. To support this work, the IDP office has started extracting all of the data and monitoring and reporting requirements across municipality against the SDGs 169 indicators over a 3- to 4-year period using annual reports and other existing local government reporting mechanisms which are audited independently. This would provide an important basis for a Voluntary Local Review on the SDGs at the level of the city of eThekwini. On the other hand, it would also contribute to more structured efforts to streamline municipal government reporting to provincial and national government and feed into national government reporting on the SDGs.

Key to eThekwini taking this next step has been the establishment of an SDG Institutionalization Committee that is chaired by the deputy city manager and assisted by the other two SDG Champions. The eThekwini SDG Institutionalization Committee serves as a vehicle for localization, implementation, monitoring, and reporting of SDGs in the municipality. It is noted that some of the indicators by the UN may not be relevant in the eThekwini Municipality context, and therefore the localization process will serve the purpose of making amendments to these indicators so that they address the city context, thereby assisting in monitoring the progress of the municipality in meeting the targets. The mandate of the committee is to be a coordinating instrument for all SDG efforts which includes, but is not limited to, facilitating partnerships with external stakeholders such as civil society, private sector organizations, academia, and local communities, with the aim of driving and measuring local action towards the attainment of SDG targets. One of the key aims of the committee will be to assess if there are critical indicators where the city is not collecting data, so that relevant sectors can be mobilized to explore ways to collect the required data for monitoring purposes.

Ultimately, the deliverables and outputs of the committee are as follows:

- Indicator Rationalization Framework
- SDG Localization and Stakeholder Engagement Plan
- SDG Monitoring Tool, allowing the city to track progress with regards to the implementation of SDGs and monitoring progress towards sustainability
- Open SDG Platform
- SDG VLR Reporting Template, Preparation, and Submission Plan

The committee has made progress on the Indicator Rationalization Framework by hosting a series of internal meetings to align the UN indicators with the relevant performance indicators at a municipal level. The performance indicators for the city are derived from various sources such as the SDBIP, the National Treasury Outcome Indicators (commonly referred to as Circular 88 Indicators), as well as Provincial Back to Basics quarterly reporting. Key to the exercise is not to duplicate reporting but to align existing performance reporting mechanisms to actively contribute towards achieving the SDGs in the city. Through this rationalization process, the city is also able to identify reporting gaps that might exist. The city has also made progress with regards to the UN-Habitat-designed SDG Monitoring Tool and the Open SDG Platform, which is currently being piloted internally with the Water and Sanitation Department. Once the pilot has been completed, these tools would be expanded to reflect the other sectors that respond to the SDGs.

12.7 Conclusion

Internal support and awareness raising have been crucial for effective SDG localization in the metropolitan city of eThekwini. Another important factor to be taken into account are the city's partnerships with a range of organizations and networks, such as UN agencies UN-Habitat and UNDP, as well as UCLG and the European Union, which has enabled access to opportunities for cross-city learning and engagement, thereby assisting the further embedding of the SDGs at a city level.

Importantly, these partnerships have been enabled by the active presence of a growing group of key city leaders and officials at the global stage, which have contributed to the city's global exposure and credibility and the development of a coherent narrative on sustainable development at the city level. This in turn has opened up doors to additional contacts and relationships into networks beyond the initial focus on climate change and resilience, as well as strengthened relationships with other cities and levels of government within South Africa.

These global partnerships and engagements have given the city of eThekwini an added advantage over other less well-connected or resourced cities within the country, as well as across the continent. However, Durban also shows what is possible when there is an active leadership with a clear and consistent vision that supports local sustainability champions. eThekwini's approach to SDG localization should not be seen as a required template, but merely as one of many different entry points and approaches. More than 5 years into the adoption of the SDGs there is now sufficient information on the many good practices that can help ramp up SDG localization quickly. To share and access that information, there is a need to concentrate efforts and include opportunities for those local governments that are not part of international networks so that they can also benefit from the experience of cities such as eThekwini.

References

Acuto M, Parnell S, Seto KC (2018) Building a global urban science. Nature Sustainability 1(1):2-4

Beall J, Parnell S, Albertyn C (2015) Elite compacts in Africa: the role of area-based management in the new governmentality of the Durban city-region. International Journal of Urban and Regional Research 39(2):390-406

Binns T, Nel E (2002) Devolving development: integrated development planning and developmental local government in post-apartheid South Africa. Regional Studies 36(8):921-932

Boon R, Cockburn J, Govender N, Ground L, Slotow R, Mclean C, Douwes E, Rouget M et al. (2016) Managing a threatened savanna ecosystem (KwaZulu-Natal Sandstone Sourveld) in an urban biodiversity hotspot: Durban, South Africa. Bothalia-African Biodiversity & Conservation 46(2):1-12

Cameron R (2001) The upliftment of South African local government. Local Government Studies 27(3):97-118

Cockburn J, Rouget M, Slotow R, Roberts D, Boon R, Douwes E, O'Donoghue S, Downs CT et al. (2016) How to build science-action partnerships for local land-use planning and management: lessons from Durban, South Africa. Ecology and Society 21(1):28

Cousins B (2021) What landmark Kwazulu-Natal court ruling means for land reform in South Africa. The Conversation, 22 June. https://theconversation.com/what-landmark-kwazulu-natal-court-ruling-means-for-land-reform-in-south-africa-162969. Accessed 20 July 2021

Croese S, Oloko M, Simon D, Valencia S (2021) Bringing the global to the local: the challenges of multi-level governance for global policy implementation in Africa. International Journal of Urban Sustainable Development 13 (3):435–447

Duminy J, Parnell S (2020) Supporting city futures: the cities support programme and the urban challenge in South Africa. African Centre for Cities at the University of Cape Town, Cape Town

eThekwini Municipality (2021) Draft Integrated Development Plan: 2021/2022 review. http://www.durban.gov.za/City_Government/City_Vision/IDP/Documents/2021_22_IDP.pdf. Accessed 6 July 2021

eThekwini Municipality (2017a) Resilience strategy 2017. http://www.durban.gov.za/City_Services/development_planning_management/environmental_planning_climate_protection/Documents/DurbansResilienceStrategy(ShortVersion)CouncilApproved.pdf. Accessed 20 July 2021

eThekwini Municipality (2017b) 2017/2018 Final Integrated Development Plan. http://www.durban.gov.za/City_Government/City_Vision/IDP/Documents/IDP%202017%202018.PDF. Accessed 20 July 2021

eThekwini Municipality (2017c) Service delivery and budget implementation plan: SDBIP report as at 31 December 2021. http://www.durban.gov.za/City_Government/Administration/city_manager/performance_management_unit/reports/SBIP/2017-2018%20Audited%20Quarter%202.pdf. Accessed 20 July 2021

eThekwini Municipality (2015) City Planning Commission: charting sustainable growth for Durban—overview. http://www.durban.gov.za/City_Government/City%20Planning%20Commission/Documents/Brochure.pdf. Accessed 20 July 2021

eThekwini Municipality (2001) Long term development framework. an overview of the key developmental challenges, unicity vision, outcomes and city strategy. http://www.durban.gov.za/Documents/City_Government/IDP_Policy/17%20Long%20Term%20Devpt%20Framework.pdf. Accessed 20 July 2021

Govender N, Reddy PS (2019) Performance budgeting in local government: a case study of eThekwini Municipality in South Africa. In: de Vries M, Nemec J, Špaček D (eds) Performance-based budgeting in the public sector. Governance and public management. Palgrave Macmillan, Cham, pp 143-59

Harrison P (2001) The genealogy of South Africa's Integrated Development Plan. Third World Planning Review 23(2):175-193

Lund F, Skinner C (2004) Integrating the informal economy in urban planning and governance. International Development Planning Review 26(4):431-456

Madzivhandila TS, Asha AA (2012) Integrated development planning process and service delivery challenges for South Africa's local municipalities. Journal of Public Administration 47(si-1):369-378

Marrengane N, Sawyer L, Tevera D (2021) Traditional authorities in African cities: setting the scene. African Studies 80(2):125-133

McDonald DA, Smith L (2004) Privatising Cape Town: from apartheid to neo-liberalism in the Mother City. Urban Studies 41(8):1461–1484

Moodley S (2016) The road towards Habitat 3—Durban urban breakfast. Municipal Institute of Learning, 5 October. http://www.mile.org.za/QuickLinks/News/Pages/news_20161005.aspx. Accessed 6 July 2021

Moyo I, Gumbo T (2021) The changing shape of economic informality in Durban. In: Moyo I, Gumbo T (eds) Urban informality in South Africa and Zimbabwe: on growth, trajectory and aftermath. Springer Publishing, Cham, pp 113-120

Mthembu DE, Nhamo G (2021) Domestication of the UN Sustainable Development Goals in South Africa. South African Journal of International Affairs 28(1):1-28

Palmer I, Moodley N, Parnell S (2017) Building a capable state: service delivery in post-apartheid South Africa. Zed Books, London

Parnell S, Pieterse E, Swilling M (2002) Democratising local government: the South African experiment. University of Cape Town Press, Cape Town

Patel K (2013) A successful slum upgrade in Durban: a case of formal change and informal continuity. Habitat International 40:211-217

Phakathi M (2020) The land question in South Africa and the re-emergence of the nationalism of AmaZulu. Journal of Nation-Building & Policy Studies 4(2):103-125

Provincial Planning Commission (PPC) Kwazulu-Natal (2012) Condensed version of the Kwazulu-Natal provincial growth and development strategy and plan: 2012 to 2030. http://www.kznppc.gov.za/images/downloads/12-05-21%20PGDPPrinted%20Booklet%20V8.pdf. Accessed 20 July 2021

Republic of South Africa (2019) South Africa's Voluntary National Review (VNR) report 2019. https://sustainabledevelopment.un.org/content/documents/23402RSA_Voluntary_National_Review_Report__9_July_2019.pdf. Accessed 20 July 2021

Republic of South Africa (2012) National Development Plan 2030: our future—make it work. National Planning Commission. https://www.gov.za/sites/default/files/gcis_document/201409/ndp-2030-our-future-make-it-workr.pdf. Accessed 20 July 2021

Resnick D (2021) The politics of urban governance in sub-Saharan Africa. Regional & Federal Studies 31(1):139-161

Roberts D, O'Donoghue S (2013) Urban environmental challenges and climate change action in Durban, South Africa. Environment and Urbanization 25(2):299-319

Roberts D (2010) Prioritizing climate change adaptation and local level resilience in Durban, South Africa. Environment and Urbanization 22(2):397-413

Roberts D (2008) Thinking globally, acting locally—institutionalizing climate change at the local government level in Durban, South Africa. Environment and Urbanization 20(2):521-537

Sim V, Sutherland C, Buthelezi S, Khumalo D (2018) Possibilities for a hybrid approach to planning and governance at the interface of the administrative and traditional authority systems in Durban. Urban Forum 29:351-368

Todes A (2004) Regional planning and sustainability: limits and potentials of South Africa's Integrated Development Plans. Journal of Environmental Planning and Management 47(6):843-861

United Cities and Local Governments (UCLG) (2020) Learning module 4: localizing the SDGs through decentralized cooperation. The trainer's guide. https://www.uclg.org/sites/default/files/module4.pdf. Accessed 20 July 2021

United Cities and Local Governments (UCLG) (2019a) Learning module 2: territorial planning to achieve the SDGs. The trainer's guide. https://learning.uclg.org/wp-content/uploads/2021/02/m2_en_web.pdf. Accessed 20 July 2021

United Cities and Local Governments (UCLG) (2019b) Learning module 3: reporting to national and local reviews—the role of local government associations in the Voluntary National Review elaboration process. https://learning.uclg.org/wp-content/uploads/2021/02/m3_en_web.pdf. Accessed 20 July 2021

United Cities and Local Governments (UCLG) (2017) Learning module 1: localizing the SDGs (introduction). The trainer's guide. https://learning.uclg.org/wp-content/uploads/2021/02/m1_en_web.pdf. Accessed 20 July 2021

United Nations (UN) (2016) New Urban Agenda: Quito declaration on sustainable cities and human settlements for all (71/256). United Nations General Assembly, New York

UN-Habitat (2016) Enhancing the culture of community crime prevention: 20 years of Safer Cities experience in Africa. Conference working paper. African Forum for Urban Safety. http://www.mile.org.za/AFUS2018/DocumentsAFUS2017/20%20Years%20of%20Safer%20Cities%20in%20Africa%20-%20Report.pdf. Accessed 6 July 2021

13 Leave No Agenda Behind: Participatory Approaches to Supporting Local Governments in Africa to Implement the New Urban Agenda and the SDGs

Thomaz M. T. Ramalho and Marcella O. Guarneri

Abstract

This chapter presents UN-Habitat's pilot experience in the localization of the 2030 Agenda and the New Urban Agenda (NUA) in selected cities in sub-Saharan Africa. The pilot was conducted by UN-Habitat's Urban Planning and Design Lab and the Regional Office for Africa, later consolidated in the Participatory Incremental Urban Planning (PIUP) toolbox. The PIUP toolbox was launched at the tenth session of the World Urban Forum in Abu Dhabi in February 2020 and it was rebranded in October 2021 as the Our City Plans toolbox. The selected cities were characterized by lack of data, resources, and local capacity. The chapter starts by contextualizing how urbanization is currently addressed in sustainable development and describes the pilot methodology to localize the SDGs and the NUA. It considers the main achievements, giving concrete examples of how the process was conducted in the cities of Bissau (Guinea-Bissau), Príncipe Island (São Tomé e Príncipe), and Hawassa and Bahir Dar (Ethiopia). It finalizes by touching on the main challenges and possible ways forward in the process of the localization of the 2030 Agenda and the New Urban Agenda in African cities.

Keywords

UN-Habitat · New Urban Agenda · Agenda 2030 · SDGs · Participatory planning · Sustainable urban development · Small cities · Africa

13.1 Introduction

In 2015, the 2030 Agenda for Sustainable Development was established by member states (UN 2015). Among the 17 Sustainable Development Goals (SDGs), one was specifically designed to tackle the challenges of urbanization: SDG 11 on sustainable cities and communities. Aiming to make cities and human settlements inclusive, safe, resilient, and sustainable, the urban SDG represented a big leap from the way in which urban issues had been framed before. For instance, the Millennium Development Goals (MDGs) only had two targets that were directly connected to urbanization: one on drinking water and sanitation and another on slum upgrading (UN 2000).

T. M. T. Ramalho (✉)
Regional Office for Africa, UN-Habitat, Maputo, Mozambique
e-mail: tramalho@gmail.com

M. O. Guarneri
Urban Planning & Design Lab, UN-Habitat, Nairobi, Kenya
e-mail: marcella.guarneri@un.org

S. Croese, S. Parnell (eds.), *Localizing the SDGs in African Cities*, Sustainable Development Goals Series, https://doi.org/10.1007/978-3-030-95979-1_13

The United Nations Conference on Housing and Sustainable Urban Development in October 2016, also known as Habitat III, was the first major UN conference after the launching of the 2030 Agenda. This explains why the Quito Declaration on Sustainable Cities and Human Settlements for All, also known as the New Urban Agenda (NUA), was so influenced by sustainable development concepts, which were clearly translated in the transformative commitments for sustainable urban development in the NUA (UN-Habitat 2016). The New Urban Agenda is the first international document in which the 193 member states of the UN General Assembly agree on principles, tools, and methodologies related to sustainable urbanization. Since Habitat III, discussions on how to localize, design, define, and implement the SDGs and the NUA at the local level have been at center stage in the United Nations and among urban development stakeholders. Many urban frameworks have been formulated and operationalized, also taking global agendas and regional, national, and subnational strategies into consideration (Local 2030). Figure 13.1 shows how all these different frameworks are interlinked, informing sustainable urban development planning in the case of Cabo Verde.

The consensus in the international community regarding the role of urbanization in sustainable development is timely. By 2030, 60% of the world's population will be living in urban areas (UNDESA 2018). In the African continent, the urban population is currently 43.5% of the total population and will increase to 48.4% in 2030 and to 58.9% in 2050. Africa has also, by far, the highest average annual rate of change of the urban population of 3.58%, against 1.90% of the world's rate. If only sub-Saharan Africa is considered, this number increases to 3.98% (UNDESA 2018). The urbanization challenge in Africa is characterized by fast growth and at the same time by a lack of capacity at local level in terms of data, resources, planning, and management which leads to informal urbanization, with 61.7% of the region's urban population estimated to live in slums in 2010 (Africa Renewal 2012).

13.2 Planning Sustainable Urban Development

In its paragraph 15, the NUA defines the four fundamental drivers of change for sustainable urban development as (1) policy and legislation, (2) urban governance, (3) urban planning and design, and (4) municipal finance and local fiscal systems (UN-Habitat 2016).

To support local governments in localizing the NUA and 2030 Agenda in their local spatial development plans, UN-Habitat's Regional Office for Africa and the Urban Planning and Design Lab have been piloting the Participatory Incremental Urban Planning (PIUP) toolbox in selected African cities (UN-Habitat 2020a). The focus of this toolbox is on local governments in developing countries, Small Island Development States (SIDS), least developed countries (LDC), and landlocked countries, and as such the pilot projects were conducted in Bissau (Guinea-Bissau), Hawassa and Bahir Dar (Ethiopia), Príncipe Island (São Tomé e Príncipe), Moroni

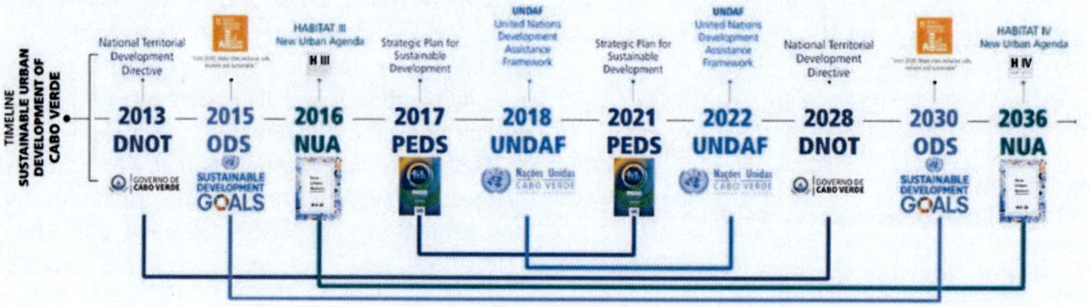

Fig. 13.1 Timeline for sustainable urban development of Cabo Verde. (Source: authors)

(Comoros), and at policy level in Cabo Verde and Angola. This experience was not only limited to the African continent, as it also included pilots in Saudi Arabia, Haiti, and Mexico, among others. The first edition of this methodological approach was published and presented during the tenth session of the World Urban Forum in Abu Dhabi on 12 February 2020 by UN-Habitat's team and city leaders from Bissau and Hawassa.

The approach is outlined in a publication, named *Participatory Incremental Urban Planning: a toolbox to support local governments in developing countries to implement the New Urban Agenda and the Sustainable Development Goals—edition for fast growing small cities* (UN-Habitat 2020a), which was expanded and revised for publication in October 2021, officially launched as *Our City Plans* (UN-Habitat 2021). The PIUP toolbox is process-based, understanding that planning practice is a "process framework rather than a design framework" (UN-Habitat 2020a, p. 7). It is fit-for-purpose, providing a "simplified roadmap for urban planning process based on best practices for contexts with limited resources and capacity" (ibid.), which is usually the case in Africa at the local level. The PIUP toolbox is "incremental and flexible, being a step-by-step methodology structured in phases, blocks and activities, supported by innovative tools to gradually develop and implement the plan" (ibid.). It is impact-oriented, through "smart mechanisms and incentives, operationalization tools, coherent policy implementation linked to land management, city finance, urban governance and legislation, also stressing the importance of action planning to put projects on the ground" (ibid.). The PIUP is also participatory and people-centered, placing local authorities, key stakeholders, and "citizens at the driver's seat of sustainable urban development" (ibid.). Finally, it is open source and modular, "creating a dynamic feedback and continuous improvement process between national and sub-national urban policies and urban planning practices at the local level. It is designed to be constantly improved and to absorb external contributions and new tools" (ibid.).

The key block within the PIUP toolbox for the localization of the SDGs is the Strategic Plan block.[1] This block "aims at defining a participatory vision, goals, targets and indicators" (UN-Habitat 2020a, p. 55) and "should be developed following the methodology for the localization of the urban SDG targets at municipal level" (ibid.) which "has been successfully applied in over 10 countries and 20 cities, to support the alignment of national priorities and plans with the local priorities, as well as international agendas reflected by the SDGs" (ibid.). The block comprises of four activities: Strategic Development Scenarios (recommended), Visioning Workshop (mandatory), Spatial Strategy (recommended), and Monitoring and Evaluation Plan (mandatory). The focus of this chapter will be on the two mandatory activities of this first edition (UN-Habitat 2020a), which are the activities under the Strategic Development Plan block of the second edition.

In practical terms, the Strategic Plan block is implemented on the ground through a Sustainable Urban Development Visioning Workshop, which is a 1-week activity led by local authorities to gather key local stakeholders and build the results framework for the city plan, including the vision, goals, targets, and selection of indicators to the monitoring framework. A typical calendar of activities for a City Sustainable Development Visioning Workshop comprises a series of intertwined activities which are flexible and adapted to the level of engagement of local stakeholders. The workshop can take from 3 to 5 days, depending on the intensity and commitment of participants and their knowledge of the city issues. It includes at least the following activities (Fig. 13.2):

The workshop starts with a training session on the methodology, the strategic planning approach of the 2030 Agenda, and selected global agendas and national and subnational strategies to be considered by the plan. The rest of the first day is dedicated to presenting and

[1]The name of the block (a section of the PIUP publication) is "Strategic Plan" in the first edition. In the second edition, it was renamed to "Strategic Development Plan".

	1st Day	2nd Day	3rd Day
	Understanding relevant global and national agendas, defining a shared VISION and identifying key issues for the city	***Transforming the vision into enabling GOALS and linking them to national strategies and global agendas***	***Setting up SMART TARGETS, a MONITORING FRAMEWORK and identifying the NEXT STEPS for the plan***
09:00 – 10:30	Methodological presentation on Strategic Planning principles, Sustainable Development Goals, New Urban Agenda and selected national and sub-national strategies	How could my Sustainable Development working group support the achievement of this shared VISION?	Based on selected issues and by aggregating indicators, how could your GOAL be structured into TARGETS which relate to national and global development targets?
10:50 – 12:20	Which are the main issues (SWOT) related to your Sustainable Development working group that should be addressed by the plan?	Defining GOALS based on the VISION and main issues	Setting SMART TARGETS to support the achievement of the GOALS.
13:40 – 15:10	Which other issues should also be addressed by the Strategic Framework?	Defining the GOALS to enable the VISION achievement based on main issues	Setting BASELINE, FREQUENCY and MEANS OF VERIFICATION for each indicator and identifying strategies to deal with data gaps
15:30 – 17:00	Which VISION could be extracted from National Strategy, SDGs and other global agendas for the city's Sustainable Development framework?	Selecting relevant INDICATORS from national strategies and global agendas, especially the NUA and SDGs to monitor development GOALs	WAY FORWARD — Presenting what was done during this workshop and which are the gaps to finalize the strategic framework for Sustainable development

Fig. 13.2 City sustainable development visioning workshop program (UN-Habitat 2020a)

agreeing on the current diagnostic of the city, the key issues, and how a development vision can be built to integrate global agendas and national and subnational strategies. This is done through six perspectives which should always be represented throughout the workshop, which are the thematic 5 Ps for Sustainable Development: People, Prosperity, Planet, Peace, and Partnership (SDG Services n.d.), complemented by the enabling P of Planning (also grouped as Policy/Planning/Programming) (Fig. 13.3).

The stakeholders are divided into groups representing the 6Ps, after which they have to agree on a common and shared vision in a plenary session as a way to ensure all perspectives from Sustainable Development are considered by stakeholders. Social issues will be brought up by the People group, while environmental challenges will be addressed by the Planet group, for instance, while economic concerns are mainstreamed by the Prosperity group. Leaving no issue and no sustainable development perspective behind is key to ensuring that the sustainable development vision is inclusive and that, later, all relevant SDGs for the local context are properly localized and integrated into the plan.

The vision which comes out of this exercise is usually a very long statement which can later be shortened to be more impactful and catchier for communication and marketing purposes. However, keeping it as long as it was originally formulated is important at this stage, since the success of the exercise depends more on being inclusive rather than on being focused. As an example, Bahir Dar's vision statement was originally defined in the workshop as "By 2035, Bahir Dar will be a sustainable, inclusive, resilient and vibrant leading cluster of responsible tourism, innovative industry and tertiary services." This vision statement was later modified by city leaders in some aspects and shortened for communication purposes to read "Bahir Dar: the first water-wise resilient forest city of Ethiopia and Africa."

Fig. 13.3 The 6Ps (or 5Ps +1) for sustainable urban development (UN-Habitat 2020a)

The vision statement is among the most dynamic outputs of the planning exercise and can be improved, modified, and shortened throughout the planning process. However, it is of critical importance to establish a comprehensive, yet simplified vision to match local capabilities and results frameworks for the localization of the SDGs and the NUA. It is from the vision statement, agreed by all stakeholders as the guiding aspiration for the plan, that the vision goals are extracted. In the case of Bahir Dar, for instance, it is clear that Goal 1–6 is directly extracted from the vision statement (Fig. 13.4).

The same example can be seen from the Príncipe 2030 Regional Sustainable Development Plan (UN-Habitat São Tomé and Príncipe 2019). The vision statement "By 2030, the Island of Príncipe, an international biosphere reserve, will become a global reference of biodiversity conservation and sustainable development that is inclusive and resilient to climate change, through the integration of responsible tourism and the green and blue economy, ensuring quality of life for all" is translated in Goal 1–5 (Fig. 13.5).

The second day of the workshop, after the Sustainable Development Vision Statement is defined, is dedicated to translating this shared vision into a framework which can guide and be integrated into local development plans. As per the examples of Bahir Dar and Príncipe Island, this is initially done by translating the vision statement into vision goals. These goals are inspired by the 2030 Agenda format and shall be short statements which quickly communicate the thematic area covered by the goal. For instance, Goal 1 of the Príncipe 2030 plan on Environmental Sustainability translates the vision of Príncipe as a global reference for biodiversity, and it is also

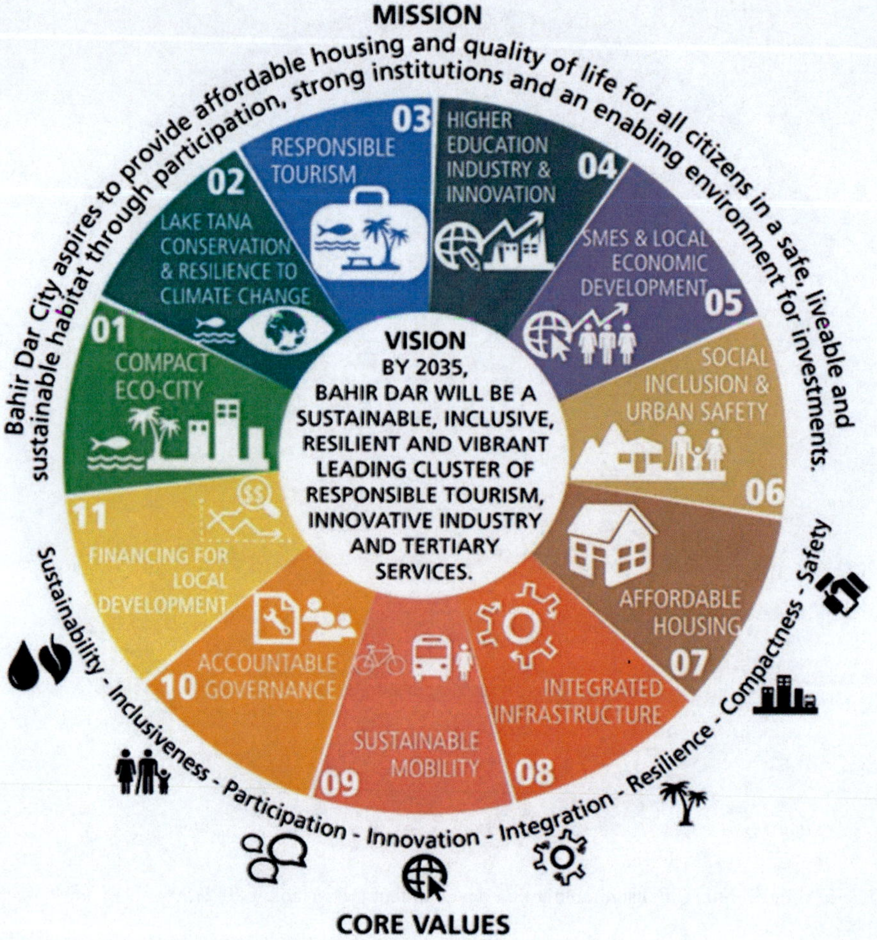

Fig. 13.4 The wheel of sustainable urban development of Bahir Dar 2035 (First version, source: authors)

formulated into a Goal Statement, as per the SDGs, which is "Protect, restore and promote the sustainable and responsible management and use of terrestrial and marine resources, halting and reversing ecosystem degradation and biodiversity loss" (UN 2015). This Goal Statement was critically discussed by the stakeholders during the workshop and then refined and approved by the plenary. It encompasses content from SDG 14 to 15, but also from the Nagoya Protocol on Access and Benefit Sharing, a 2010 supplementary agreement to the 1992 Convention on Biological Diversity and Príncipe's specific commitments as a UNESCO Biosphere Reserve.

In parallel to the designing of vision goals, the results framework at goal level will be also complemented by enabling goals, which usually come from the main issues pointed out in the first day of the workshop. Experience has shown that usually these issues are related to governance, finance, planning, innovation, capacity building, infrastructure, mobility, housing, and so forth. While the vision goals are usually linked to the Ps of People, Prosperity, and Planet, the enabling goals are associated with the P of Peace, Partnership and Planning, Policy, and Programming. As soon as the goals are defined, one important strategy to ensure engagement of the stakeholders is elaborating and sharing the Wheel of Sustainable Urban Development of the city (see examples above in Fig. 13.4 of Bahir, Fig. 13.5 of Dar and Príncipe, and Fig. 13.6 below of Bissau). The wheel is projected on a wall or printed and distributed to focal groups

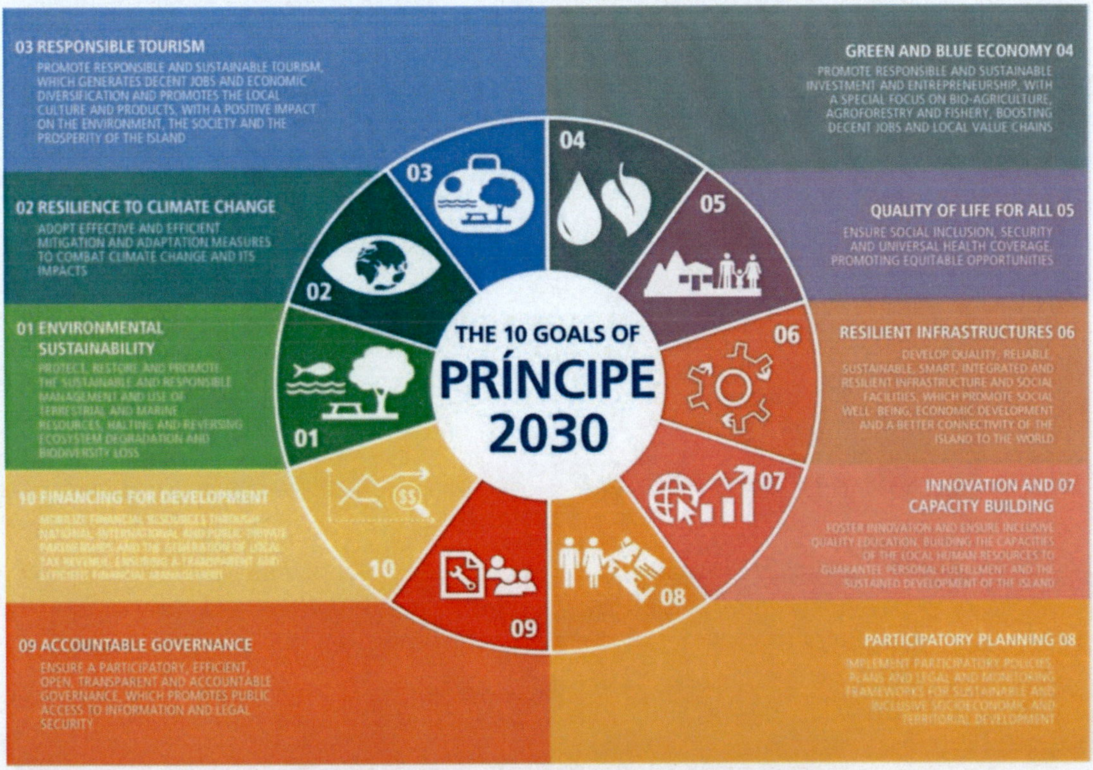

Fig. 13.5 Príncipe 2030 regional sustainable development plan (UN-Habitat São Tomé and Príncipe 2019)

and allows all participants to see the progress made so far and already communicates the centrality of the SDGs in the city planning process. This energizes the group since quick progress can already be seen by participants, and it will be critical in the future communication strategy of the plan.

The third day of the workshop is then dedicated to building a simplified operational framework for the plan, and it is when the localization of SDGs and the NUA principles happens in practical terms. Some of the critical challenges in developing countries, and particularly in Africa, for the localization of the SDGs in cities are the lack of local capacity to (1) manage and monitor data, (2) build baselines, and (3) promote effective participation of key stakeholders in the planning and governance process (Valencia et al. 2019; Croese et al. 2021). For that reason, applying an effective, simplified, and engaging methodology is key to ensuring the ownership of the plan by local government and civil society. During the implementation of the catalytic projects, it was identified that the most complex issue in the framework to operationalize the city plan is setting up the monitoring system and making it effective, especially on building and selecting indicators and setting up baselines. The steps taken to arrive at the simplified operational framework will be detailed in the next section.

13.3 Monitoring Sustainable Urban Development

The draft version of the NUA Monitoring Framework was launched on 17 September 2020 with 77 indicators, 28 of them also being SDG indicators, covering SDG 11, SDG 1, SDG 3, SDG 5, SDG 6, SDG 7, SDG 8, SDG 9, SDG 12, SDG 16, and SDG 17 (UN-Habitat 2020b). Some NUA indicators are quite straightforward and are immediately available for utilization, while some are still being discussed in terms of

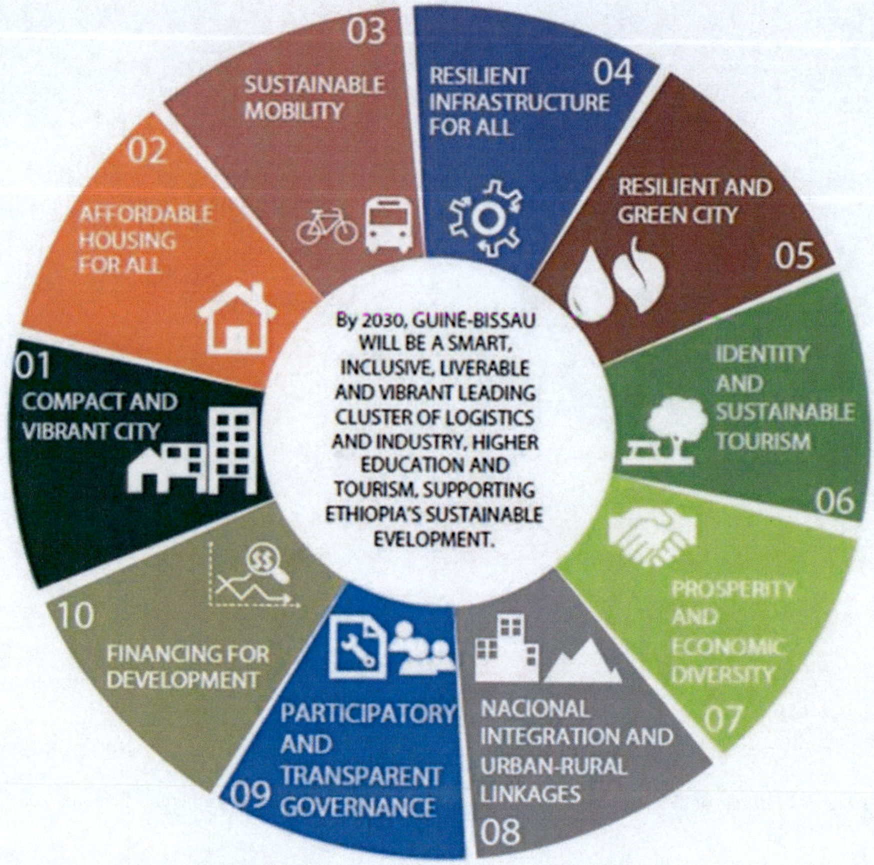

Fig. 13.6 Bissau 2030 sustainable development plan (UN-Habitat Bissau 2019)

their methodological aspects. Similarly, the so-called SDG Tier III indicators lacked an internationally established methodology, which was only developed following a comprehensive review in 2020, 5 years after their adoption (UN IAEG-SDGs 2022).

Until this review, despite representing important progress on the sustainable urban development agenda, SDG 11 contained two Tier III indicators (SDG 11.a.1 and SDG 11.c.1), a majority of Tier II indicators (11 out of 15), meaning those that have an established methodology but no regular data collection, and only two Tier I indicators, those with an established methodology as well as regular data collection: SDG 11.1.1 and 11.6.2 (UN IAEG-SDGs 2022). Up until a subsequent review of the SDG tier classification in February 2022, the situation was still worrisome with SDG 11.c.1 remaining as a Tier III indicator, 9 Tier II indicators and 5 Tier I indicators, with the inclusion of SDG 11.5.1, 11.a.1 and 11.b.1 in the list. This has created enormous challenges for the effective localization and integration of the SDG 11 indicators in the monitoring framework of urban and territorial plans. A similar lack of methodological standards applies to other SDG targets with a clear urban dimension. However, this challenge should not impede SDG 11 targets to translate into urban policies and plans at local level.

Based on these challenges and on the lack of capacity to develop their own city-based indicators, the selected approach to localizing the SDGs at city level is through the selection of established indicators in existing plans and strategies, especially those already being monitored. Alternatively, it is possible to consider an important indicator which is not measured at local level and create its baseline and monitoring system based on perception surveys.

For example, in Bahir Dar, the planning team wanted to build an indicator on affordable transportation to measure the implementation of a specific target of their Goal 9 on Sustainable Mobility (UN-Habitat 2020a). The local experts were elaborating a very complex set of indicators in which data would not be easily collected and processed. After discussions with UN-Habitat experts, they opted for considering SDG indicator 11.2.1 (proportion of population that has convenient access to public transport, by sex, age, and persons with disabilities), still a Tier II indicator, and establish its baseline through perception, since data was not yet available at the city level.

The application of the PIUP methodology on the third day of workshop then starts with the mapping, listing, and selection of all relevant indicators from the selected global agendas and national and subnational strategies. For instance, for Ethiopian cities, all indicators were extracted from the National Urban Development Spatial Plan (NUDSP), the Second Growth and Transformation Plan (GTP II), the subnationals and sectoral (urban development and housing) plans linked to GTP II, the New Urban Agenda, and the 2030 Agenda. Many of these indicators already have national or regional baselines set up by the plans, and at city level, the effort was domesticating them and spatializing their implementation.

These selected indicators already provided a time frame for the city plan, and the indicators from NUDSP 2035 and the NUA 2036 functioned as frameworks for monitoring long-term targets, while indicators from the 2030 Agenda were used for monitoring mid-term targets, and the indicators appropriated from the GTP II 2025 were linked to short-term targets (see Fig. 13.7).

After the selection of relevant indicators, they were connected with the drafted goals, and whenever possible, indicators were associated to provide inputs for drafting a SMART target, which should be Specific, Measurable, Achievable, Realistic, and Time bound. The SMART principles are currently a mainstream method for developing strategic plans (Bjerke and Renger 2016). As an example from Hawassa Structure Plan, target 1.7 was "by 2030, reduce the adverse per capita environmental impact of cities, including by paying special attention to air quality, sound pollution and waste generation" (UN-Habitat 2020a) and encompassed indicators from the

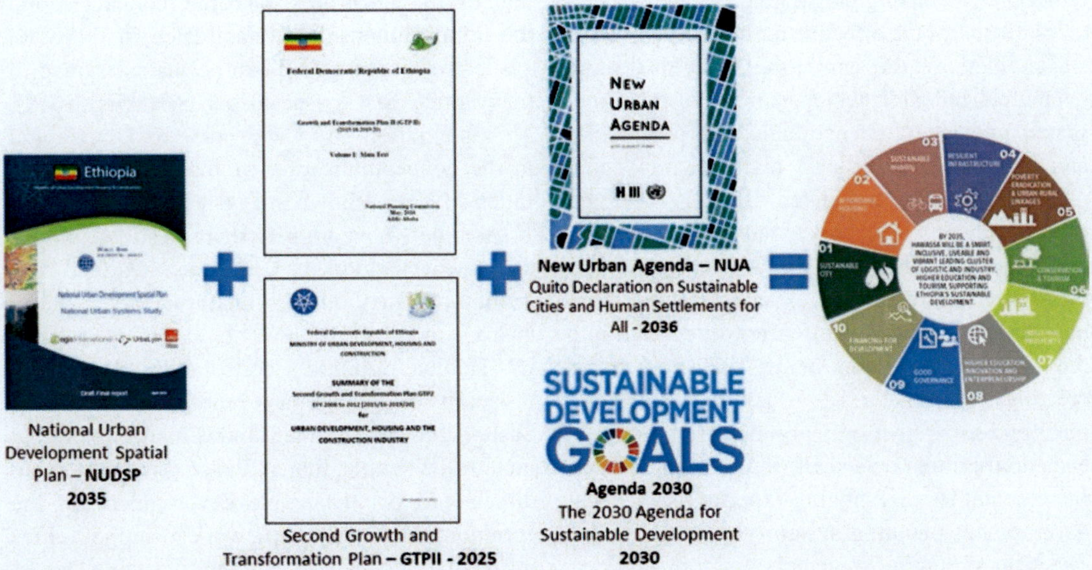

Fig. 13.7 Integrating local targets with global and national development agendas (UN-Habitat 2018)

GTP II (number of cases of noise pollution reported to Environmental Authority disaggregated by neighborhood), SDG 11.6 (annual mean levels of fine particulate matter—e.g., PM 2.5 and PM 10—population weighted), and SDG 12.5 (city recycling rate, tons of material recycled).

Even though some of these indicators are not yet being calculated by the Central Statistics Agency in Ethiopia, community perception surveys can fill the gap, while government bodies dealing with statistics advance with national data collection. In this sense, there are a number of advantages in using existing indicators from global, national, and subnational agendas and strategies. First, facilitating fundraising for city plan implementation, since subnational, national, and global stakeholders have a framework to measure the efficiency of resources they are allocating to the city; second, avoiding the creation of more indicators in a context already characterized by a lack of data, capacity, and resources; and, finally, establishing a localization process which effectively integrates global agendas into the local development planning culture and practice.

This experience in multiple cities has shown that localizing specific indicators and linking them to goals that are tailored to local realities rather than localizing global goals provides a better platform for the effective participation of local stakeholders. At the same time, it builds a more contextualized plan which can be further integrated into local and national budgetary cycles and function as a strategic platform for spatial plans. In many experiences, SDG localization processes at city level are characterized by bringing relevant goals or targets down to the local context. Even though this is an easier and more automatic process, it is often not integrated into the way planning is done at city level, sometimes creating a plan that exists in parallel to existing plans and sometimes only producing a report that ends up forgotten on a shelf. In this Participatory Incremental Urban Planning experience, the local development planning system is respected, and rather than bringing an agenda down from top to bottom, local stakeholders are requested to build their city goals from the bottom up through the lens of the 5Ps + 1. This ensures that no agenda, no issues, and no sustainable development perspectives are left behind and that an effective discussion between experts and stakeholders from different sectoral areas takes place throughout the planning process.

The charter below, for instance (see Fig. 13.8), has been guiding many SDG localization processes, trying to fit the SDG framework as a whole into a box. However, a deeper analysis at SDG target level shows that many SDGs have targets that can be linked to different Ps. For instance, SDG target 11.1 on Housing and Slum Upgrading is clearly linked to People, while SDG target 11.2 on Sustainable Transportation can have a Planet, but also a Prosperity linkage. The same applies to SDG 11.a on Rural-Urban Linkages and the SDG 11.c on Resilient Construction. This is of course not limited to SDG 11, and it is clear that SDG 6, boxed under People, has targets linked to both Planet and Peace dimensions, and SDG 7 has a Planet dimension, while SDG 10 also has a People dimension. Also, all means-based SDG targets finishing with letters (e.g., SDG 1.b, 15.a) are usually also linked to both Partnership and Planning/Policy dimensions.

Rather than part of different boxes linked to the 5Ps, the SDGs are clearly interconnected, and the urban dimension, consolidated in the SDG 11, is cross-cutting and can be mainstreamed in many other SDGs (see also UN-Habitat 2018). This integration effort was made by UN-Habitat in the conceptualization of the City Prosperity Index (CPI), as shown in Fig. 13.9.

As others have argued, there is still scope for the improvement of the CPI as an integrated conceptual and methodological framework for city-based monitoring (Wong 2015). Ideally, UN-Habitat should consolidate the City Prosperity Index, perhaps rebranded as the City Sustainable Development Index to also expressly include the People, Planet, Peace, and Partnership dimensions of sustainable development, as the common monitoring framework to support cities in localizing SDGs, by integrating the SDG 11 with relevant SDG urban targets and the NUA

Fig. 13.8 Scheme for the localization of SDGs with the 5Ps for sustainable development (UN-Habitat 2020a)

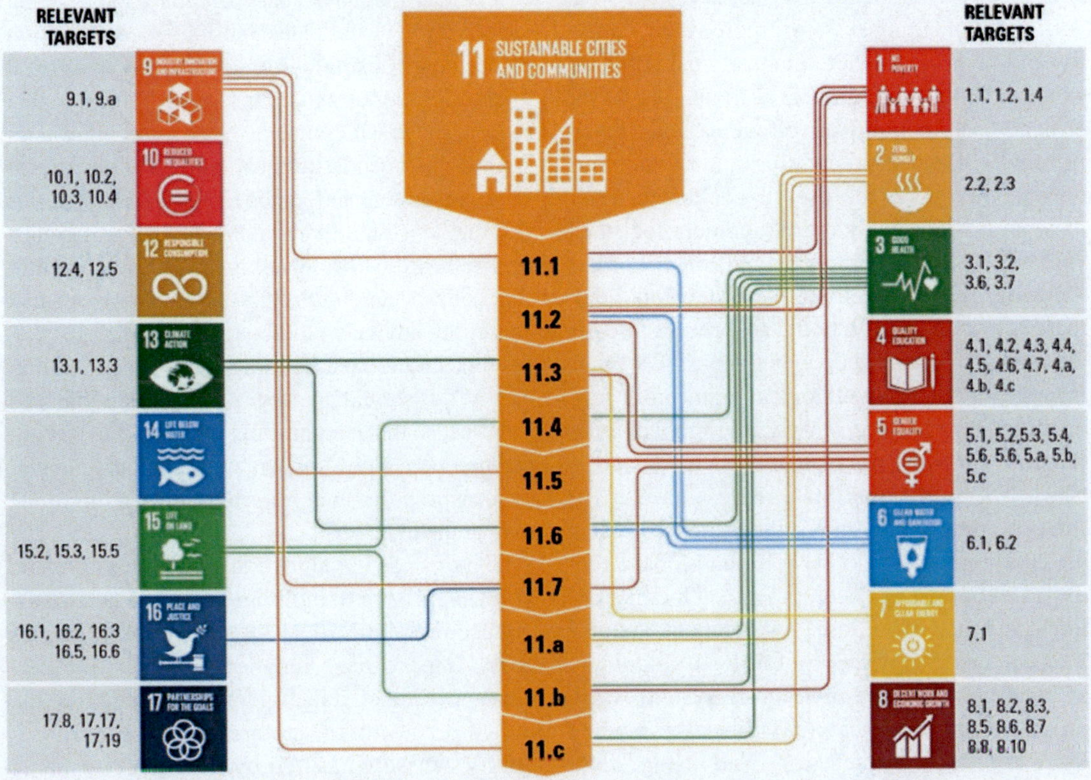

Fig. 13.9 Integration of SDG 11 with other SDGs (UN-Habitat 2018)

Monitoring Framework. This would greatly facilitate city leaders and planners to effectively integrate global agendas in their current planning system. This can also be associated with other existing initiatives, such as the Local 2030 platform, a network for the sharing of tools, experiences, solutions, and guides to support SDG localization.

13.4 Challenges Related to This Experience

Among the challenges and lessons learnt through this pilot experience, one critical issue relates to the Wheel of Sustainable Urban Development design and how this communicates very similarly to the original SDGs. Some criticism has been expressed regarding the fact that the locally tailored goals, based on the vision and issues, are not the SDGs themselves and the graphic similarity can create confusion with local stakeholders. The fact is that in areas with weak planning systems, scarce data, capacity, and resources, where plans are usually nonexistent or not operational, participation is the critical tool to build an effective and implementable plan. Participation and ownership, however, is much more easily achieved when local authorities and key stakeholders feel their voices and ideas are reflected in the plan. Allowing local stakeholders to develop goals that reflect a locally built vision and locally identified critical issues using the SDGs as an inspiration and methodological framework may be more effective than simply asking local communities to prioritize global goals discussed in New York by Member State representatives. As such, a compromise may be reached by using the 2030 Agenda and other relevant global agendas to support the development of locally tailored and fit-for-purpose development visions. This can be done as support for the development of local planning and monitoring system, since proposing new indicators is usually beyond local capacity. By selecting and using these existing global indicators, the SDGs, NUA, and strategic national plans can be effectively integrated into local plans, even though there is no local effective monitoring capacity.

Another critical challenge derives from the fact that 5 years after the adoption of the 2030 Agenda, many SDG indicators are still not fully operational with most data not being produced by National Statistics Offices, and when available, data are usually not disaggregated to the city level. This can be a very strong argument to integrate city plans and global agendas at the goal and target level, rather than still not operational global indicators. However, the results framework of the 2030 Agenda is indivisible and that means exactly the contrary, that the indicator level is the best entry point for localization, since one goal will never be measured until all indicators are operational. Also, a target with two indicators can have a measurable and relevant indicator at the city level, while a Tier III indicator may not be relevant to the local context (see also Simon et al. 2016).

Besides that, the 2030 Agenda should be an aspiration for local plans and influence the vision of the city. Raising awareness around the global agendas is important in this regard, and for that reason, the crash course on SDGs and other relevant global agendas and national and subnational strategies is done before the identification of issues and the division of participants in six focal groups inspired by the 5Ps of Sustainable Development + the Policy/Planning/Programming group. In that sense, the advice would be that if the local stakeholders find a Tier III indicator relevant to their reality, this should be integrated into their plan and baseline, so that a monitoring database can be built through perception surveys with the local population, again promoting a stronger ownership of the plan at the local level.

Experience has shown that an alternative and viable approach to building baselines and assuring proper monitoring of locally tailored targets is relying on local knowledge and on the perception of local stakeholders on issues that the indicators want to capture. Transport affordability can be measured by hard data on transport fees, government subsidies, service

Fig. 13.10 Bissau 2030 sustainable development plan (UN-Habitat Bissau 2019)

provision, and local salaries, but also by explaining to local stakeholders what affordability means and how affordable their daily commuting is. Such examples were presented from Brazil during the First UN-Habitat Assembly in 2019. In a partnership between UN-Habitat and COLAB, data was collected through a mobile app on the perception of citizens of SDG 11. Citizens could not only grade but also provide comments and upload georeferenced photos to illustrate certain issues, thereby showing how innovative and interactive tools can be used to support a vibrant and engaging planning, implementation, and monitoring process of city plans (UN-Habitat n.d.).

One critical lesson learnt from this pilot experience is the need to link the SDG and NUA localization with spatial planning (Watson 2016). Identifying land to accommodate a landfill for the achievement of SDG 11.6 and SDG 12.4 and 12.5 is particularly crucial for effective implementation at city level. Spatializing the SDGs, such as in the example below of Bissau, allows local government, stakeholders, investors, and development partners to predict land conflicts and promote sustainable and inclusive city development. This is a crucial issue for African cities growing at a fast rate and with very low levels of planning capacity.

For the spatialization of goals to be more effective, targets should be translated into spatial strategies and catalytic projects. These pilot projects should be classified and prioritized and be immediately integrated into existing budgeting cycles and other funding mechanisms. The experience of Bissau, in that sense, is very positive. By being the first spatial plan developed for the city since 1996 (see Fig. 13.10), the Bissau 2030 Sustainable Development Plan was embraced by critical local partners, such as the World Bank and the African Development Bank, and some pilot projects are already in the funding pipeline of these institutions. The same happened in Príncipe Island, where private investors, which participated enthusiastically in the planning process, could immediately support the local government on critical issues related to island development.

13.5 Way Forward

The localization of the SDGs at city level will certainly follow many methodologies, and the most important lesson from this pilot experience is that this exercise should not be detached of local planning practice already developed by cities. In this sense, the strategic phase of city plans represents a good entry point for both SDG and NUA localization at city level. This calls for a new generation of master plans, and since usually these spatial planning tools have a timeframe of 10 years, the city master or structure plans can be adequate frameworks for the operationalization of the UN Secretary-General's Decade of Action towards 2030 at city level.

These new-generation master plans would integrate Sustainable Development and NUA principles into the strategic visioning exercise by recognizing all dimensions of sustainable development embedded in the 5Ps + 1 (People, Prosperity, Planet, Peace and Partnership + Policy/Planning/Programming) and ensuring no agendas, no issues, and no sustainable development perspective are left behind. The integration of the SDG and NUA principles into these new-generation master plans should also be done by prioritizing measurable indicators at the local level from the relevant global agendas, and by using perception tools and others forms of local data collection for indicators that are still not collected at the city level. Using citizen perceptions to build baselines for global indicators at the local level should be further explored in developing contexts where data are still scarce and expensive, through the utilization of apps and other IT tools.

At the global level, relevant institutions should discuss a global monitoring framework based on SDG 11, urban SDG targets and the NUA which can be applied to cities, including those with limited data, resources, and local capacity. UN-Habitat has been piloting the City Prosperity Index (CPI) which could be the basis for this integrated monitoring framework and could be rebranded as the City Sustainable Development Index (CSDI).[2]

Lastly, it is important to promote capacity building and financing mechanisms to ensure effective implementation after the SDG and NUA localization is done at local level lest localization exercises be seen locally as another frustrating planning initiative with no actual impact. SDG project accreditation and financing mechanisms, such as UN-Habitat's City Investment Facility, can be a global example of this type of mechanism and inspire similar regional, national, and subnational funding schemes.

References

Africa Renewal (2012) Towards African cities without slums: governments set course towards improving poor urban areas. https://www.un.org/africarenewal/magazine/april-2012/towards-african-cities-without-slums. Accessed 14 December 2020

Bjerke MB, Renger R (2016) Being smart about writing SMART objectives—evaluation and program planning. https://linkinghub.elsevier.com/retrieve/pii/S0149718916302580. Accessed 7 February 2021

Croese S, Dominique M, Raimundo IM (2021) Co-producing urban knowledge in Angola and Mozambique: towards meeting SDG 11. npj Urban Sustainability. https://doi.org/10.1038/s42949-020-00006-6

SDG Services (n.d.) The main principle of sustainability is the common good. https://www.sdg.services/principles.html. Accessed 2 February 2021

Simon D, Arfvidsson H, Anand G, Bazaz A, Fenna G, Foster K, Jain G, Hansson S et al. (2016) Developing and testing the urban Sustainable Development Goal's targets and indicators—a five-city study. Environment & Urbanization 28:49–63

United Nations (UN) (2000) Millennium Development Goals. https://www.un.org/millenniumgoals. Accessed on 30 January 2021

United Nations (UN) (2015) Transforming our world: the 2030 Agenda for Sustainable Development (70/1). United Nations General Assembly, New York. https://www.un.org/ga/search/view_doc.asp?symbol=A/RES/70/1&Lang=E. Accessed 30 January 2021

United Nations Department of Economic and Social Affairs (UNDESA) Population Division (2018) World urbanization prospects, online edn, 2018 revision.

[2] See the UN-Habitat website for more information on the CPI: https://unhabitat.org/programme/city-prosperity-initiative.

https://population.un.org/wpp/Download/Standard/Population. Accessed 27 November 2020

UN-Habitat (2020a) Participatory incremental urban planning: a toolbox to support local governments in developing countries to implement the New Urban Agenda and the Sustainable Development Goals—edition for fast growing small cities. https://unhabitat.org/participatory-incremental-urban-planning-toolbox-a-toolbox-to-support-local-governments-in. Accessed 27 November 2020

UN-Habitat (2020b) Data analytics—the New Urban Agenda monitoring framework. https://www.urbanagendaplatform.org/data_analytics. Accessed 14 December 2020

UN-Habitat (2021) Our City Plans: A Global Toolbox for Customized Planning Processes. https://urbanoctober.unhabitat.org/event/our-city-plans-global-toolbox-customized-planning-processes-0. Accessed 20 March 2022

UN-Habitat (n.d.) UN-Habitat and Colab launch a mobile app for open public consultation on living conditions in Brazilian cities. United Nations Sustainable Development Goals Partnerships Platform. https://sustainabledevelopment.un.org/partnership/?p=29756. Accessed 22 July 2021

UN-Habitat (2018) Tracking progress towards inclusive, safe, resilient and sustainable cities and human settlements. SDG 11 synthesis report—High Level Political Forum. United Nations-Habitat, Nairobi

UN-Habitat (2016) New Urban Agenda. https://uploads.habitat3.org/hb3/NUA-English.pdf. Accessed 1 February 2021

UN-Habitat Bissau (2019) Bissau 2030: sustainable development plan. https://unhabitat.org/node/143079. Accessed 30 January 2021

UN-Habitat São Tomé & Príncipe (2019) Príncipe 2030: sustainable development plan for the autonomous region of Príncipe. https://en.principe2030.com. Accessed 30 January 2021

United Nations Inter-Agency and Expert Group on Sustainable Development Goal Indicators (UN IAEG-SDGs) (2022) Tier classification for global SDG indicators. https://unstats.un.org/sdgs/iaeg-sdgs/tier-classification/. Accessed 20 March 2022

Valencia SC, Simon D, Croese S, Nordqvist J, Oloko M, Sharma T, Taylor Buck N, Versace I (2019) Adapting the Sustainable Development Goals and the New Urban Agenda to the city level: initial reflections from a comparative research project. International Journal of Urban Sustainable Development. https://doi.org/10.1080/19463138.2019.1573172

Watson V (2016) Locating planning in the New Urban Agenda of the urban Sustainable Development Goal. Planning Theory 15(4):435-448

Wong C (2015) A framework for 'City Prosperity Index': linking indicators, analysis and policy. Habitat International 45(1):3-9

14 SDG Localization in African Cities: The Crucible of the 2030 Agenda

Susan Parnell

Abstract

African cities are the crucible of SDG implementation, as they have the most to gain from advancing sustainability aspirations. Unsurprisingly therefore, some of the most interesting examples of SDG localization and implementation are now emerging from cities and towns from Cairo to Cape Town. This conclusion reflects on the contributions in this book and distills their learnings about sustainable practice and thought when viewed from the urban African frontier. The collection offers a range of local African experiences to nuance the way forward on the road to global sustainable development based on local ownership and adaptation of the 2030 Agenda to local priorities. As such, one of the most positive things the SDG process has made legible is the extent to which Africa matters in the overall urban conversation. If cities matter to sustainability, African cities matter more than most to our collective chances of realizing the SDGs globally.

Keywords

SDG localization · African cities · Agenda 2030 · Agenda 2063

14.1 Introduction

Our hope in curating this volume was to set out new ways of learning about Africa's transformation through the urban implementation of the SDGs. We also sought to offer fresh modes of exploring what has happened in African cities under the SDG rubric and to debate how collective knowledge might usefully be expanded or amended to drive SDG implementation in and through African cities. The intention is, through both the writers and readers of the volume, to build an interdisciplinary community of intellectual practice that is centrally involved in making sustainability change happen locally and nationally. At the global scale, our aspiration is to ensure that African voices, both from academia and practice, are at the forefront of forging an urban community of practice that has broad capacity to engage urban policy based on deep local knowledge and the bigger issues surrounding urban transformation. Ideally, Africans will be leaders in supra-local forums and debates about urban sustainability like Local 2030 (a collaborative platform to advance the local

S. Parnell (✉)
Human Geography, University of Bristol, Bristol, UK

African Centre for Cities, University of Cape Town, Cape Town, South Africa
e-mail: susan.parnell@bristol.ac.uk

S. Croese, S. Parnell (eds.), *Localizing the SDGs in African Cities*, Sustainable Development Goals Series, https://doi.org/10.1007/978-3-030-95979-1_14

implementation of the SDGs[1]) as well as the development of key tools and processes for urban SDG planning and reporting, such as National Urban Policies and Voluntary Reviews.

Contributing to the large-scale reconfiguration of the urban process requires action on multiple fronts and across varied scales. There is no use signing up to a long-term vision, as all African member states of the United Nations did when they endorsed *Transforming Our World: The Sustainable Development Goals* (UN 2015), if there is no commitment to honestly take stock of progress and adjust the course of action in a particular location to realize the intended outcomes. The nature of such large-scale reconfiguration and adaptation to the normal way of doing things requires continual adjustment, especially on the ground, as the implications of implementing the ambitions of the SDGs becomes apparent. These local experiences of SDG implementation are not simply an endpoint of the global agenda, but a hub in the feedback loops for revisions to the global agenda.

While it is difficult to make robust assessments of progress less than halfway through a large-scale multi-year program of change (monitoring implementation of the SDGs is a task that is set to run until 2030 and after), it has to be done. It also has to be done everywhere, or the integrity of the international process is put at risk. Regular formal evaluation is thus built into global programs like the SDGs, the New Urban Agenda, and the United Nations Intergovernmental Panel on Climate Change (IPCC). Although it is far from clear how the local scale can and should engage with multi-lateral mechanisms that are designed primarily for national to global reporting, there can be no doubt that it is local implementation that is the litmus test of global change (Dellas et al. 2018). The critical importance of the local scale for putting global aspirations into practice makes it essential to generate knowledge about how the SDGs are working, or not, in particular places, most especially those cities that have the most to gain from advancing sustainability aspirations. It is no overstatement therefore to assert that rapidly changing African cities, who currently perform poorly against the SDG metrics, are the crucible of SDG implementation.

African cities and nations have engaged actively in both structured and informal review processes associated with the global urban agenda, including SDG 11 and SDG localization (c.f. Chaps. 11–13; Kundu et al. 2020). There is also a strong African engagement with the SDG Agenda more generally (Ramutsindela and Mickler 2020). However, there is still so much more to say about Africa—the region whose progress is central to the international success or failure of the SDGs. Turmoil, like the COVID-19 pandemic that has disproportionately disrupted the economies of low- and middle-income contexts, makes localized and urbanized African reflections of the direction of travel in the SDG process even more important than usual.

As early as 2018, just 3 years after the endorsement of the SDGs, the Committee for Development Policy (United Nations Economic and Social Council 2018) reflected that to achieve the UN's pledge to "leave no one behind" would mean a special focus on low-income nations and the sub-Saharan region. Curiously, that committee expected the tricky parts of ensuring progress on the global 2030 vision to lie only in rural areas, which they argued should be prioritized and, in so doing, ignored almost entirely the challenges and opportunities that African cities would pose for SDG implementation. By contrast, African leaders at the regional, national, and local level, alert to the urbanizing character of the region and the significance of cities to social, economic, and environmental integrity and large-scale change, have not been as short-sighted or anti-urban (AU 2015). The positive urban position of the Africans within the UN has been bolstered by the wider recognition that addressing challenges of resource use, climate change, and biodiversity would mean understanding the place of cities in the global system and getting to grips with African urbanization.

Increasingly, cities are seen as key to implementing transformative practices that will con-

[1]See the online Local 2030 platform: https://sustainabledevelopment.un.org/partnerships/local2030.

tribute to and benefit from ideas about sustainability. It was on this basis that member states endorsed a dedicated city SDG and from which the move to urbanize the SDG process has been forged. The recognition of the importance of spatial and locational dynamics to achieving sustainable development outcomes also pervades Africa 2063 (AU 2015). Although that landmark policy from the African Union preceded publication of the 2030 position of the UN (UN 2015) and did not focus overtly on urban areas, it is a city-sensitive approach to sustainable development for Africa that authors of several chapters point to as facilitating the new ethos of urban management that SDG implementation demands. Unsurprisingly therefore, some of the most interesting examples of SDG localization and implementation are now emerging from cities and towns, and we have seen that there is much to learn about sustainable practice and thought when viewed from the urban African frontier.

An overarching theme of the book is that ensuring local actors are integrally involved in efforts to reflect and revise the direction of travel on the urban implementation of the 2030 vision contributes to the legitimacy of SDG process. By extension, African adoption of and commitment to the SDGs will permeate the regional processes and inform the global SDG program. With these pointers in mind, the preparation of this volume brought together a range of African activists, specialists, and scholars. Unlike governments (city, regional, or national), our reflections illuminate rather than assess the SDG implementation efforts. They should be read alongside, not in place of or in critique of, SDG self-assessments from African member states. Our cases are in no way representative of the whole continent. We are methodologically eclectic and work on urban issues at varied scales. The different conceptual entry points we deploy all underscore the problems of making comparison and of aggregation, especially for Africa—a large and hugely diverse region. The paradox we navigate is that while it is politically important that the global community foreground the African urban question, it is imperative that there is no flattening of the urban condition in Africa.

There are further limits to the book. Most contributors seek to highlight gains and opportunities from the innovations we track, rather than dwell on the dark conundrums of implementing an overly ambitious global program in places that lack the necessary resources. While this Afro-optimism is essential to energizing the SDG implementation effort, it should not detract from the very serious structural barriers African cities face in raising the requisite finance to scale up innovations. African urban SDG implementation success must be celebrated at the same as these fragile and often burdened places and people are presented for heightened attention and additional support.

Taken as a whole, the 13 chapters in the book are relatively soft on the barriers to SDG realization, presenting a largely optimistic take on the uptake of the SDGs, based on local ownership of the pathways to change. There is a collective confirmation of the overarching value of the 2030 vision, for Africa and for its cities. Support for the SDGs is especially strong where urban change interventions for sustainable development are embedded in wholly owned local interventions. Areas of concern, especially capacity and the lack of finance, are regularly flagged. Rather than turning their backs on the SDGs, the contributors to this book record problems of implementation. Barriers to action range from the paucity of relevant ideas, the difficulty prioritizing in a context that needs multiple interventions for complex problems, the persistence of violence, the lack of full political support, and the drudgery of navigating bureaucracy. They furthermore endorse the value of multi-lateral action to spur change and nudge the direction and form of local action in Africa. Reading across the contributions, the collection offers a range of local African experiences to nuance the way forward on the road to global sustainable development. Most exciting of all is the emergence of a cadre of young engaged African urbanists who know and love their cities and who want to have a powerful voice in recrafting urban society and systems, aligning their aspirations with the normative base of sustainability.

14.2 What Can We Learn About SDG Localization From Africa?

It is very clear from each of the contributions to this volume that the SDG Agenda, with all of its shortcomings, has gained significant local traction in less than a decade. There may not be many academic texts on the SDGs in African cities, but there is already widespread acceptance of and buy-in to the vision. The extent of progress is, however, both slow and uneven, and the barriers to implementation experienced in African cities and towns are varied. For African cities, especially the smaller and less well-resourced centers, the value of the universal construction of the SDG indicator is sometimes a poor fit or it cannot be measured without available data. Crucially, the implied focus of the SDGs sometimes lacks potency either because of the imperatives of extreme poverty or because the varied configuration of political authority in African settlements does not readily translate into the multi-scalar notions of the state that most UN members understand. As one might expect from such a large continent, which at the time of producing this text was in the midst of a dramatic urban transition and a city altering pandemic, it is hard to generalize. Issues of African exceptionalism aside, there is no reason to abandon the 2030 project. For African cities to maximize their role, more can be done. Working from what is already being done well, there is much scope to scale up SDG implementation, to expand support to more African cities, and even to refine some of the detail and timeframes of the global and national policy processes to allow cities to consolidate their implementation plans.

Crucial support to Africa must come in establishing robust knowledge and data capacity, and any urban African support for implementing the SDGs must address the well-understood fiscal gap for cities. There is no need to start from scratch—there are already scalable lessons on SDG implementation to be drawn from African SDG experiences. First is the value of having African urbanists who are generalists and not just specialists and people who understand local articulations of community and government; who are literate in the way the local and the national interact; who have city-based experiences of the interactions of the water, energy, and food systems; and who can see beyond their narrow area of expertise. Second is the imperative of getting younger people into positions of authority and influence. The urban skills shortage in Africa can mean rapid promotion for younger professionals in the field, creating upward mobility that comes with the risk of compromising experience, but it also brings the dividend of energy and innovation into the urban management system. The third lesson from Africa is the merit in sharing accounts of SDG implementation among close neighbors in order to build a common narrative about the African urban challenge and how these cases fit into global stories about sustainability. The book thus speaks to the importance of getting the auxiliary urbanites—scholars, donors, development banks, and leaders of community-based organizations—into structured learning opportunities to generate authoritative urban vocabularies from which to shift international debates about the future of cities.

Although the SDGs set out a universal agenda, there is simply no blueprint for implementation of the goals in African cities. Thankfully, there are no obvious examples of leaders seeking to mold their cities to some idealized norm, as has been the case in the past with world city aspirations. Instead, perhaps reflecting the confidence of a new generation, African cities and towns seem to have taken different elements of the 2030 Agenda and fused them with their own priorities, often articulating local versions of the SDG values rather than invoking any international format. There is also no Africa-wide uniformity to the monitoring, governance, or financing of the SDGs, though the importance of generating both better and more comparable data systems was a common and frequent lament.

Cities that find value in associating with the ideas and monitoring systems of the 2030 Agenda sometimes, but not always, work in conjunction with SDG-related processes such as National

Urban Policies, the New Urban Agenda, or Voluntary National and Local Reviews. The problems of populating information for and from SDG implementation highlight the paucity of African data and the more general issue of the absence of public open access subnational or geospatial information. Our failure to present a composite picture of SDG implementation progress is but one example of the patchy, incomplete, misconfigured, and generally dysfunctional format of city data on Africa.

On the positive side, the multiplicity of policy formats and reporting lines seen in the large and small African cities we have covered just in this volume speaks to a healthy diversity of local practices and the absence of formulaic implementation. All chapters in the fourth part of the volume that cover collaborative experiences from the frontier of practice highlight the importance of the embeddedness of the global ideas in local planning processes and agendas. Exactly how these ideas are absorbed differs greatly from one place to another. Assembling local energy in places as different as Bissau, Cairo, Hawassa, or Lusaka, the book's authors reflect on barriers to and progress in the collective project of transforming urban areas to achieve the ambitions of sustainability. Time and again we read that cities and towns have taken from the global and national policy visions what they find works for them—but they have also (tentatively) managed to feed back into the bigger national and international narratives what has been working on the ground. In the chicken and egg narrative of sustainable urban development, it matters less where innovative ideas started than that they are relevant to local conditions and that they are gaining traction in Africa.

Ideas matter—at both the high-level conceptual stage and once there is physical or institutional operation of projects and programs. Ideas that eschewed single entry points of monolithic solutions have the most local African support. Across the continent, the urban leadership on the SDGs has rejected single projects and focused on multiscale operations, complexity, interdisciplinarity, and partnerships. There is an acute awareness that change takes time and demands widespread support. Pragmatism is very evident with many of the chapters highlighting that, in order to advance the SDGs in an African urban environment, we need to understand the objectives and aspirations of the global agenda. But far more importantly, what is needed is a grounded understanding of how cities actually work. A deep knowledge of government, local history, and structures, and a deep sensitivity to political possibilities, helps make a difference to how well the SDGs "land" in a place. Grounded community engagement and a deep understanding of politics were seen to be imperative at all stages of urban change. Possibly a surprise to some, across the chapters there is a recognition and endorsement of the role of expert knowledge and of specialization for city governance. Alongside ongoing community mobilization, the importance of building a culture of cooperation in technical and professional processes within government emerges as an indicator of likely SDG success.

In truth, the time available to implement the SDGs in Africa's cities is too short. Yet the thread that emerged from all the cities is the importance of endurance and the will to complete programs that might advance the SDGs. Implementation must go on—even in the face of a pandemic, a climate catastrophe, and systemic injustices.

14.3 Conclusion

Looking to 2030, the date by which the SDGs are supposed to be realized from the vantage point of an African city is daunting. We will not make it. We were never going to. Even before the ravages of the COVID-19 pandemic, the goals were unattainable. Aspirations for building in social, economic, and environmental integrity into the DNA of African cities remain a hope and a goal.

But just because the 2030 timeframe is so tight, and the targets and indicators were never designed with the realities of vast informality, extreme and deep poverty, and minimal public investment in mind, it does not mean that the

SDG path set out is the wrong one— even for this special continent. While there are undoubtedly particular African SDG obstacles, there are also African opportunities, not least of which is the energetic and aspirational commitment of residents and across local civil society. In local and national government too, there is a drive to improve African cities and forge more sustainable patterns of urban living.

One of the most positive things about the SDG process is that it has made legible the extent to which Africa matters in the overall global conversation. If cities matter to sustainability, African cities matter more than most to our collective chances of realizing the SDGs. Perhaps, the African Union timeline of 2063 is more realistic than the timeline of 2030, but at least the SDGs help guide, rather than distort, urban development for Africa.

References

African Union (AU) (2015) Agenda 2063: report of the commission on the African Union Agenda 2063. African Union, Addis Ababa

Dellas E, Carius A, Beisheim M, Parnell S, Messner D (2018) Realising synergies in follow-up and review: the role of local and regional governments and their partners in the follow-up and review of global sustainability agendas. Adelphi consult GmBH, Berlin; Cities Alliance, Brussels

Kundu D, Sietchiping R, Kinyanjui M (2020) Developing National Urban Policies. Springer Publishing, Singapore

Ramutsindela M, Mickler D (eds) (2020) Africa and the Sustainable Development Goals. Sustainable Development Goals series. Springer Publishing, Singapore

United Nations (UN) (2015) Transforming our world: the 2030 Agenda for Sustainable Development (70/1). United Nations General Assembly, New York

United Nations Economic and Social Council (ECOSOC) (2018) Committee for Development Policy report on the twentieth session, 12-16 March 2018 (E/2018/33, supplement no. 13). United Nations Economic and Social Council, New York

Index

S. Croese, S. Parnell (eds.), *Localizing the SDGs in African Cities*, Sustainable Development Goals Series, https://doi.org/10.1007/978-3-030-95979-1

R

S